Inverse Theory for Petroleum Reservoir Characterization and History Matching

This book is a guide to the use of inverse theory for estimation and conditional simulation of flow and transport parameters in porous media. It describes the theory and practice of estimating properties of underground petroleum reservoirs from measurements of flow in wells, and it explains how to characterize the uncertainty in such estimates.

Early chapters present the reader with the necessary background in inverse theory, probability, and spatial statistics. The book then goes on to develop physical explanations for the sensitivity of well data to rock or flow properties, and demonstrates how to calculate sensitivity coefficients and the linearized relationship between models and production data. It also shows how to develop iterative methods for generating estimates and conditional realizations. Characterization of uncertainty for highly nonlinear inverse problems, and the methods of sampling from high-dimensional probability density functions, are discussed. The book then ends with a chapter on the development and application of methods for sequentially assimilating data into reservoir models.

This volume is aimed at graduate students and researchers in petroleum engineering and groundwater hydrology and can be used as a textbook for advanced courses on inverse theory in petroleum engineering. It includes many worked examples to demonstrate the methodologies, an extensive bibliography, and a selection of exercises.

Color figures that further illustrate the data in this book are available at www.cambridge.org/9780521881517

Dean Oliver is the Mewbourne Chair Professor in the Mewbourne School of Petroleum and Geological Engineering at the University of Oklahoma, where he was the Director for four years. Prior to joining the University of Oklahoma, he worked for seventeen years as a research geophysicist and staff reservoir engineer for Chevron USA, and for Saudi Aramco as a research scientist in reservoir characterization. He also spent six years as a professor in the Petroleum Engineering Department at the University of Tulsa. Professor Oliver has been awarded 'best paper of the year' awards from two journals and received the Society of Petroleum Engineers (SPE) Reservoir Description and Dynamics award in 2004. He is currently the Executive Editor of *SPE Journal*. His research interests are in inverse theory, reservoir characterization, uncertainty quantification, and optimization.

Albert Reynolds is Professor of Petroleum Engineering and Mathematics, holder of the McMan chair in Petroleum Engineering, and Director of the TUPREP Research Consortium at the University of Tulsa. He has published over 100 technical articles and one previous book, and is well known for his contributions to pressure transient analysis and history matching. Professor Reynolds has won the SPE Distinguished Achievement Award for Petroleum Engineering Faculty, the SPE Reservoir Description and Dynamics Award and the SPE Formation Award. He became an SPE Distinguished Member in 1999.

Ning Liu holds a Ph.D. from the University of Oklahoma in petroleum engineering and now works as a Reservoir Simulation Consultant at Chevron Energy Technology Company. Dr Liu is a recipient of the Outstanding Ph.D. Scholarship Award at the University of Oklahoma and the Student Research Award from the International Association for Mathematical Geology (IAMG). Her areas of interest are history matching, uncertainty forecasting, production optimization, and reservoir management.

Inverse Theory for Petroleum Reservoir Characterization and History Matching

Dean S. Oliver

Albert C. Reynolds

Ning Liu

CAMBRIDGE
UNIVERSITY PRESS

CAMBRIDGE
UNIVERSITY PRESS

University Printing House, Cambridge CB2 8BS, United Kingdom

One Liberty Plaza, 20th Floor, New York, NY 10006, USA

477 Williamstown Road, Port Melbourne, VIC 3207, Australia

314-321, 3rd Floor, Plot 3, Splendor Forum, Jasola District Centre, New Delhi - 110025, India

79 Anson Road, #06-04/06, Singapore 079906

Cambridge University Press is part of the University of Cambridge.

It furthers the University's mission by disseminating knowledge in the pursuit of education, learning and research at the highest international levels of excellence.

www.cambridge.org
Information on this title: www.cambridge.org/9781108462075

First published 2008
First paperback edition 2018

A catalogue record for this publication is available from the British Library

ISBN 978-0-521-88151-7 Hardback
ISBN 978-1-108-46207-5 Paperback

Cambridge University Press has no responsibility for the persistence or accuracy of URLs for external or third-party internet websites referred to in this publication, and does not guarantee that any content on such websites is, or will remain, accurate or appropriate.

Al Reynolds dedicates the book to Anne, his wife and partner in life.
Ning Liu dedicates the book to her parents and teachers.
Dean Oliver dedicates the book to his wife Mary
and daughters Sarah and Beth.

Contents

10   Quantifying uncertainty   269

11   Recursive methods   347

Preface

The intent of this book is to provide a rather broad overview of inverse theory as it might be applied to petroleum reservoir engineering and specifically to what has, in the past, been called history matching. It has been strongly influenced by the geophysicists' approach to inverse problems as opposed to that of mathematicians. In particular, we emphasize that measurements have errors, that the quantity of data are always limited, and that the dimension of the model space is usually infinite, so inverse problems are always underdetermined. The approach that we take to inverse theory is governed by the following philosophy.

1. All inverse problems are characterized by large numbers of parameters (conceptually infinite). We only limit the number of parameters in order to solve the forward problem.
2. The number of data is always finite, and the data always contain measurement errors.
3. It is impossible to correctly estimate all the parameters of a model from inaccurate, insufficient, and inconsistent data,[1] but reducing the number of parameters in order to get low levels of uncertainty is misleading.
4. On the other hand, we almost always have some prior information about the plausibility of models. This information might include positivity constraints (for density, permeability, and temperature), bounds (porosity between 0 and 1), or smoothness.
5. Most petroleum inverse problems related to fluid flow are nonlinear. The calculation of gradients is an important and expensive part of the problem; it must be done efficiently.
6. Because of the large cost of computing the output of a reservoir simulation model, trial and error approaches to inverting data are impractical.
7. Probabilisitic estimates or bounds are often the most meaningful. For nonlinear problems, this is usually best accomplished using Monte Carlo methods.
8. The ultimate goal of inverse theory (and history matching) is to make informed decisions on investments, data aquisition, and reservoir management. Good decisions can only be made if the uncertainty in future performance, and the consequences of actions can be accurately characterized.

---

[1] This is part of the title of a famous paper by Jackson [1]: "Interpretation of inaccurate, insufficient, and inconsistent data."

Other general references

Several good books on geophysical inverse theory are available. Menke [2] provides good introductory information on the probabilistic interpretation of an answer to an inverse problem, and much good material on the discrete inverse problem. Parker [3] contains good material on Hilbert space, norms, inner products, functionals, existence and uniqueness (for linear problems), resolution and inference, and functional differentiation. He does not, however, get very deeply into nonlinear problems or stochastic approaches. Tarantola [4] comes closest to covering the material on linear inverse problems, but has very little material on calculation of sensitivities. Sun [5] focusses on problems related to flow in porous media, and contains useful material on the calculation of sensitivities for flow and transport problems. A highly relevant free source of information on inverse theory is the book by John Scales [6].

No single book contains a thorough description of the nonlinear developments in inverse theory or the applications to petroleum engineering. Most of the material that is specifically related to petroleum engineering is based on our publications.

The choice of material for these notes is based on the observation that while many scientists and engineers have good intuition for the outcome of an experiment, they often have poor intuition regarding inverse problems. This is not to say that they can not estimate some parameter values that might result in a specified response, but that they have little feel for the degree of nonuniqueness of the answer, or of the relationship of their answer to other answers or to the true parameters. We feel that this intuition is best developed through a study of linear theory and that the method of Backus and Gilbert is good for promoting understanding of many important concepts at a fundamental level. On the other hand, the Backus and Gilbert method can produce solutions that are not *plausible* because they are too erratic or too smooth. We, therefore, introduce methods for incorporating prior information on smoothness and variability. One of the principal uses of these methods is to investigate risk and to make informed decisions regarding investment. For many petroleum engineering problems, evaluation of uncertainty requires the ability to generate a meaningful distribution multiple of models. Characterization of uncertainty for highly nonlinear inverse problems, and the methods of sampling from high-dimensional probability density functions are discussed in Chapter 10.

Most history-matching problems in petroleum engineering are strongly nonlinear. Efficient incorporation of production-type data (e.g. pressure, concentration, water-oil ratio, etc.) requires the calculation of sensitivity coefficients or the linearized relationship between model and data. This is the topic of Chapter 9.

Although history matching has typically been a "batch process" in which all data are assimilated simultaneously, the installation of permanent sensors in wells has increased the need for methods of updating reservoir models by sequentially assimilating data as it becomes available. A method for doing this is described in Chapter 11.

1 Introduction

If it were possible for geoscientists and engineers to know the locations of oil and gas, the locations and transmissivity of faults, the porosity, the permeability, and the multi-phase flow properties such as relative permeability and capillary pressure at all locations in a reservoir, it would be conceptually possible to develop a mathematical model that could be used to predict the outcome of any action. The relationship of the *model* variables, m, describing the system to observable variables or *data*, d, is denoted

$$g(m) = d.$$

If the model variables are known, outcomes can be predicted, usually by running a numerical reservoir simulator that solves a discretized approximation to a set of partial differential equations. This is termed the *forward problem*.

Most oil and gas reservoirs are inconveniently buried beneath thousands of feet of overburden. Direct observations of the reservoir are available only at well locations that are often hundreds of meters apart. Indirect observations are typically made at the surface, either at the well-head (production rates and pressures) or at distributed locations (e.g. seismic). In the *inverse problem*, the observations are used to determine the variables that describe the system. Real observations are contaminated with errors, ϵ, so the inverse problem is to "solve" the set of equations

$$d_{obs} = g(m) + \epsilon$$

for the model variables, with the goal of making accurate predictions of future performance.

1.1 The forward problem

In a forward problem, the physical properties of some system (system or model parameters) are known, and a deterministic method is available for calculating the response or outcome of the system to a known stimulus. The physical properties are referred to as system or model parameters. A typical forward problem is represented by a differential equation with specified initial and/or boundary conditions. A simple example of a forward problem of interest to petroleum engineers is the following steady-state problem for a one-dimensional flow in a porous medium:

$$\frac{d}{dx}\left(\frac{k(x)A}{\mu}\frac{dp(x)}{dx}\right) = 0,$$

(1.1)

for $0 < x < L$, and

$$\frac{dp}{dx}\bigg|_{x=L} = -\frac{q\mu}{k(L)A},$$

(1.2)

$$p(0) = p_e$$

(1.3)

where A (cross sectional area to flow in cm^2), μ (viscosity in cp), q (flow rate in cm^3/s), and pressure p_e (atm) are assumed to be constant. The length of the system in cm is represented by L. The function $k(x)$ represents the permeability field in Darcies. This steady-state problem could describe linear flow in either a core or a reservoir. For this forward problem, the model parameters, which are assumed to be known, are A, L, μ, and $k(x)$. The stimulus for the system (reservoir or core) is provided by prescribing q (the flow rate out the right-hand end) and $p(0)$ (the pressure at the left-hand end), for example, by the boundary conditions, which are assumed to be known exactly. The system output or response is the pressure field, which can be determined by solving the boundary-value problem. The solution of this steady-state boundary-value problem is given by

$$p(x) = p_e - \frac{q\mu}{A}\int_0^x \frac{1}{k(\xi)}\,d\xi.$$

(1.4)

If the emphasis is on the relationship between the permeability field and the pressure, we might formally write the relationship between pressure, p_i, at a location, x_i, and the permeability field as $p_i = g_i(k)$. This expression indicates that the function g_i specifies the relation between the permeability field and pressure at the point x_i.

Forward problems of interest to us can usually be represented by a differential equation or system of differential equations together with initial and/or boundary conditions. Most such forward problems are well posed, or can be made to be well posed by imposing natural physical constraints on the coefficients of the differential equation(s) and the auxiliary conditions. Here, auxiliary conditions refer to the initial and boundary conditions. A boundary-value problem, or initial boundary-value problem, is said to be *well posed* in the sense of Hadamard [7], if the following three criteria are satisfied:
(a) the problem has a solution,
(b) the solution is unique, and
(c) the solution is a continuous function of the *problem data*.
It is important to note that the *problem data* include the functions defining the initial and boundary conditions and the coefficients in the differential equation. Thus, for the

boundary-value problem of Eqs. (1.1)–(1.3), the problem data refers to $p_e, q\mu/k(L)A$ and $k(x)$.

If $k(x)$ were zero in some part of the core, then we can not obtain steady-state flow through the core and the pressure solution of Eq. (1.4) is not defined, i.e. the boundary-value problem of Eqs. (1.1)–(1.3) does not have a solution for $q > 0$. However, if we impose the restriction that $k(x) \geq \delta > 0$ for any arbitrarily small δ then the boundary-value problem is well posed.

If a problem is not well posed, it is said to be *ill posed*. At one time, most mathematicians believed that ill-posed problems were incorrectly formulated and nonphysical. We know now that this is incorrect and that a great deal of useful information can be obtained from ill-posed problems. If this were not so, there would be little reason to study inverse problems, as almost all inverse problems are ill posed.

1.2   The inverse problem

In its most general form, an inverse problem refers to the determination of the plausible physical properties of the system, or information about these properties, given the observed response of the system to some stimulus. The observed response will be referred to as observed data. For example, for the steady-state problem considered above, an inverse problem could represent the problem of determining the permeability field from pressure data measured at points in the interval $[0, L]$. Note that measured or observed data is different from the problem data introduced in the definition of a well-posed problem.

In both forward and inverse problems, the physical system is characterized by a set of model parameters, where here, a model parameter is allowed to be either a function or a scalar. For the steady single-phase flow problem, the model parameters can be chosen as the inverse permeability ($m(x) = 1/k(x)$), fluid viscosity, cross sectional area A and length L. Note, however, the model parameters could also be chosen as $(k(x)A)/\mu$ and L. If we were to attempt to solve Eq. (1.1) numerically, we might discretize the permeability function, and choose $k_i = k(x_i)$ for a limited number of integers i as our parameters. The choice of model parameters is referred to as a parameterization of the physical system. Observable parameters refer to those that can be observed or measured, and will simply be referred to as observed data. For the above steady-state problem, forcing fluid to flow through the porous medium at the specified rate q provides the stimulus and measured values of pressure at certain locations that represent observed data. Pressure can be measured only at a well location, or in the case where the system represents a core, at locations where pressure transducers are situated. Although the relation between observed data and model parameters is often referred to as the model, we will refer to this relationship as the (assumed) theoretical model, because we wish to refer to any feasible set of specific model parameters as a model. In the continuous inverse problem, the model or model parameters may represent a function or set of functions rather than simply a discrete set of parameters. For the steady-state problem of Eqs. (1.1)–(1.3), the boundary-value problem implicitly defines the theoretical model with the explicit relation between observable parameters and the model or model parameters given by Eq. (1.4).

The inverse problem is almost never well posed. In the cases of most interest to petroleum reservoir engineers and hydrogeologists, an infinite number of equally good solutions exist. For the steady-state problem, the general inverse problem represents the determination of information about model parameters (e.g. $1/k(x)$, μ, A, and L) from pressure measurements. As pressure measurements are subject to noise, measured pressure data will not, in general, be exact. The assumed theoretical model may also not be exact. For the example problem considered earlier, the theoretical model assumes constant viscosity and steady-state flow. If these assumptions are invalid, then we are using an approximate theoretical model and these modeling errors should be accounted for when generating inverse solutions.

For now, we state the general inverse problem as follows: determine plausible values of model parameters given inexact (uncertain) data and an assumed theoretical model relating the observed data to the model. For problems of interest to petroleum engineers, the theoretical model always represents an approximation to the true physical relation between physical and/or geometric properties and data. Left unsaid at this point is what is meant by plausible values (solutions) of the inverse problems. A plausible solution must of course be consistent with the observed data and physical constraints (permeability and porosity can not be negative), but for problems of interest in petroleum reservoir characterization, there will normally be an infinite number of models satisfying this criterion. Do we want to choose just one estimate? If so, which one? Do we want to determine several solutions? If so, how, why, and which ones? As readers will see, we have a very definite philosophical approach to inverse problems, one that is grounded in a Bayesian viewpoint of probability and assumes that prior information on model parameters is available. This prior information could be as simple as a geologist's statement that he or she believes that permeability is 200 md plus or minus 50. To obtain a mathematically tractable inverse problem, the prior information will always be encapsulated in a prior probability density function. Our general philosophy of the inverse problem can then be stated as follows: given prior information on some model parameters, inexact measurements of some observable parameters, and an uncertain relation between the data and the model parameters, how should one modify the prior probability density function (PDF) to include the information provided by the inexact measurements? The modified PDF is referred to as the a posteriori probability density function. In a sense, the construction of the a posteriori PDF represents the solution to the inverse problem. However, in a practical sense, one wishes to construct an estimate of the model (often, the maximum a posteriori estimate) or realizations of the model by sampling the a posteriori PDF. The process of constructing a particular estimate of the model will be referred to as estimation; the process of constructing a suite of realizations will be referred to as simulation.

Here, our emphasis is on estimating and simulating permeability and porosity fields. Our approach to the application of inverse problem theory to petroleum reservoir characterization problems may be summarized as follows.

1. Postulate a prior PDF for the model parameters from analog fields, core, logs, and seismic data. We will often assume that the prior PDF is multi-variate Gaussian, in which case the means and the covariance fully define the stochastic model.
2. Formulate the a posteriori PDF conditioned to all observed data. Data could include both production data and "hard" data (direct measurements of the variables to be estimated) for the rock property fields.
3. Construct a suite of realizations of the rock property fields by sampling the a posteriori PDF.
4. Generate a reservoir performance prediction under proposed operating conditions for each realization. This step is done using a reservoir simulator.
5. Construct statistics (e.g. histogram, mean, variance) from the set of predicted outcomes for each performance variable (e.g. cumulative oil production, water–oil ratio, breakthrough time). Determine the uncertainty in predicted performance from the statistics.

In our view, steps 2 and 3 are both vital, albeit difficult, and most of our research effort has focussed either on step 3 or on issues related to computational efficiency including the development of methods to efficiently generate sensitivity coefficients. Note that if one simply generates a set of rock property fields consistent with all observed data, but the set does not characterize the true uncertainty in the rock property fields (in our language, does not represent a correct sampling of the a posteriori PDF), steps 4 and 5 can not be expected to yield a meaningful characterization of the uncertainty in predicted reservoir performance.

2 Examples of inverse problems

The inverse problems examples presented in this chapter illustrate the concepts of data, model, uniqueness, and sensitivity. Each of these concepts will be developed in much greater depth in subsequent chapters. The examples are all quite simple to describe and understand, but several are difficult to solve.

2.1 Density of the Earth

The mass, M, and moment of inertia, I, of the Earth are related to the density distribution, $\rho(r)$, (assuming mass density is only a function of radius) by the following formulas:

$$M = 4\pi \int_0^a r^2 \rho(r)\, dr, \tag{2.1}$$

$$I = \frac{8\pi}{3} \int_0^a r^4 \rho(r)\, dr, \tag{2.2}$$

where a is the radius of the Earth. If the true density is known for all r, then it is easy to compute the mass and the moment of inertia. In reality, the mass and moment of inertia can be estimated from measurements of the precession of the axis of rotation and the gravitational constant; the density distribution must be estimated. The data vector consists of the "observed" mass and moment of inertia of the Earth:

$$d = [M \quad I]^{\mathrm{T}} \tag{2.3}$$

and the model variable, $m = \rho(r)$, is the density. (Throughout this book, the superscript T on a matrix or vector denotes its transpose.) The relationship between the model variable and the theoretical data is

$$d = \int_0^a \begin{bmatrix} 4\pi r^2 \\ \frac{8\pi}{3} r^4 \end{bmatrix} m\, dr. \tag{2.4}$$

Note that, in this example, the dimension of the model to be estimated is infinite, while the dimension of the data space is just 2. Prior information might be a lower

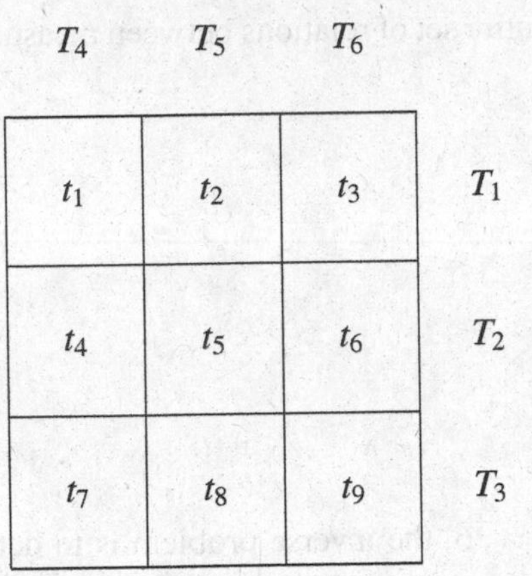

$$T_4 \quad\quad T_5 \quad\quad T_6$$

t_1	t_2	t_3	T_1
t_4	t_5	t_6	T_2
t_7	t_8	t_9	T_3

Figure 2.1. The array of nine blocks with traveltime parameters, t_i, and the six measurement locations for total traveltime, T_i, across the array.

bound on the density. A loose lower bound would be that density is positive. A reasonable lower bound with more information is that density is greater than or equal to 2250 kg/m^3. Although it is easy to generate a model that fits the data exactly, unless other information is available, the uncertainty in the estimated density at a point or a radius is unbounded.

Note also that the theoretical relationship between the density and the data in this example is only approximate as the Earth is not exactly spherical, and there is no a priori reason to believe that the density is only a function of radius.

2.2   Acoustic tomography

One of the simplest examples that demonstrates the concepts of sensitivity, nonuniqueness, and inconsistency is the problem of estimation of the spatial distribution of acoustic slowness (1/velocity) from measurements of traveltime along several ray paths through a solid body. For simplicity, we assume that the material properties are uniform within each of the nine blocks (Fig. 2.1) and we only consider paths that are orthogonal to the block boundaries so that refraction can be ignored and the paths remain straight. If t denotes the acoustic slowness of a homogeneous block, and T denotes the time required to travel a distance D within or across a block, then $T = tD$. Consider a 3 × 3 array of blocks of various materials shown in Fig. 2.1. Each homogeneous block is 1 unit in width by 1 unit in height. Measurements of traveltime have been made for each column and each row of blocks. If the slowness of the (1, 1) block is t_1, the slowness of the (1, 2) block is t_2, and the slowness of the (1, 3) block is t_3, then T_1, the total traveltime for a sound wave to travel across the first row of blocks, is given by $T_1 = t_1 + t_2 + t_3$. Similar relations hold for the other rows and columns. If the measurements of traveltime are exact, the entire set of relations between measurements and slowness in each block is

$$T_1 = t_1 + t_2 + t_3$$
$$T_2 = t_4 + t_5 + t_6$$
$$T_3 = t_7 + t_8 + t_9$$
$$T_4 = t_1 + t_4 + t_7$$
$$T_5 = t_2 + t_5 + t_8$$
$$T_6 = t_3 + t_6 + t_9.$$

(2.5)

Given measured values of T_i, $i = 1, 2, \ldots, 6$, the inverse problem is to determine information about the acoustic slownesses, t_j, $j = 1, 2, \ldots, 9$. More specifically, we may wish to determine the set of all solutions of Eq. (2.6)

$$
\begin{bmatrix} T_1 \\ T_2 \\ T_3 \\ T_4 \\ T_5 \\ T_6 \end{bmatrix} =
\begin{bmatrix}
1 & 1 & 1 & 0 & 0 & 0 & 0 & 0 & 0 \\
0 & 0 & 0 & 1 & 1 & 1 & 0 & 0 & 0 \\
0 & 0 & 0 & 0 & 0 & 0 & 1 & 1 & 1 \\
1 & 0 & 0 & 1 & 0 & 0 & 1 & 0 & 0 \\
0 & 1 & 0 & 0 & 1 & 0 & 0 & 1 & 0 \\
0 & 0 & 1 & 0 & 0 & 1 & 0 & 0 & 1
\end{bmatrix}
\begin{bmatrix} t_1 \\ t_2 \\ t_3 \\ t_4 \\ t_5 \\ t_6 \\ t_7 \\ t_8 \\ t_9 \end{bmatrix}.
$$

(2.6)

With the notation commonly used in this book, Eq. (2.6) is written as

$$d = Gm,$$

(2.7)

where the data, d, is the vector of traveltime measurements, i.e.

$$d = [T_1 \quad T_2 \quad T_3 \quad T_4 \quad T_5 \quad T_6]^{\mathrm{T}},$$

(2.8)

the model, m, is the vector of slowness values given by

$$m = [t_1 \quad t_2 \quad \cdots \quad t_9]^{\mathrm{T}}$$

(2.9)

and the sensitivity matrix, G, is the matrix that relates the data to the model variables and is given by

$$
G =
\begin{bmatrix}
1 & 1 & 1 & 0 & 0 & 0 & 0 & 0 & 0 \\
0 & 0 & 0 & 1 & 1 & 1 & 0 & 0 & 0 \\
0 & 0 & 0 & 0 & 0 & 0 & 1 & 1 & 1 \\
1 & 0 & 0 & 1 & 0 & 0 & 1 & 0 & 0 \\
0 & 1 & 0 & 0 & 1 & 0 & 0 & 1 & 0 \\
0 & 0 & 1 & 0 & 0 & 1 & 0 & 0 & 1
\end{bmatrix}.
$$

(2.10)

The reason for calling G the sensitivity matrix is easily understood by examining the particular row of G associated with a particular measurement. Note that there are as many rows as there are measurements. Each row has nine elements in this example, one for each model variable. The element in the ith row and jth column of G gives the "sensitivity" ($\partial T_i/\partial t_j$) of the ith measurement to a change in the jth model variable. So, for example, the fourth measurement is only sensitive to t_1, t_4, and t_7. As can be seen easily from Eq. (2.5) or (2.6), $\partial T_4/\partial t_j = 1$ for $j = 1, 4, 7$ and $\partial T_4/\partial t_j = 0$ otherwise. Note when $\partial T_i/\partial t_j = 0$, a change in the acoustic slowness t_j will not affect the value of the traveltime T_i, thus we can find no information on the value of t_j from the measured value of T_i.

When we want to visualize the sensitivity for a particular measurement, we often display the row in a natural ordering, one that corresponds to the spatial distribution of model parameters; see Fig. 2.1. Here, we let G_i denote the ith row of G and display G_2

as:

0	0	0
1	1	1
0	0	0

. This display is convenient as it indicates that the second traveltime measurement only depends on the slowness values in the second row. Similarly, G_4 can be displayed as:

1	0	0
1	0	0
1	0	0

, which, when compared to Fig. 2.1 shows clearly that the fourth traveltime measurement is only sensitive to the slowness values of the first column of blocks. Of course, when the models become very large, we will not display all of the numbers. Instead we will use a shading scheme that shows the strength of the sensitivity by the darkness of the grayscale.

Solutions

Suppose that the values of acoustic slowness are such that the exact measurement of one-way traveltime in each of the columns and rows is equal to 6 units (i.e. $T_i = 6$ for all i). Clearly, a homogeneous model for which the slowness of each block is 2 will satisfy this data exactly, i.e. with all $t_j = 2$ and all $T_i = 6$, Eq. (2.6) is satisfied. Similarly, it is easy to see that

$$\hat{m} = [2 \quad 2 \quad 2 \quad 2+b \quad 2-b \quad 2 \quad 2-b \quad 2+b \quad 2]^{\text{T}}. \tag{2.11}$$

is a solution of Eq. (2.6), for any real constant b, when all entries of the data vector are equal to 6. A little examination shows that the following models also satisfy the data exactly:

1	2	3
2	2	2
3	2	1

-2	0	8
-2	6	2
10	0	-4

$2+a$	2	$2-a$
2	2	2
$2-a$	2	$2+a$

$2+b$	$2-b$	2
$2-b$	$2+b$	2
2	2	2

,

Box 1. Nonuniqueness

The null space of G is the set of all real, nine-dimensional column vectors m such that $Gm = 0$. It is easy to verify that each of the following models represent vectors in the null space of G,

0	1	−1		1	−1	0		0	0	0		0	0	0
0	−1	1		−1	1	0		0	1	−1		1	−1	0
0	0	0		0	0	0		0	−1	1		−1	1	0

In fact, the four vectors represented by these four models represent a basis for the null space of G, so any vector in the null space of G can be written as a unique linear combination of these four vectors. If v is any vector in the null space of G and m is a vector of acoustic slownesses that satisfies $Gm = d$ where d is the vector of measured traveltimes, then the model $m + v$ also satisfies the data because

$$G(m + v) = Gm + Gv = d. \tag{2.12}$$

Thus, we can add any linear combination of models (vectors) in the null space of G to a model that satisfies the traveltime data and obtain another model which also satisfies the data.

This acoustic tomography problem has an infinite number of models that satisfy the data exactly for certain data. As there are fewer traveltime data than model variables, this is not surprising. We show next, however, that for other values of the traveltime data, there are no values of acoustic slowness that satisfy Eq. (2.6).

No solution

As measurements are always noisy, let us assume that because of the inaccuracy of the timing, the following measurements were made:

$$T_{obs} = [6.07 \quad 6.07 \quad 5.77 \quad 5.93 \quad 5.93 \quad 6.03]^T. \tag{2.13}$$

Interestingly, despite the fact that there are fewer data than model parameters, there are *no* models that satisfy this data. Eq. (2.5) indicates that T_1 should be the sum of the slowness values in the first row, T_2 should be the sum of slowness values in the second row, and T_3 should be the sum of the slowness values in the third row. Thus

$$T_1 + T_2 + T_3 = t_1 + t_2 + \cdots + t_9. \tag{2.14}$$

But T_4 is the sum of slowness values in column one, and similarly for T_5 and T_6 so if there are values of the model parameters that satisfy these data, we must also have

$$T_4 + T_5 + T_6 = t_1 + t_2 + \cdots + t_9. \tag{2.15}$$

From these results, it is clear that in order for a solution to exist, we must have $T_1 + T_2 + T_3 = T_4 + T_5 + T_6$, but when the data contain noise this is extremely unlikely. For the data of Eq. (2.13), $T_1 + T_2 + T_3 = 17.91$ and $T_4 + T_5 + T_6 = 17.89$, so that with these data, Eq. (2.6) has no solution. Generally, in this case one seeks a solution that comes as close as possible to satisfying the data. A reasonable measure of the goodness of fit is the sum of the squared errors,

$$O(m) = \sum_{j=1}^{6} (d_{\text{obs},j} - d_j(m))^2 = (d_{\text{obs}} - Gm)^T (d_{\text{obs}} - Gm).$$ (2.16)

Here, we have introduced notation that will be used throughout this book. Specifically, $d_{\text{obs},j}$ denotes the j component of the vector of measured or observed data (traveltimes in this example), and d_j denotes the corresponding data that would be calculated (predicted) from the assumed theoretical model relationship (Eq. (2.7) in this example) for a given model variable, m. $O(m)$ denotes an objective function to be minimized and is defined by the first equality of Eq. (2.16). The second equality of Eq. (2.16) follows from standard matrix vector algebra. One solution that has the minimum data mismatch is

2.011	2.011	2.044
2.011	2.011	2.044
1.911	1.911	1.944

, or equivalently,

$$\hat{m} = [2.011 \ 2.011 \ 2.044 \ 2.011 \ 2.011 \ 2.044 \ 1.911 \ 1.911 \ 1.944]^T.$$ (2.17)

From the last equality of Eq. (2.16), it is clear that if m is a least-squares solution then so is $m + v$ where v is a solution in the null space of G. Thus, similar to the case where data are exact, an infinite number of solutions satisfy the data equally well in the least-squares sense.

2.3 Steady-state 1D flow in porous media

Here, the steady-state flow problem introduced in Section 1.1 is formulated as a linear inverse problem. It is assumed that the cross sectional area A, the viscosity μ, the flow rate q, and the end pressure p_e in Eq. (1.4) are known exactly. Although many other characteristics of the porous medium are also unknown (e.g. color, mineralogy, grain size, porosity), we will treat the permeability field as the only unknown. Let

$$d(x) = p_e - p(x)$$ (2.18)

and

$$d_i = d(x_i),$$ (2.19)

$$p_1 \qquad p_2 \qquad\qquad\qquad\qquad p_{N_d}$$

$$p_e \;\boxed{\quad \bullet \qquad \bullet \qquad\qquad\qquad\qquad\qquad \bullet \;}\, q$$

Figure 2.2. A porous medium with constant pressure p_e at the left-hand end, constant production rate q at the right-hand end, and N_d measurements of pressure at various locations along the medium.

for $i = 1, 2, \ldots, N_d$ where the x_is define N_d distinct locations in the interval $[0, L]$ at which pressure measurements are recorded. If the inverse problem under consideration involves linear flow in a reservoir, the x_is would correspond to points at which wells are located. However, the steady-state problem could also represent flow through a core with the x_is representing locations of pressure transducers. The d_is now represent pressure drops, or more generally, pressure changes. However, for simplicity, the data d_i for this problem will be referred to simply as pressure data.

For linear flow problems, it is convenient to define the model variable, $m(x)$, as inverse permeability

$$m(x) = \frac{1}{k(x)}. \tag{2.20}$$

With this notation, Eq. (1.4) can be written as

$$d(x) = \int_0^L G(x, \xi) m(\xi) \, d\xi, \tag{2.21}$$

where

$$G(x, \xi) = \begin{cases} q\mu/A & \text{for } \xi \le x, \\ 0 & \text{for } \xi > x. \end{cases} \tag{2.22}$$

Note that G is only nonzero in the region between the constant pressure boundary location and the measurement location, so the data (pressure drop) are only sensitive to the permeability in that region; changing the permeability beyond the measurement location would have no effect on the measurement.

Assuming pressure data, $d_i = d(x_i)$, are recorded at $x_1 < x_2 < \cdots < x_{N_d}$, Eq. (2.21) is replaced by the inverse problem

$$d_i = \frac{q\mu}{A} \int_0^{x_i} m(\xi) \, d\xi = \int_0^L G_i(\xi) m(\xi) \, d\xi, \tag{2.23}$$

for $i = 1, 2, \ldots, N_d$ where

$$G_i(\xi) = G(x_i, \xi). \tag{2.24}$$

In a general sense, solving this inverse problem means determining the set of functions that satisfy Eq. (2.23) given the values of the d_is.

If only a single pressure drop measurement, $d_1 = d(L)$ is recorded at the right-hand end ($x = L$) of the system, the problem is to solve

$$d_1 = p_e - p(L) = \frac{q\mu}{A} \int_0^L m(x)\,dx, \tag{2.25}$$

for $m(x)$. Clearly there is not a unique function that satisfies Eq. (2.25) since if $m(x)$ satisfies this equation, and $u(x)$ is any function such that

$$\int_0^L u(\xi)\,d\xi = 0, \tag{2.26}$$

then the function $m(x) + u(x)$ also satisfies Eq. (2.25).

Discretization

A discrete inverse problem for the estimation of permeability in steady-state flow can be formulated in more than one way. By approximating the integral in Eq. (2.23) or (2.25) using numerical quadrature, a discrete inverse problem can be obtained. A second procedure for obtaining a discrete inverse problem would be to discretize the differential equation, i.e. write down a finite-difference scheme for the steady-state flow problem of Eqs. (1.1)–(1.3). There is no guarantee that these two approaches are equivalent. Most work on petroleum reservoir characterization is focussed on the second approach, i.e. when observed and predicted data correspond to production data, the forward problem is represented by a reservoir simulator. Here, however, we consider the general continuous inverse problem, Eq. (2.23), and use a numerical quadrature formula to obtain a discrete inverse problem.

In many cases, the best choice of a numerical integration procedure would be a Gauss–Legendre formula (see, for example, chapter 18 of Press *et al.* [8]). But, since our purpose is only illustrative, a midpoint rectangular rule is applied here to perform numerical integration. Let M be a positive integer,

$$x_{1/2} = 0 \tag{2.27}$$

and

$$\Delta x = \frac{L}{M}. \tag{2.28}$$

Then let

$$x_{j+1/2} = x_{j-1/2} + \Delta x \tag{2.29}$$

and

$$x_j = \frac{x_{j-1/2} + x_{j+1/2}}{2}, \tag{2.30}$$

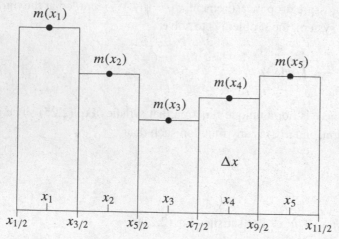

Figure 2.3. Discretization of the porous medium for integration using the midpoint rectangular method. In this figure, $m(x_i)$ is the value of $m(x)$ in the middle of the interval that extends from $x_{i-1/2}$ to $x_{i+1/2}$.

for $j = 1, 2, \ldots, M$. Using the preceding partitioning of the interval $[0, L]$, defining the constant β by

$$\beta = \frac{q\mu}{A}\Delta x, \tag{2.31}$$

and applying the midpoint rectangular rule [9] for integration, Eq. (2.25) can be approximated by

$$d_1 = \beta \sum_{j=1}^{M} m(x_j). \tag{2.32}$$

For simplicity in notation, it is again assumed that pressure data are measured at $x_{r_i+1/2}, i = 1, 2, \ldots, N_d$, where the r_is are a subset of $\{i\}_{i=1}^{M}$ and $r_1 < r_2 < \cdots < r_{N_d}$. The pressure change data at $x_{r_i+1/2}$ is denoted by $d_{\text{obs},i}$ with corresponding calculated data represented by d_i for $i = 1, 2, \ldots, N_d$. With this notation, applying the midpoint rectangular integration rule to Eq. (2.23) (with i replaced by $r_i + 1/2$) gives the approximation

$$d_i = \beta \sum_{j=1}^{r_i} m(x_j), \tag{2.33}$$

for $i = 1, 2, \ldots, N_d$.

Now let d denote the vector of calculated data given by

$$d = [d_1, d_2, \ldots, d_{N_d}]^{\text{T}}, \tag{2.34}$$

and let d_{obs} denote the corresponding vector of observed (measured) pressure drop data. Also let $G = [g_{i,j}]$ be the $N_d \times M$ matrix with the entry in the ith row and jth column defined by

$$g_{i,j} = \beta, \tag{2.35}$$

for $j \leq r_i$ and

$$g_{i,j} = 0, \tag{2.36}$$

for $j > r_i$. Then defining $m_i = m(x_i)$ for all i, and using the notation of Eqs. (3.4) and (3.5), Eq. (2.33) can be written as

$$d = Gm, \tag{2.37}$$

where G is an $N_d \times M$ matrix. Solutions of Eq. (2.23) are functions and as such represent elements of an infinite-dimensional linear space $L^2[0, L]$, whereas, "solutions" of Eq. (2.37) are vectors and are elements of a finite-dimensional linear space. In replacing $m(x)$ by its values at discrete points, the model has been reparameterized. To approximate $m(x)$ from its values at discrete points would require interpolation. Alternately, one could set $m(x) = m_i$, for $x_{i-1/2} < x_{i+1/2}$, i.e. $k_i = 1/m_i$, which corresponds to defining one permeability for each "gridblock" in the interval $[0, L]$.

For the problem under consideration, the discrete inverse problem is specified as

$$d_{obs} = Gm + \varepsilon, \tag{2.38}$$

where d_{obs} is the vector of observed "pressure drop data" and ε represents measurement errors. The objective is to characterize the set of vectors m that in some sense satisfy or are consistent with Eq. (2.38).

In the case where pressure drop data is available at $x_{i+1/2}$ for $i = 1, 2, \ldots, N_d = M$, G is a square $N_d \times N_d$ matrix which can be written as

$$G = \beta \begin{bmatrix} 1 & 0 & \cdots & 0 \\ 1 & 1 & \cdots & 0 \\ \vdots & \vdots & \ddots & \vdots \\ 1 & 1 & \cdots & 1 \end{bmatrix}. \tag{2.39}$$

Note that G is a lower triangular matrix with all diagonal elements nonzero. Thus, G is nonsingular and the unique solution of Eq. (2.38) is $m = G^{-1}d_{obs}$.

If the number of data is fewer than the number of model parameters (components of m), $N_d < M$, then Eq. (2.38) represents N_d equations in M unknowns. As the number of equations is fewer than the number of unknowns, the system of equations is said to be underdetermined. Similarly, if the number of equations is greater than the number of unknowns, $N_d > M$, the problem is said to be overdetermined. A detailed classification of underdetermined, overdetermined and mixed determined problems is presented later.

Underdetermined problem

Suppose the interval $[0, L]$ is partitioned into five gridblocks of equal size and pressure drop data $d_{obs,1}$ is observed at $x_{7/2}$ and $d_{obs,2}$ is observed at $x_{11/2} = L$. Then Eq. (2.38) becomes

$$\begin{bmatrix} d_{obs,1} \\ d_{obs,2} \end{bmatrix} = \beta \begin{bmatrix} 1 & 1 & 1 & 0 & 0 \\ 1 & 1 & 1 & 1 & 1 \end{bmatrix} \begin{bmatrix} m_1 \\ m_2 \\ m_3 \\ m_4 \\ m_5 \end{bmatrix}, \tag{2.40}$$

or, equivalently,

$$\frac{m_1 + m_2 + m_3}{3} = \frac{d_{obs,1}}{3\beta} \tag{2.41}$$

and

$$\frac{m_1 + m_2 + m_3 + m_4 + m_5}{5} = \frac{d_{obs,2}}{5\beta}. \tag{2.42}$$

Clearly the preceding two equations uniquely determine the average value of the first three model parameters and the average value of all five model parameters, but do not uniquely determine the values of the individual m_is. There are, in fact, an infinite number of vectors m that satisfy Eq. (2.40).

Integral equation

Many inverse problems are naturally formulated as integral equations, instead of matrix equations. In Chapter 1, we considered a boundary-value problem for one-dimensional, single-phase, steady-state flow; see Eqs. (1.1)–(1.3). Here we assume that the constant flow rate q, viscosity μ and cross sectional area A are known exactly, and rewrite Eq. (1.4) as

$$p_e - p(x) = C \int_0^x \frac{1}{k(\xi)} d\xi, \tag{2.43}$$

where the constant C is defined by $C = (q\mu)/A$, q is the volumetric flow rate, μ is the viscosity, and A is the cross sectional area to flow. If the function $p_e - p(x)$ is also known at a measurement location x_0, then Eq. (1.4) represents a Fredholm integral equation of the first kind [10]. The inverse problem is then to find a solution, or characterize the solutions, of the integral equation, i.e. to find a model $m(x) = k(x)$ which satisfies Eq. (2.43). Stated this way the integral equation, and hence the inverse problem, is nonlinear. This particular problem is somewhat atypical as it is possible to reformulate the problem as a linear inverse problem by defining the model as

$$m(x) = 1/k(x) \tag{2.44}$$

and rewrite the integral equation as

$$p_e - p(x) = C \int_0^x m(\xi)\,d\xi. \tag{2.45}$$

Although for the physical problem under consideration, $m(x)$ must be positive for $k(x) = 1/m(x)$ to represent a plausible permeability field, here it is convenient to define the inverse problem as the problem of finding piecewise continuous real functions, $m(x)$, defined on $[0, L]$ which satisfy Eq. (2.43) and to define the model space M as the set of all positive piecewise continuous functions defined on $[0, L]$. (M is a real vector space, whereas the subset of M consisting of all positive real-valued functions defined on $[0, L]$ is not a vector space.) The operator G defined on the model space by

$$[Gm](x) = C \int_0^x m(\xi)\,d\xi, \tag{2.46}$$

is a linear operator, i.e. for any constants α and β and any two models $m_1(x)$ and $m_2(x)$

$$G(\alpha m_1 + \beta m_2) = \alpha G m_1 + \beta G m_2. \tag{2.47}$$

Thus, by replacing the parameter $k^{-1}(x)$ by $m(x)$, we have converted the original nonlinear inverse problem (nonlinear integral equation) to a linear inverse problem. Also note Gm is a continuous function of x. Defining

$$d(x) = p_e - p(x), \tag{2.48}$$

Eq. (2.45) can be written as

$$d(x) = [Gm](x). \tag{2.49}$$

Note the similarity to Eq. (2.7).

If the pressure change across the core, $d(L) = p_e - p(L) = p(0) - p(L)$, is measured, the inverse problem becomes to find models $m(x)$ such that

$$d(L) = [Gm](L), \tag{2.50}$$

where the linear operator G is now defined by

$$[Gm](L) = C \int_0^L m(\xi)\,d\xi = \frac{q\mu}{A} \int_0^L m(\xi)\,d\xi. \tag{2.51}$$

Note that G defined by Eq. (2.51) maps functions $m(x)$ in the model space into the set of real numbers.

2.4   History matching in reservoir simulation

A major inverse problem of interest to reservoir engineers is the estimation of rock property fields by history-matching production data. Here, we introduce the complexities, using a single-phase, flow problem.

The finite-difference equations for one-dimensional single-phase flow can be obtained from the differential equation,

$$C_1 \frac{\partial}{\partial x}\left(\frac{k(x)A}{\mu}\frac{\partial p(x,t)}{\partial x}\right) - q\delta(x - x_0) = C_2\phi(x)c_t A\frac{\partial p(x,t)}{\partial t},$$
(2.52)

$$\text{for } 0 < x < L \text{ and } t > 0,$$

$$\frac{\partial p(0,t)}{\partial x} = \frac{\partial p(L,t)}{\partial x} = 0, \qquad \text{for all } t > 0$$
(2.53)

and

$$p(x,0) = p_{\text{in}}, \qquad \text{for all } t > 0,$$
(2.54)

where p_{in} is the initial pressure which is assumed to be uniform. The constants C_1 and C_2 which appear in Eq. (2.52) depend on the system of units. In SI units, both constants are equal to unity. Here, we use oil field units in which case, $C_1 = 1.127 \times 10^{-3}$ and $C_2 = 5.615$. Eq. (2.53) represents no flow boundaries at the ends of the system. In Eq. (2.52), A has units of ft^2 and represents the cross sectional area to flow which we assume to be uniform; μ in centipoise represents the fluid viscosity which we assume to be constant; $k(x)$ in millidarcies represents a heterogeneous permeability field; $\phi(x)$ represents a heterogeneous porosity field; c_t is the total compressibility in psi^{-1} and is assumed to be constant. In Eq. (2.52), the Dirac delta function, $\delta(x - x_0)$, is used to model a production well at x_0 produced at a rate q. The units of the delta function are ft^{-1}. For simplicity, we partition the reservoir into N uniform gridblocks of width Δx in the x direction, let x_i denote the center of the ith gridblock, let $x_{i+1/2}$ and $x_{i-1/2}$, respectively, denote the right- and left-hand boundaries of gridblock i. The grid system is shown in Fig. 2.4, where the circles represent the gridblock centers.

We assume that a single producing well is located in gridblock k. Integrating Eq. (2.52) with respect to x over the ith gridblock, i.e. from $x_{i-1/2}$ to $x_{i+1/2}$, and using the fact that the resulting integral of the Dirac delta function is equal to 1 gives

$$C_1\left(\frac{k(x)A}{\mu}\frac{\partial p}{\partial x}\right)_{(x_{i+1/2},t)} - C_1\left(\frac{k(x)A}{\mu}\frac{\partial p}{\partial x}\right)_{(x_{i-1/2},t)} - q\delta_{i,k}$$

$$= C_2 \int_{x_{i-1/2}}^{x_{i+1/2}}\left(\phi(x)c_t A\frac{\partial p(x,t)}{\partial t}\right)dx$$

$$= \phi_i c_t A\Delta x\left(\frac{\partial p}{\partial t}\right)_{(x_i,t)},$$
(2.55)

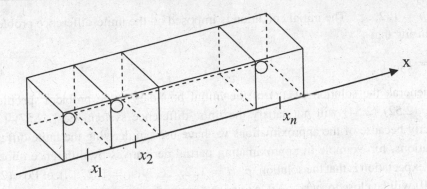

Figure 2.4. One-dimensional grid system.

for $i = 1, 2, \ldots, N$ and $t > 0$. In Eq. (2.55), the last equality assumes $\phi(x)$ and the time derivative of pressure are constant on the interval $(x_{i-1/2}, x_{i+1/2})$ and equal to their values at the gridblock center. If this assumption is invalid then Eq. (2.55) represents an approximation of Eq. (2.52). Throughout, $\delta_{i,k}$ denotes the Kronecker delta function defined by

$$\delta_{i,k} = \begin{cases} 0 & \text{for } k \neq i, \\ 1 & \text{for } k = i. \end{cases} \tag{2.56}$$

Note that Eq. (2.55) applies at any value of time. A sequence of discrete times is defined using a constant time step, Δt, by $t_n = t_{n-1} + \Delta t$ for $n = 0, 1, 2, \ldots$, where $t_0 = 0$. If we consider Eq. (2.55) at any $t = t_n > 0$ and use standard Taylor series approximations for the spatial and time derivatives, we obtain the following finite-difference equation:

$$C_1 \frac{k_{i+1/2}A}{\mu} \left(\frac{p_{i+1}^n - p_i^n}{\Delta x} \right) - C_1 \frac{k_{i-1/2}A}{\mu} \left(\frac{p_i^n - p_{i-1}^n}{\Delta x} \right) - q\delta_{i,k}$$
$$= \phi_i c_t A \Delta x \left(\frac{p_i^n - p_i^{n-1}}{\Delta t} \right) \tag{2.57}$$

for $i = 2, 3, \ldots, N - 1$ and $n = 1, 2, \ldots$. At $i = 1$ and $i = N$, respectively, we impose the no flow boundary conditions of Eq. (2.53) and obtain instead of Eq. (2.57), the following two equations:

$$C_1 \frac{k_{3/2}A}{\mu} \frac{p_2^n - p_1^n}{\Delta x} - q\delta_{1,k} = \phi_1 c_t A \Delta x \left(\frac{p_1^n - p_1^{n-1}}{\Delta t} \right), \tag{2.58}$$

and

$$-C_1 \frac{k_{N-1/2}A}{\mu} \frac{p_N^n - p_{N-1}^n}{\Delta x} - q\delta_{N,k} = \phi_N c_t A \Delta x \left(\frac{p_N^n - p_N^{n-1}}{\Delta t} \right), \tag{2.59}$$

for $n = 1, 2, \ldots$. The initial condition is imposed on the finite-difference problem by requiring that

$$p_i^0 = p_{\text{in}}. \tag{2.60}$$

In general, the solution $p(x, t)$ of the initial boundary-value problem specified by Eqs. (2.52)–(2.54) will not satisfy the finite-difference system, Eqs. (2.57)–(2.59), exactly because of the approximations we have used in deriving the finite-difference equations, for example in approximating partial derivatives by difference quotients. The expectation is that the solution, p_i^n ($i = 1, 2, \ldots, N, n = 1, 2, \ldots$), of Eqs. (2.57)–(2.60) will be close to $p(x_i, t_n)$ if Δt and Δx are sufficiently small.

Given the cross sectional area to flow, rock and fluid properties, the initial pressure and the flow rate, the forward problem is to solve the system of finite-difference equations (Eqs. (2.57)–(2.59)) for p_i^n, $i = 1, 2, \ldots, N$, given p_i^{n-1}, $i = 1, 2, \ldots, N$. At the first time step, $n = 1$ and $p_i^{n-1} = p_i^0 = p_{\text{in}}$.

As is usually done in reservoir simulation, we now assume that permeability is constant on each gridblock, $x_{i-1/2} < x < x_{i+1/2}$, with $k(x) = k_i$ for $i = 1, 2, \ldots, N$. Using the standard harmonic average to relate the permeabilities at a gridblock boundary to the permeabilities of the two adjacent gridblocks gives

$$k_{i+1/2} = \frac{2k_i k_{i+1}}{k_i + k_{i+1}}, \tag{2.61}$$

for $i = 1, 2, \ldots, N - 1$. A typical history-matching problem would be to estimate the permeability and porosity fields given the value of flow rate, A, μ, c_t and observations of gridblock pressure at a few locations.

Multiple solutions

Using a numerical reservoir simulator, we have generated a solution of the system of finite-difference equations given by Eqs. (2.57)–(2.59) for parameter values given in Table 2.1.

Table 2.1. *Reservoir data.*

Number of gridblocks, N	9
Cross sectional area, A, ft^2	2500
Porosity, ϕ	0.25
Permeability, k, md	150
Δx, ft	500
Well location	$i = 9$
Well production rate, q, RB/D	250
System compressibility, c_t, psi^{-1}	10^{-5}
Fluid viscosity, μ, cp	0.5
Initial pressure, p_{in}, psi	3500

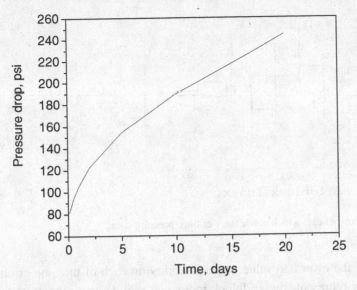

Figure 2.5. Pressure drop for one-dimensional single-phase flow example.

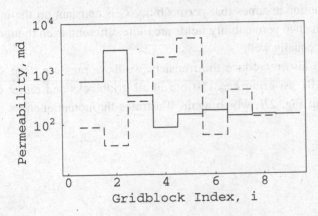

Figure 2.6. Two permeability fields which honor the wellbore pressure data.

Note that the "true" reservoir is homogeneous. Also note that the reservoir is produced by a single well located in gridblock 9. The wellbore pressure at the well in gridblock 9 was obtained by using a Peaceman [11] type equation to relate gridblock and flowing bottomhole wellbore pressure, $p_{wf}(t)$. A plot of the wellbore pressure drop, $\Delta p = p_{in} - p_{wf}$, versus time for twenty days of production is shown in Fig. 2.5.

Figure 2.6 shows two different permeability fields that were obtained as solutions to the history-matching problem, assuming that $\phi = 0.25$ in all gridblocks. Both solutions match the wellbore pressure data of Fig. 2.5 to within 0.01 psi. This example illustrates clearly that the inverse problem of determining the gridblock porosities and permeabilities from flowing wellbore pressure will not have a unique solution when the data are inaccurate and measurements are obtained at only a few locations. In Fig. 2.6,

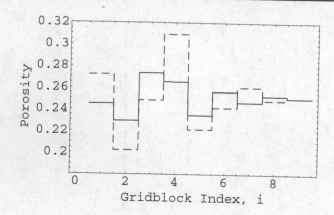

Figure 2.7. Two porosity fields which honor the wellbore pressure data.

we have plotted the estimated value of permeability on each of the nine gridblocks, versus i where i represents the gridblock index. The solid curve represents the first permeability field estimate and the dashed curve represents a second permeability field estimate. Each solution assumes that permeability k_i is constant on the interval $(x_{i-1/2}, x_{i+1/2})$. Note, the two permeability fields are quite different, even though both honor the pressure data equally well.

Interestingly, we can also reproduce the transient wellbore pressure drop shown in Fig. 2.5 to within 0.01 psi using $k = 150$ md in all gridblocks and either of the porosity fields shown in Fig. 2.7, which again illustrates the nonuniqueness of the inverse problem.

2.5   Summary

The examples in this chapter would all have been infinite dimensional in their parameterization, if a natural parameterization had been chosen. It was often necessary, however, to discretize the system in order to solve the forward problem. That is typical for systems that are described mathematically by differential equations. Even with a reduced parameterization, however, the inverse solutions were not unique. When the measurements contain noise (which is always the case), there may be no solutions to the problem that match the data exactly. In the acoustic tomography example, there were no solutions that honored the noisy data exactly, but infinitely many that approximately honored the data equally well.

The relationships of the data to the model variables varied from very simple weighted integrals for the relationship between mass of the Earth (data) and the mass density distribution (model), to a highly complex, nonlinear relationship between pressure (data) and permeability (model) for transient flow in a heterogeneous porous medium.

One of the difficult features of petroleum inverse problems is that the relationship between measurements (water-cut, pressure, seismic amplitude) and variables to be estimated (permeability, porosity, fault transmissibility) is difficult to compute.

For those cases where the solutions are nonunique or no exact solutions exist, it is useful to relax the definition of a solution. It will sometimes be useful to identify a "best estimate" after carefully specifying the meaning of best. In some cases it might be the estimate with the fewest features not required by the data, or the smoothest estimate. In any case, it is also useful to provide an estimate of uncertainty, either in the parameters or in some function of the parameters.

3 Estimation for linear inverse problems

In this chapter, the notions of underdetermined problems, overdetermined problems, mixed determined problems, the null space, the generalized inverse, methods of constructing estimates, sensitivities and resolution are explored for linear finite-dimensional inverse problems. In petroleum reservoir characterization, neither permeabilities nor pressure data are available at every point in the reservoir. Thus, it is assumed that a discrete set of measured data, d_i, $i = 1, 2, \ldots, N_d$, are available and the solution of the inverse problem means the construction of estimates or realizations of the model conditional to these data. The concepts are illustrated by considering the steady-state flow problem introduced in Section 1.1.

Linear inverse problems are those for which the theoretical relation between data and the model can be represented by

$$d_i = (G_i, m),\tag{3.1}$$

where m represents the model variables, d_i represents the data predicted by a particular model m, and $(\cdot, \cdot)$ represents an inner product on some suitably chosen inner product space which contains all feasible models. In many continuous inverse problems of interest, Eq. (3.1) can be represented by the equation,

$$d_i = \int_a^b G_i(x)m(x)\,dx,\tag{3.2}$$

for $i = 1, 2, \ldots, N_d$, where a and b are constants and the inner product is defined on the space of functions which are square integrable on the interval $[a, b]$, i.e. $L^2[a, b]$. Any $m(x)$ in $L^2[a, b]$ which satisfies Eq. (3.2) is a solution of the inverse problem. In general, it is not required that $m(x)$ exactly satisfy this equation to qualify as a solution, if for no other reason than the measured data are corrupted by noise. Classical least-squares fitting of data and nonlinear regression are commonly used in pressure transient analysis to generate models that "honor" data, but do not exactly reproduce the measured data. The continuous inverse problem of Eq. (3.2) is said to be linear because for any $u(x)$ and $v(x)$ in $L^2[a, b]$ and any constants α and β,

$$(G_i, \alpha u + \beta v) = \alpha(G_i, u) + \beta(G_i, v), \tag{3.3}$$

for all i.

Although most natural systems are best modeled using continuous functional representations, we primarily consider discrete inverse problems. Discrete problems refer to those where the physical system under consideration is characterized by a finite number of model variables, say $m_1, m_2, \ldots, m_M$. For discrete inverse problems, it is often convenient to describe a model by the vector of model variables

$$m = [m_1 \quad m_2 \quad \cdots \quad m_M]^T, \tag{3.4}$$

where the superscript T on a matrix or vector denotes its transpose. It is convenient to also include all calculated data in a vector

$$d = [d_1 \quad d_2 \quad \cdots \quad d_{N_d}]^T. \tag{3.5}$$

Then, a discrete linear inverse problem can be represented by

$$d = Gm, \tag{3.6}$$

where G is an $N_d \times M$ matrix representing the sensitivity of data to model variables.

Eq. (3.2) (or more generally, Eq. (3.1)) and Eq. (3.6) predict the data that will be calculated given a model m. Thus, d is referred to as the calculated or theoretical data. If measured data are exact (zero measurement error) and m is the true (actual) physical model, then d will be identical to measured data. In general, however, the observed data will be corrupted by measurement error. The vector of measured or observed data is denoted by d_{obs}.

3.1   Characterization of discrete linear inverse problems

Throughout this section, m denotes a real M-dimensional column vector of model variables; d_{obs} denotes an N_d-dimensional vector of observed data; d denotes the associated calculated data; G denotes an $N_d \times M$ matrix and the inverse problem is formally represented by

$$d_{obs} = Gm. \tag{3.7}$$

The relation between calculated data and any model m is given by

$$d = Gm. \tag{3.8}$$

We assume that the vectors d and m and the matrix G have been normalized so that the entries of each of them are dimensionless. This will be important when eigenvalues and eigenvectors are discussed.

3.1.1   The null space and range

Assume that the unknown vector of model variables is an element of a linear vector space. We refer to this vector space as the model space. Since all solutions of Eq. (3.7) must be M-dimensional vectors, it is convenient to assume that the model space is R^M for this particular inverse problem. The set of all possible vectors of observed data is referred to as the data space. Since such data vectors are N_d dimensional, it is convenient to assume that the data space is R^{N_d}.

Two definitions are useful in characterizing the matrix G in the linear inverse problem. The **null space** of G is defined to be the set of all vectors in the model space that satisfy $Gm = 0$. The **range** of G is the set of all vectors d such that there is at least one m which satisfies $Gm = d$.

The dimension of the range of G is called the **rank** of G. Two important results from linear algebra, relating the rank of G and the dimension of the null space of G, are
 (i) the rank of G is equal to the rank of G^T and
 (ii) the sum of the dimension of the null space of G and the rank of G is equal to M, where G is an $N_d \times M$ matrix [see, for example, 12].

For $i = 1, 2, \ldots, M$, let g_i denote the ith column of the $N_d \times M$ matrix G so

$$G = [g_1 \quad g_2 \quad \cdots \quad g_M].\tag{3.9}$$

Note that for any M-dimensional vector x

$$Gx = [g_1 \quad g_2 \quad \cdots \quad g_M]\begin{bmatrix} x_1 \\ x_2 \\ \vdots \\ x_M \end{bmatrix} = \sum_{j=1}^{M} x_j g_j.\tag{3.10}$$

Then, it is clear that every vector in the range of G is a linear combination of the columns of G. It can be shown that if the columns of G are linearly independent then there are no nontrivial solutions to $Gm = 0$ and vice versa.

If the number of linearly independent columns of G is equal to N_d, then there is at least one vector m which satisfies $Gm = d_{obs}$. On the other hand, if $M < N_d$, then there exist data vectors d_{obs} such that the equation $Gm = d_{obs}$ has **no solution**.

We can also show that $Gm = d_{obs}$ has a **unique solution** for every d_{obs} in the data space if all of the rows of G are independent and the number of model variables M is equal to the number of data N_d. This result indicates that Eq. (3.7) has a unique solution for **every** $d_{obs} \in R(N_d)$ if and only if G is a nonsingular square matrix. However, if G is not a square matrix, there may exist a particular vector d_{obs} such that $Gm = d_{obs}$ has a unique solution. This could occur if some individual equations represented in $Gm = d_{obs}$ are linear combinations of other equations. For example, if rows k and l are identical, then the kth and lth components of the calculated data would be identical, and assuming no measurement error, $d_{obs,k}$ and $d_{obs,l}$ will be equal. In general, however, observed data will be corrupted by measurement errors and $d_{obs,k}$ and $d_{obs,l}$ will not be equal. For this reason, all results on uniqueness and existence are phrased in terms of arbitrary d_{obs}.

Examples

Let us illustrate these concepts with the following simple examples. In each case, we will consider the discrete inverse problem for estimation of permeability from measurements of pressure at various locations along a porous medium with one-dimensional steady flow. For simplicity, we will assume that $q\mu\Delta x/A = 1$ and the core has been subdivided into five intervals ($M = 5$).

One measurement at $x = L$

If only one measurement of pressure drop at $x = L$ is available, the relationship between theoretical pressure drop and model variables ($m_i = 1/k_i$) is

$$Gm = [1 \quad 1 \quad 1 \quad 1 \quad 1] \begin{bmatrix} m_1 \\ m_2 \\ m_3 \\ m_4 \\ m_5 \end{bmatrix} = d_1. \tag{3.11}$$

The rank of G is the number of independent rows or columns (the two are equivalent). There is only one row so the rank is 1. The dimension of the null space must then be 4 since the sum of the dimension of the null space and the rank must be 5. The null space consists of all vectors in R^5 that satisfy $Gm = 0$. This is just the set of all 5-vectors whose elements sum to 0. A basis for the null space is

$$b_1 = \begin{bmatrix} 1 \\ -1 \\ 0 \\ 0 \\ 0 \end{bmatrix}, b_2 = \begin{bmatrix} 1 \\ 0 \\ -1 \\ 0 \\ 0 \end{bmatrix}, b_3 = \begin{bmatrix} 1 \\ 0 \\ 0 \\ -1 \\ 0 \end{bmatrix}, b_4 = \begin{bmatrix} 1 \\ 0 \\ 0 \\ 0 \\ -1 \end{bmatrix}.$$

Any solution of Eq. (3.11) can be written in the form

$$m = \begin{bmatrix} d_1 \\ 0 \\ 0 \\ 0 \\ 0 \end{bmatrix} + a_1 \begin{bmatrix} 1 \\ -1 \\ 0 \\ 0 \\ 0 \end{bmatrix} + a_2 \begin{bmatrix} 1 \\ 0 \\ -1 \\ 0 \\ 0 \end{bmatrix} + a_3 \begin{bmatrix} 1 \\ 0 \\ 0 \\ -1 \\ 0 \end{bmatrix} + a_4 \begin{bmatrix} 1 \\ 0 \\ 0 \\ 0 \\ -1 \end{bmatrix},$$

where the a_i are arbitrary real scalars.

Two independent measurements

If one measurement of pressure drop is available at $x = L$ and another at $x_{7/2}$, the relationship between theoretical pressure drop and model variables is

$$Gm = \begin{bmatrix} 1 & 1 & 1 & 0 & 0 \\ 1 & 1 & 1 & 1 & 1 \end{bmatrix} \begin{bmatrix} m_1 \\ m_2 \\ m_3 \\ m_4 \\ m_5 \end{bmatrix} = \begin{bmatrix} d_1 \\ d_2 \end{bmatrix}. \tag{3.12}$$

There are two independent rows in G so the rank of G is 2. The dimension of the null space must then be 3 since the sum of the dimension of the null space and the rank must be 5. The null space consists of all vectors in R^5 that satisfy $Gm = 0$. The null space must consist of all 5-element vectors whose first three elements sum to zero and whose last two elements sum to 0. A basis for the null space is

$$b_1 = \begin{bmatrix} 1 \\ -1 \\ 0 \\ 0 \\ 0 \end{bmatrix}, b_2 = \begin{bmatrix} 1 \\ 0 \\ -1 \\ 0 \\ 0 \end{bmatrix}, b_3 = \begin{bmatrix} 0 \\ 0 \\ 0 \\ 1 \\ -1 \end{bmatrix}.$$

Any solution of Eq. (3.12) can be written in the form

$$m = \begin{bmatrix} d_1 \\ 0 \\ 0 \\ d_2 - d_1 \\ 0 \end{bmatrix} + a_1 \begin{bmatrix} 1 \\ -1 \\ 0 \\ 0 \\ 0 \end{bmatrix} + a_2 \begin{bmatrix} 1 \\ 0 \\ -1 \\ 0 \\ 0 \end{bmatrix} + a_3 \begin{bmatrix} 0 \\ 0 \\ 0 \\ 1 \\ -1 \end{bmatrix}.$$

Two redundant measurements

If two measurement of pressure drop are available at $x = L$, the relationship between theoretical pressure drop and model variables is

$$Gm = \begin{bmatrix} 1 & 1 & 1 & 1 & 1 \\ 1 & 1 & 1 & 1 & 1 \end{bmatrix} \begin{bmatrix} m_1 \\ m_2 \\ m_3 \\ m_4 \\ m_5 \end{bmatrix} = \begin{bmatrix} d_1 \\ d_2 \end{bmatrix}. \tag{3.13}$$

There is only one independent row in G so the rank of G is 1. The dimension of the null space must then be 4 since the sum of the dimension of the null space and the rank must be 5. Note that the number of independent columns is the same as the number of independent rows.

3.1.2   Underdetermined problems

Eq. (3.7) has *more than one solution* for every d_{obs} if and only if all of the rows of G are linearly independent (the data are independent) *and* there are more model variables than data.

Now consider the real-symmetric $N_d \times N_d$ matrix GG^T where the entries of the $N_d \times M$ matrix G are dimensionless. It is often important in the computation of solutions to know if the product GG^T is nonsingular. Since GG^T is real symmetric, all eigenvalues are real numbers. Let λ and x be an eigenvalue–eigenvector pair of GG^T so

$$GG^Tx = \lambda x. \tag{3.14}$$

Premultiplying by x^T gives

$$x^TGG^Tx = (G^Tx)^TG^Tx = \lambda x^Tx. \tag{3.15}$$

From Eq. (3.15), it follows that

$$\lambda = \frac{\|G^Tx\|^2}{\|x\|^2}, \tag{3.16}$$

so $\lambda \geq 0$ and $\lambda = 0$ if and only if $G^Tx = 0$, i.e. the nonzero eigenvector x is in the null space of G^T. It then follows that all eigenvalues of GG^T are positive, i.e. GG^T is positive definite, if the dimension of the null space of G^T is 0. Conversely, if the dimension of the null space of G^T is 0, then the null space of G^T contains only the zero vector and hence for any nonzero N_d-dimensional column vector x,

$$x^TGG^Tx = \|G^Tx\|^2 > 0, \tag{3.17}$$

thus, GG^T is positive definite. In summary, we have shown that GG^T is positive definite if and only if the dimension of the null space of G^T is 0. If G is a real $N_d \times M$ matrix and the rank of G is N_d, then GG^T is a nonsingular positive-definite matrix.

The linear inverse problem of Eq. (3.7) is said to be **underdetermined** if $M > N_d$, i.e. the number of unknowns is greater than the number of equations. However, for our purposes the stricter definition of a **purely underdetermined** problem is required.

Definition 3.1. *The inverse problem of $Gm = d_{obs}$ is said to be purely underdetermined if for every d_{obs} in the data space, there exist more than one vector m in the model space which satisfies the equation.*

The inverse problem represented by $Gm = d_{obs}$ is purely underdetermined if and only if the rank of G is equal to N_d, and the dimension of the null space of G is greater than 0. In this case, GG^T is a nonsingular positive-definite matrix and $M > N_d$.

3.1.3   Overdetermined and mixed determined problems

For the case where the rank of the $N_d \times M$ matrix G is less than N_d, there exist vectors d_{obs} in the data space which are not in the range of G. For any such d_{obs}, there is no m which satisfies Eq. (3.7). However, it is still desirable to find "solutions" m of the inverse problem, i.e. vectors m which in some sense approximately satisfy Eq. (3.7). For such problems, it will be shown that the classical least-squares solution can be calculated provided $G^T G$ is nonsingular.

If the rank of G is less than N_d, then there exists at least one d_{obs} such that $Gm = d_{obs}$ has no solution. It can be shown that the following three conditions are equivalent:

(i) the $M \times M$ real-symmetric matrix $G^T G$ is a nonsingular positive-definite matrix;
(ii) the rank of G is M;
(iii) the dimension of the null space of G is 0.

Definition 3.2. *The inverse problem $Gm = d_{obs}$ is said to be a purely overdetermined problem if the dimension of the null space of G is 0 and there exist d_{obs} for which the equation has no solution.*

An alternative definition is that the inverse problem of Eq. (3.7) is purely overdetermined if and only if $G^T G$ is a nonsingular positive-definite matrix and $M < N_d$.

Definition 3.3. *The inverse problem of Eq. (3.7) is said to be a mixed determined problem if it is neither purely undetermined nor purely overdetermined.*

Mixed determined problems represent those where some of the model variables are overdetermined and some are underdetermined. Suppose for the steady-state flow problem considered earlier, pressure had been measured five times at $x_{5/2}$ but only once at $x_{11/2} = L$. Then the inverse problem is to solve the following equation for m:

$$
\begin{bmatrix} d_{obs,1} \\ d_{obs,2} \\ d_{obs,3} \\ d_{obs,4} \\ d_{obs,5} \\ d_{obs,6} \end{bmatrix}
=
\begin{bmatrix} 1 & 1 & 0 & 0 & 0 \\ 1 & 1 & 0 & 0 & 0 \\ 1 & 1 & 0 & 0 & 0 \\ 1 & 1 & 0 & 0 & 0 \\ 1 & 1 & 0 & 0 & 0 \\ 1 & 1 & 1 & 1 & 1 \end{bmatrix}
\begin{bmatrix} m_1 \\ m_2 \\ m_3 \\ m_4 \\ m_5 \end{bmatrix},
\tag{3.18}
$$

or equivalently

$$
m_1 + m_2 = d_{obs,i},
\tag{3.19}
$$

for $i = 1, 2, \ldots, 5$ and

$$
m_1 + m_2 + m_3 + m_4 + m_5 = d_{obs,6}.
\tag{3.20}
$$

The observed data, $d_{obs,i}$, $i = 1, 2, \ldots, 5$, represent five different "measurements" of the same pressure change, but due to measurement errors all of these $d_{obs,i}$ values will generally be different. In this case, Eq. (3.19) represents five distinct equations in two unknowns and will have no solution. Thus it is reasonable to say that the sum of variables m_1 and m_2 is overdetermined. On the other hand, Eq. (3.20) is one equation in five unknowns. Regardless of the values assigned to m_1 and m_2, there are infinitely many values for m_3, m_4, and m_5 which satisfy the equation and it makes sense to say the variables m_3, m_4, and m_5 are underdetermined. Thus, intuitively, one expects that Eq. (3.18) should be classified as a mixed determined problem. To show formally that this problem is mixed determined in the sense of Definition 3.3, we simply need to apply the theoretical results established previously. The rank of G is the dimension of the range of G, which is equal to the number of linearly independent columns of G, which is clearly equal to 2. Since $N_d = 6$, the statement following Definition 3.1 indicates that the problem is not purely underdetermined. Since $M = 5$, the dimension of the null space of G is 3. Thus, by Definition 3.2, the problem is not purely overdetermined. By Definition 3.3, it follows that Eq. (3.18) is a mixed determined problem.

To further illustrate the basic concepts, we determine a basis for the null space of G and the range of G. A vector m is in the null space of G if and only if $Gm = 0$, i.e.

$$m_1 + m_2 = 0, \tag{3.21}$$

for $i = 1, 2, \ldots, 5$ and

$$m_1 + m_2 + m_3 + m_4 + m_5 = 0. \tag{3.22}$$

Eq. (3.21) holds if and only if

$$m_2 = -m_1, \tag{3.23}$$

in which case Eq. (3.22) reduces to

$$m_5 = -m_4 - m_3. \tag{3.24}$$

Now by inspection it is clear that any m whose components satisfy Eqs. (3.23) and (3.24) can be written as

$$m = m_1[1, -1, 0, 0, 0]^T + m_3[0, 0, 1, 0, -1]^T + m_4[0, 0, 0, 1, -1]^T, \tag{3.25}$$

i.e. the three five-dimensional column vectors, $x_1 = [1, -1, 0, 0, 0]^T$, $x_2 = [0, 0, 1, 0, -1]^T$ and $x_3 = [0, 0, 0, 1, -1]^T$, form a basis for the null space of G.

If d is in the range of G, then there exist m in $S(m)$ such that $d_6 = \beta(m_1 + m_2 + \cdots + m_5)$ and $d_i = \beta(m_1 + m_2)$ for $i = 1, 2, \ldots, 5$, i.e. $d_1 = d_2 = \cdots = d_5$. Any such d can be written as

$$d = d_1[1, 1, 1, 1, 1, 0]^T + d_6[0, 0, 0, 0, 0, 1]^T. \tag{3.26}$$

Thus, the two vectors $y_1 = [1, 1, 1, 1, 1, 0]^T$ and $y_2 = [0, 0, 0, 0, 0, 1]^T$ form a basis for the range of G.

Now consider the problem described by

$$m_1 + m_2 + m_3 + m_4 + m_5 = d_{\text{obs},i},$$ (3.27)

for $i = 1, 2, \ldots, 6$, i.e.

$$
\begin{bmatrix}
d_{\text{obs},1} \\
d_{\text{obs},2} \\
d_{\text{obs},3} \\
d_{\text{obs},4} \\
d_{\text{obs},5} \\
d_{\text{obs},6}
\end{bmatrix}
=
\begin{bmatrix}
1 & 1 & 1 & 1 & 1 \\
1 & 1 & 1 & 1 & 1 \\
1 & 1 & 1 & 1 & 1 \\
1 & 1 & 1 & 1 & 1 \\
1 & 1 & 1 & 1 & 1 \\
1 & 1 & 1 & 1 & 1
\end{bmatrix}
\begin{bmatrix}
m_1 \\
m_2 \\
m_3 \\
m_4 \\
m_5
\end{bmatrix}.
$$ (3.28)

Eq. (3.28) can be described by Eq. (3.7) where

$$
G =
\begin{bmatrix}
1 & 1 & 1 & 1 & 1 \\
1 & 1 & 1 & 1 & 1 \\
1 & 1 & 1 & 1 & 1 \\
1 & 1 & 1 & 1 & 1 \\
1 & 1 & 1 & 1 & 1 \\
1 & 1 & 1 & 1 & 1
\end{bmatrix}.
$$ (3.29)

Note that for this problem $N_d = 6$ and $M = 5$. It is easy to see that if d is in the range of G then all components of d are equal and thus $d = d_1[1, 1, 1, 1, 1, 1]^{\text{T}}$. This shows that the rank of G is 1 which is less than $N_d = 6$ and that the single vector $[1, 1, 1, 1, 1, 1]^{\text{T}}$ is a basis for the range of G. Because the sum of the rank of G and the dimension of the null space of G must be equal to the dimension of the model space ($M = 5$), we see that the dimension of the null space of G is equal to 4. Again, from Definition 3.2 and the discussion following Definition 3.1, Eq. (3.29) represents a mixed determined problem.

It is fairly common to informally classify Eqs. (3.18) and (3.29) as overdetermined problems because the number of equations is greater than the number of unknowns. More generally, it is common to classify Eq. (3.7) as overdetermined if the number of equations is greater than the number of unknowns ($N_d > M$) and to classify the problem as underdetermined if the number of unknowns is greater than the number of equations $M > N_d$. Although this classification often seems reasonable, it is useful to subscribe to the more accurate terminology of purely overdetermined because a purely overdetermined problem can be "solved" by least squares since $G^{\text{T}}G$ is nonsingular. Similarly, for a purely underdetermined problem, GG^{T} is nonsingular and one reasonable solution to choose among the infinite set of solutions is the one closest to a preconceived estimate of the true model. As will be seen, this solution involves the inverse of GG^{T}. The assumption $M > N_d$ by itself does not guarantee that GG^{T} is nonsingular, however, as can be verified by considering Eq. (3.7) with

$$G = \begin{bmatrix} 1 & 0 & 0 \\ 1 & 0 & 0 \end{bmatrix}, \tag{3.30}$$

so

$$GG^{\mathrm{T}} = \begin{bmatrix} 1 & 1 \\ 1 & 1 \end{bmatrix}, \tag{3.31}$$

which is singular.

3.2   Solutions of discrete linear inverse problems

The inverse problem is again described by

$$d_{\mathrm{obs}} = Gm, \tag{3.32}$$

where d_{obs} corresponds to measured data in the N_d-dimensional "data space" $S(d)$, G is an $N_d \times M$ matrix, and m is in the M-dimensional "model space" $S(m)$. We have seen that some problems of this form have no solutions while other problems have solutions that are not unique. In the remainder of this chapter, we will develop methods for solutions of both types of problems. To do this, we will often need to compute model vectors, m, that minimize matrix products of the form $(Gm - d_{\mathrm{obs}})^{\mathrm{T}} W_D (Gm - d_{\mathrm{obs}})$ and $(m - \mu)^{\mathrm{T}} W_M (m - \mu)$ with respect to m. As most practical methods of finding the minimum of continuous functions make extensive use of gradients, we begin by reviewing basic material on gradient operators.

3.2.1   Gradient operators

Given a vector

$$m = \begin{bmatrix} m_1 \\ m_2 \\ \vdots \\ m_M \end{bmatrix}, \tag{3.33}$$

the gradient operator ∇_m is defined by

$$\nabla_m = \begin{bmatrix} \frac{\partial}{\partial m_1} \\ \frac{\partial}{\partial m_2} \\ \vdots \\ \frac{\partial}{\partial m_M} \end{bmatrix}. \tag{3.34}$$

The subscript m has been attached to indicate that the gradient is with respect to the vector variable m. However, in most cases, it is clear which derivatives are meant by the gradient operator and in such cases, one normally suppresses the subscript on the gradient operator.

If $f(m)$ is a real-valued function of the vector m, the gradient with respect to m of $f(m)$ is defined by

$$\nabla_m f(m) = \begin{bmatrix} \frac{\partial f(m)}{\partial m_1} \\ \frac{\partial f(m)}{\partial m_2} \\ \vdots \\ \frac{\partial f(m)}{\partial m_M} \end{bmatrix}. \tag{3.35}$$

Although one can formally take the dot product of ∇_m and m, $\nabla_m m$ does not make sense, however,

$$\nabla_m m^\mathrm{T} = \begin{bmatrix} \frac{\partial}{\partial m_1} \\ \frac{\partial}{\partial m_2} \\ \vdots \\ \frac{\partial}{\partial m_M} \end{bmatrix} \begin{bmatrix} m_1 & m_2 & \cdots & m_M \end{bmatrix} = I_\mathrm{M}, \tag{3.36}$$

where I_M is the $M \times M$ identity matrix. Normally, the order of the identity matrix considered is clear and the subscript M is suppressed.

In general, ∇_m acts like an $M \times 1$ vector operator that can be applied to any $1 \times N$ matrix $g(m)$. If

$$g(m) = \begin{bmatrix} g_1(m) & g_2(m) & \cdots & g_N(m) \end{bmatrix}, \tag{3.37}$$

where each g_i is a real-valued function of m, then the gradient of the $1 \times N$ vector-valued function g is defined as the $M \times N$ matrix given by

$$\nabla_m g(m) = \begin{bmatrix} \frac{\partial g_1}{\partial m_1} & \frac{\partial g_2}{\partial m_1} & \cdots & \frac{\partial g_N}{\partial m_1} \\ \frac{\partial g_1}{\partial m_2} & \frac{\partial g_2}{\partial m_2} & \cdots & \frac{\partial g_N}{\partial m_2} \\ \vdots & \vdots & \ddots & \vdots \\ \frac{\partial g_1}{\partial m_M} & \frac{\partial g_2}{\partial m_M} & \cdots & \frac{\partial g_N}{\partial m_M} \end{bmatrix}. \tag{3.38}$$

If A is a $1 \times N$ matrix with elements dependent on m and B is an $N \times 1$ matrix whose elements are also functions of m, then it is easy to show that

$$\nabla_m(AB) = (\nabla_m A)B + (\nabla_m B^\mathrm{T})A^\mathrm{T}. \tag{3.39}$$

If B is an $M \times 1$ matrix with elements independent of m, then applying Eqs. (3.36) and (3.39),

$$\nabla_m(B^T m) = (\nabla_m B^T)m + (\nabla_m m^T)B = I_M B = B \tag{3.40}$$

and

$$\nabla_m(m^T B) = (\nabla_m m^T)B + (\nabla_m B^T)m = B. \tag{3.41}$$

In a similar way, we could show that

$$\nabla_m(m^T C) = C, \tag{3.42}$$

for an $M \times N$ matrix C that is independent of m.

Now suppose that m_0 is an M-dimensional column vector whose entries are independent of m, W is an $M \times M$ matrix which has entries independent of M, and define $S(m)$ by

$$S(m) = (m - m_0)^T W_M(m - m_0). \tag{3.43}$$

Then applying Eqs. (3.36), (3.39), (3.40), and (3.42) as needed gives

$$\nabla_m S(m) = (\nabla_m[(m - m_0)^T])W_M(m - m_0) + (\nabla_m[(m - m_0)^T W_M^T])(m - m_0)$$
$$= W_M(m - m_0) + W_M^T(m - m_0). \tag{3.44}$$

If W_M is a symmetric matrix, i.e. $W_M^T = W_M$, then Eq. (3.44) reduces to

$$\nabla_m S(m) = 2W_M(m - m_0). \tag{3.45}$$

Also note that if $W_M^T = W_M$, then from Eqs. (3.45) and (3.41)

$$\nabla_m[\nabla_m S(m)]^T = 2\nabla_m[(m - m_0)^T W_M^T] = 2W_M^T = 2W_M. \tag{3.46}$$

Now consider the expression

$$S(m) = [Gm - d_{\text{obs}}]^T W_D(Gm - d_{\text{obs}}), \tag{3.47}$$

where d_{obs} is a fixed N_d-dimensional column vector (independent of m), G is an $N_d \times M$ matrix and W_D is a real-symmetric $N_d \times N_d$ matrix with entries independent of m. Similar to Eqs. (3.45) and (3.46), it is straightforward to show that

$$\nabla_m S(m) = 2G^T W_D(Gm - d_{\text{obs}}) \tag{3.48}$$

and

$$\nabla_m([\nabla_m S(m)]^T) = 2G^T W_D G. \tag{3.49}$$

3.2.2   Solution of purely underdetermined problems

Suppose that the inverse problem

$$d_{\text{obs}} = Gm \tag{3.50}$$

is purely underdetermined so there exists more than one m which exactly satisfies this equation for any d_{obs}. We assume m is M dimensional, d_{obs} is of dimension $N_d < M$ and G is an $N_d \times M$ matrix.

Because Eq. (3.50) has more than one solution for any right-hand side, construction of a specific solution requires specification of additional information or constraints. This information may represent belief about features of the true solution, or it could simply represent an additional constraint such as the selection of the smallest or smoothest of all the models that honor the data.

In some case, the best solution is the one that not only satisfies the data exactly, but is also closest to m_0 in a least-squares sense. The solution can be obtained using Lagrange multipliers, i.e. by minimizing

$$S(m, \lambda) = (m - m_0)^{\text{T}}(m - m_0) + \lambda^{\text{T}}(Gm - d_{\text{obs}}), \tag{3.51}$$

where λ is the N_d-dimensional column vector of Lagrange multipliers. Taking the gradient of Eq. (3.51) with respect to m gives

$$\nabla_m S = 2(m - m_0) + G^{\text{T}}\lambda. \tag{3.52}$$

Taking the gradient of Eq. (3.51) with respect to λ gives

$$\nabla_\lambda S = Gm - d_{\text{obs}}. \tag{3.53}$$

If m and λ minimize S, it is necessary that $\nabla_m S = 0$ and $\nabla_\lambda S = 0$. As expected, setting $\nabla_\lambda S = 0$ implies that m must satisfy Eq. (3.50). Setting $\nabla_m S = 0$ and solving for m gives

$$m = m_0 - \frac{1}{2}G^{\text{T}}\lambda. \tag{3.54}$$

Using this result in Eq. (3.50) gives

$$d_{\text{obs}} = Gm_0 - \frac{1}{2}GG^{\text{T}}\lambda. \tag{3.55}$$

Since it is assumed that Eq. (3.50) is a purely underdetermined problem, $(GG^{\text{T}})^{-1}$ exists. Thus, Eq. (3.55) can be solved for λ to obtain

$$\lambda = 2(GG^{\text{T}})^{-1}(Gm_0 - d_{\text{obs}}). \tag{3.56}$$

Inserting this result into Eq. (3.54) gives

$$m = m_0 + G^{\mathrm{T}}(GG^{\mathrm{T}})^{-1}(d_{\mathrm{obs}} - Gm_0). \tag{3.57}$$

Note that the estimate of m in Eq. (3.57) is a linear combination of the columns of G^{T}, which suggests that the columns of G^{T} provide useful basis vectors for construction of estimates.

Steady-state flow underdetermined example problem

Reconsider the steady-state flow inverse problem considered in Section 2.3. For simplicity, it is assumed that $\beta = 1$, and for concreteness, we consider the specific problem in which $m_{\mathrm{true}} = [1.2, 1.3, 1.35, 1.35, 1.3]^{\mathrm{T}}$. The true pressure drop at $x_{7/2}$ and at $x_{11/2}$ would be 3.85 and 6.50, respectively. Suppose, however, that because of noise in the measurements, the observed values are 3.95 and 6.30. Thus,

$$d_{\mathrm{obs}} = \begin{bmatrix} 3.95 \\ 6.30 \end{bmatrix} = Gm = \begin{bmatrix} m_1 + m_2 + m_3 \\ m_1 + m_2 + m_3 + m_4 + m_5 \end{bmatrix}, \tag{3.58}$$

where

$$G = \begin{bmatrix} 1 & 1 & 1 & 0 & 0 \\ 1 & 1 & 1 & 1 & 1 \end{bmatrix}. \tag{3.59}$$

Assume that our prior estimate of permeability ($k_j = 1/m_j$) is approximately 0.83 md in every gridblock, so the prior estimate of the model variables is

$$m_0 = \begin{bmatrix} 1.2 & 1.2 & 1.2 & 1.2 & 1.2 \end{bmatrix}^{\mathrm{T}}. \tag{3.60}$$

Then

$$Gm_0 = \begin{bmatrix} 3.6 \\ 6.0 \end{bmatrix}, \tag{3.61}$$

$$GG^{\mathrm{T}} = \begin{bmatrix} 3 & 3 \\ 3 & 5 \end{bmatrix}, \tag{3.62}$$

$$(GG^{\mathrm{T}})^{-1} = \begin{bmatrix} 5/6 & -3/6 \\ -3/6 & 3/6 \end{bmatrix}, \tag{3.63}$$

and so Eq. (3.57) gives

$$m = m_0 + G^{\mathrm{T}}(GG^{\mathrm{T}})^{-1}(d_{\mathrm{obs}} - Gm_0) = \begin{bmatrix} 1.317 \\ 1.317 \\ 1.317 \\ 1.175 \\ 1.175 \end{bmatrix}, \tag{3.64}$$

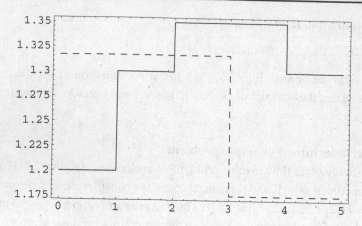

Figure 3.1. The inverse of permeability estimated from two inaccurate pressure measurements at $x = 3$ and 5. The estimate is required to honor the data exactly. The solid curve shows the true model that was used to generate the data before adding random errors.

i.e. the estimated values of permeability on the five gridblocks are

$$\begin{bmatrix} k_1 & k_2 & k_3 & k_4 & k_5 \end{bmatrix}^{\mathrm{T}} = \begin{bmatrix} 0.439 & 0.439 & 0.439 & 0.851 & 0.851 \end{bmatrix}^{\mathrm{T}}. \qquad (3.65)$$

Figure 3.1 compares the estimate with the truth.

Note the "first" equation of the system of two equations specified in Eq. (3.58) requires that $m_1 + m_2 + m_3 = 3.95$. Since there is no information available to distinguish between the individual values of the first three components of m, it is reasonable that the first three components of m_{est} are equal. Subtracting the "first equation" of Eq. (3.58) from the second gives $m_4 + m_5 = 2.35$ and the estimate of Eq. (3.64) satisfies this equation with $m_4 = m_5 = 1.175$.

Remark 3.1. The result suggests that the choice of m_0 had no effect on the estimate, which is true for the specific example considered. In fact for this problem, it is easy to show (see comment below) that if the first three components of m_0 are equal and the fourth and fifth components are equal, then

$$m_0 - G^{\mathrm{T}}(GG^{\mathrm{T}})^{-1}Gm_0 = 0 \qquad (3.66)$$

in which case, Eq. (3.64) reduces to

$$m = G^{\mathrm{T}}(GG^{\mathrm{T}})^{-1}d_{\mathrm{obs}} \qquad (3.67)$$

i.e. the m closest to m_0 that satisfies Eq. (3.50) does not depend on the specific values of the components of m_0. (It is also possible to give a geometric interpretation of this result.) On the other hand, if all components of m_0 are different, then Eqs. (3.57) and (3.67) are not equivalent.

$$[G^{\mathrm{T}}(GG^{\mathrm{T}})^{-1}G - I]m_0 = \begin{bmatrix} -\frac{2}{3} & \frac{1}{3} & \frac{1}{3} & 0 & 0 \\ \frac{1}{3} & -\frac{2}{3} & \frac{1}{3} & 0 & 0 \\ \frac{1}{3} & \frac{1}{3} & -\frac{2}{3} & 0 & 0 \\ 0 & 0 & 0 & -\frac{1}{2} & \frac{1}{2} \\ 0 & 0 & 0 & \frac{1}{2} & -\frac{1}{2} \end{bmatrix} \cdot \begin{bmatrix} m_{01} \\ m_{02} \\ m_{03} \\ m_{04} \\ m_{05} \end{bmatrix}.$$

Hence, if m_0 is in the null space of $[G^{\mathrm{T}}(GG^{\mathrm{T}})^{-1}G - I]$ then Eq. (3.66) is satisfied. It is easy to see that only two of the first three rows are independent and only one of the last two rows are independent. The dimension of the null space must therefore be two. A basis for the null space is provided by the two vectors $(1, 1, 1, 0, 0)$ and $(0, 0, 0, 1, 1)$. If m_0 is uniform, as it was in the previous example, then it is an element of the null space and satisfies Eq. (3.66).

The previous result was not very satisfying, in that it was not very close to the truth, even though this problem only has five unknowns. Suppose that another measurement of pressure is taken at $x_{5/2}$, but that this measurement also has measurement error, and

$$d_{obs} = \begin{bmatrix} 2.08 \\ 3.95 \\ 6.30 \end{bmatrix} = Gm, \tag{3.68}$$

where

$$G = \begin{bmatrix} 1 & 1 & 0 & 0 & 0 \\ 1 & 1 & 1 & 0 & 0 \\ 1 & 1 & 1 & 1 & 1 \end{bmatrix}. \tag{3.69}$$

From an examination of the columns of G we see that the rank is 3, which is the same as the number of data. Hence this problem is also purely underdetermined. Using exactly the same approach as before, we can calculate the smallest model that exactly satisfies the data. Figure 3.2 shows the estimate from three data. If we had hoped that adding data would improve the estimate, then this is a disappointing result, as it is even further from the truth.

There are of course a large variety of choices for specifying a particular model that satisfies Eq. (3.50). Two obvious choices are the "smoothest" or "flattest" models that satisfy this equation. For the continuous inverse problem

$$d_{obs,i} = \int_a^b G_i(x)m(x)\,dx, \tag{3.70}$$

the flattest solution is the function $m(x) \in L^2[a, b]$ whose derivative is square integrable and minimizes

$$F(m) = \int_a^b \left(\frac{dm(x)}{dx}\right)^2 dx, \tag{3.71}$$

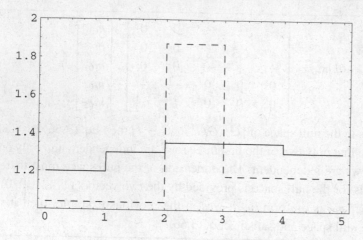

Figure 3.2. The inverse of permeability estimated from three inaccurate pressure measurements at $x = 2, 3$, and 5. The estimate is required to honor the data exactly. The solid curve shows the true model that was used to generate the data before adding random errors.

subject to the constraint that $m(x)$ must satisfy Eq. (3.70). In the discrete linear inverse problem, difference operators replace derivatives. Thus, define the $(M - 1) \times M$ matrix D by

$$
D = \begin{bmatrix}
-1 & 1 & 0 & 0 & \cdots & 0 \\
0 & -1 & 1 & 0 & \cdots & 0 \\
0 & 0 & -1 & 1 & \cdots & 0 \\
\vdots & \vdots & \ddots & \ddots & \ddots & \vdots \\
0 & 0 & 0 & \cdots & -1 & 1
\end{bmatrix}.
\tag{3.72}
$$

Letting $d_{i,j}$ denote the element in the ith row and jth column of D, Eq. (3.72) indicates that $d_{i,i} = -1$ and $d_{i,i+1} = 1$ for $i = 1, 2, \ldots, M - 1$, with all other entries of D equal to zero. Note that if $\|Dm\|^2 = 0$, then $Dm = 0$ which implies $-m_i + m_{i+1} = 0$ for $i = 1, 2, \ldots, M - 1$, i.e. all components of m must be identical.

Again Lagrange multipliers can be used to minimize $\|Dm\|^2$ subject to the constraint that Eq. (3.50) holds, i.e. we seek a minimum of

$$
S(m, \lambda) = \|Dm\|^2 + \lambda^{\mathsf{T}}(Gm - d_{\text{obs}}) = m^{\mathsf{T}} D^{\mathsf{T}} Dm + \lambda^{\mathsf{T}}(Gm - d_{\text{obs}}).
\tag{3.73}
$$

Setting the gradients of S with respect to m and λ equal to zero results in the system

$$
\begin{bmatrix}
2D^{\mathsf{T}} D & G^{\mathsf{T}} \\
G & 0
\end{bmatrix}
\begin{bmatrix}
m \\
\lambda
\end{bmatrix}
=
\begin{bmatrix}
0 \\
d_{\text{obs}}
\end{bmatrix}.
\tag{3.74}
$$

Although the matrix $D^{\mathsf{T}} D$ is clearly singular, so that the system can not be solved as in Section 3.2.2, the system of equations defined by Eq. (3.74) has a unique solution if there is at least one measurement.

Remark 3.2. It is obvious that any vector m whose elements are identical will minimize the "flatness" term $m^T D^T Dm$, so at least one measurement is required to make the solution unique.

Note that all solutions to the purely underdetermined system (without regularization) can be written in the form

$$m = V_p a + V_0 b, \tag{3.75}$$

where the coefficients of V_p are determined by the data, and the coefficients of V_0 are arbitrary. Columns of V_0 form the null space of G. When written this way, the objective function (Eq. 3.73) can be replaced by an equivalent objective function to determine the coefficients of the null space vectors,

$$S(b) = (V_p a + V_0 b)^T D^T D (V_p a + V_0 b). \tag{3.76}$$

A unique solution is obtained when $V_0^T D^T D V_0$ is nonsingular.

3.2.3 Solution of purely overdetermined problems

Again consider the inverse problem,

$$d_{obs} = Gm, \tag{3.77}$$

where G is an $N_d \times M$ matrix mapping the M-dimensional model space into the N_d-dimensional data space. Assume now that the discrete linear inverse problem is purely overdetermined. In this case, $G^T G$ is nonsingular. Since there exist d_{obs} in the data space such that Eq. (3.77) is not satisfied for any m in the model space, it is reasonable to seek a least-squares type solution, i.e. seek m that minimizes

$$S(m) = (Gm - d_{obs})^T (Gm - d_{obs}). \tag{3.78}$$

Given any model m, the vector of calculated data is $d = Gm$ so minimizing S is equivalent to minimizing the sum of squared differences between data calculated from m and the observed data, i.e. minimizing

$$S(m) = \sum_{j=1}^{N_d} (d_j(m) - d_{obs,j})^2, \tag{3.79}$$

where $d_j(m)$ denotes the jth component of $d = Gm$ and $d_{obs,j}$ is the jth entry of the observed data vector, d_{obs}.

From the results of Section 3.2.1, the gradient with respect to m of Eq. (3.78) is given by

$$\nabla S = 2G^T (Gm - d_{obs}) \tag{3.80}$$

and

$$\nabla[(\nabla S)^T] = 2G^TG. \tag{3.81}$$

Since the inverse problem is purely overdetermined, G^TG is positive definite. The least-squares solution can be found by setting $\nabla S = 0$ and solving for m to obtain

$$m_{ls} = (G^TG)^{-1}G^Td_{obs}. \tag{3.82}$$

3.2.4 Regularized least-squares solution

If G^TG is singular, the least-squares solution of Eq. (3.82) is not defined. If GG^T is singular, the solution of Eq. (3.57) is undefined. In either case, one feasible solution can be found by minimizing

$$S(m) = \frac{1}{2}(Gm - d_{obs})^T(Gm - d_{obs}) + \frac{\alpha}{2}(m - m_0)^T(m - m_0), \tag{3.83}$$

where here α is a specified real number and m_0 represents a guess for the true model. In general, the minimum of $S(m)$ is referred to as a regularized least-squares solution where the term $\alpha(m - m_0)^T(m - m_0)$ provides regularization [13]. For any fixed $\alpha > 0$, it can be shown that $S(m)$ has a unique global minimum. Note if $\alpha = 0$ then seeking a minimum of S is equivalent to seeking the classical least-squares solution. Also note that a factor of $1/2$ has been added for convenience; it has no effect on the value of m that minimizes Eq. (3.83). Using results from Section 3.2.1, the gradient of S with respect to m is given by

$$\begin{aligned}
\nabla S &= \alpha(m - m_0) + G^T(Gm - d_{obs}) \\
&= \alpha(m - m_0) + G^T(Gm - Gm_0 + Gm_0 - d_{obs}) \\
&= (\alpha I_M + G^TG)(m - m_0) + G^T(Gm_0 - d_{obs})
\end{aligned} \tag{3.84}$$

and

$$\begin{aligned}
H &= \nabla[(\nabla S)^T] \\
&= \alpha I_M + G^TG,
\end{aligned} \tag{3.85}$$

where I_M denotes the $M \times M$ identity matrix. The first equality of Eq. (3.85) defines the so called Hessian matrix H for the objective function S specified by Eq. (3.83). For $\alpha > 0$, H is positive definite. Hence, S has a unique global minimum which can be found by setting the gradient of S equal to zero. Setting $\nabla S = 0$ in Eq. (3.84) and solving for $m = m_{est}$ gives

$$m_{est} = m_0 + (\alpha I_M + G^TG)^{-1}G^T(d_{obs} - Gm_0). \tag{3.86}$$

Using matrix inversion lemmas, which are discussed in Section 7.4, Eq. (3.86) can be rewritten as

$$m_{\text{est}} = m_0 + G^{\text{T}}(\alpha I_{N_d} + GG^{\text{T}})^{-1}(d_{\text{obs}} - Gm_0), \tag{3.87}$$

where I_{N_d} is the $N_d \times N_d$ identity matrix. Analytically, the two preceding formulas are equivalent, however, the application of either requires construction of a matrix inverse, or the solution of a matrix problem. Specifically calculating m_{est} from Eq. (3.86) requires solution of the $M \times M$ matrix problem

$$(\alpha I_M + G^{\text{T}}G)x = G^{\text{T}}(d_{\text{obs}} - Gm_0). \tag{3.88}$$

Calculating m_{est} from Eq. (3.87) requires solution of the $N_d \times N_d$ matrix problem

$$(\alpha I_{N_d} + GG^{\text{T}})x = (d_{\text{obs}} - Gm_0). \tag{3.89}$$

Assuming matrix problems are to be solved numerically, it is preferable to solve Eq. (3.88) when the number of model variables, M, is significantly smaller than the number of observed data, N_d, and more computationally efficient to solve Eq. (3.89) if $N_d \ll M$.

Note, the magnitude of α influences the estimate obtained. In particular, α determines the relative weighting of the two components of the objective function of Eq. (3.83). If $\alpha \ll 1$, the data mismatch term is weighted more than the term $\|m - m_0\|^2$ and intuitively, we expect that m_{est} will "almost" satisfy $Gm_{\text{est}} = d_{\text{obs}}$. On the other hand, if $\alpha \gg 1$, then we expect that m_{est} will be "close" to m_0. Letting $\alpha \to 0$ in Eq. (3.87) formally gives

$$m_{\text{est}} = m_0 + G^{\text{T}}(GG^{\text{T}})^{-1}(d_{\text{obs}} - Gm_0), \tag{3.90}$$

which is the same estimate obtained in Eq. (3.57) for purely underdetermined problems. One should note, however, that Eq. (3.90) is applicable only if the inverse of GG^{T} exists. However if $G^{\text{T}}G$ is nonsingular, we can let $\alpha \to 0$ in Eq. (3.86) to obtain

$$\begin{aligned} m_{\text{est}} &= m_0 + (G^{\text{T}}G)^{-1}G^{\text{T}}(d_{\text{obs}} - Gm_0) \\ &= (G^{\text{T}}G)^{-1}G^{\text{T}}d_{\text{obs}}, \end{aligned} \tag{3.91}$$

which is the classical least-squares solution obtained in Eq. (3.82) for purely overdetermined problems. Even if $G^{\text{T}}G$ is singular, however, the matrix inverse involved in Eq. (3.86) exists, because a "regularization" term involving $\|m - m_0\|^2$ was added to the normal least-squares objective function; see Eq. (3.83). For this reason, the estimate obtained is referred to as a regularized or generalized least-squares solution.

The important points regarding the regularized least-squares solution of Eq. (3.86) or equivalently Eq. (3.87), can be summarized as follows.

1. As long as $\alpha > 0$, the regularized least-squares solution can be calculated, i.e. this solution exists regardless of whether the inverse problem of Eq. (3.77) is purely underdetermined, purely overdetermined or mixed determined.

2. In the limit as $\alpha \to 0$, the regularized least-squares solution approaches the classical least-squares solution if $G^T G$ is nonsingular and approaches the solution constructed previously for the purely underdetermined problem if GG^T is nonsingular.
3. The magnitude of α governs the relative size of $\|d_{obs} - Gm_{est}\|$ and $\|m_{est} - m_0\|$. As α decreases, we expect $\|d_{obs} - Gm_{est}\|$ to decrease and if α is very large, we expect $\|m_{est} - m_0\|$ to be small.

The selection of α is a problem that depends on the goal of estimation and the knowledge of the character of the true parameter distribution. If the data are measured very accurately, then one would typically choose α to be small to ensure that the estimate nearly satisfies the data. On the other hand, if the model is thought to vary little from the prior estimate and the data are inaccurate, it would be beneficial to choose a large value for α.

Regularized estimation for a steady-state flow problem

Here, the example considered in Section 3.2.2 is reconsidered. Again, all entries of m_0 are set equal to 1.2,

$$G = \begin{bmatrix} 1 & 1 & 1 & 0 & 0 \\ 1 & 1 & 1 & 1 & 1 \end{bmatrix}, \tag{3.92}$$

and

$$d_{obs} = \begin{bmatrix} 3.95 \\ 6.30 \end{bmatrix}. \tag{3.93}$$

The model estimate obtained using Eq. (3.87) is

$$m_{est} = m_0 + \frac{1}{6 + 8\alpha + \alpha^2} \left(0.301\,65 \begin{bmatrix} \alpha \\ \alpha \\ \alpha \\ 3+\alpha \\ 3+\alpha \end{bmatrix} + 0.350\,35 \begin{bmatrix} 2+\alpha \\ 2+\alpha \\ 2+\alpha \\ -3 \\ -3 \end{bmatrix} \right). \tag{3.94}$$

Solutions for three different values of α are shown in Fig. 3.3. In the limit $\alpha \to 0$ we would want to match the data exactly (either because the data *are* exact or because the uncertainty in the model variables is very large).

$$GG^T = \begin{bmatrix} 3 & 3 \\ 3 & 5 \end{bmatrix}, \tag{3.95}$$

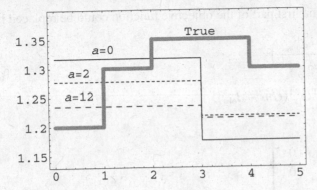

Figure 3.3. The regularized estimates with various values of the regularization parameter.

so GG^T is nonsingular. For this problem,

$$G^TG = \begin{bmatrix} 2 & 2 & 2 & 1 & 1 \\ 2 & 2 & 2 & 1 & 1 \\ 2 & 2 & 2 & 1 & 1 \\ 1 & 1 & 1 & 1 & 1 \\ 1 & 1 & 1 & 1 & 1 \end{bmatrix},$$

(3.96)

is singular. Thus, the classical least-squares solution of Eq. (3.82) does not exist. If $\alpha \to \infty$, then we obtain $m_{est} = m_0$.

3.2.5   General regularization

In considering possible methods to estimate model variables from a limited set of inaccurate data, it appeared that fitting inaccurate data exactly may not be desirable because it can give rise to estimates that are very erratic, even when the truth is known to be smooth. For steady flow the problem became worse when the data were closely spaced, since the model estimate was closely related to the derivative of the data. (That could be seen both from the discrete problem and from the differential equation that says that dp/dx is proportional to $1/k$.)

The weighted least-squares approach is an alternative to exact matching of the data. In this approach the following functional of the data is minimized.

$$S(m) = \frac{1}{2}(Gm - d_{obs})^T(Gm - d_{obs}) + \frac{1}{2}\alpha(m - m_0)^T(m - m_0).$$

(3.97)

When α is large, the estimate is forced to be close to m_0, and when α is small the data are matched very closely.

One way to use this result would be to weight the data according to their accuracy (or inversely according to their uncertainty). If the measurement error of the ith data is on the order of σ_d then the first part of the objective function could be replaced by

$$
\begin{aligned}
S_1(m) &= \frac{1}{2}\sum_i^N \left(\frac{G_i m - d_{obs,i}}{\sigma_{d,i}}\right)^2 \\
&= \frac{1}{2}(Gm - d_{obs})^{\mathrm{T}} C_{\mathrm{D}}^{-1}(Gm - d_{obs}),
\end{aligned}
\tag{3.98}
$$

where

$$
C_{\mathrm{D}} = \begin{bmatrix} \sigma_{d,1}^2 & 0 & \cdots & 0 \\ 0 & \sigma_{d,2}^2 & & 0 \\ \vdots & & \ddots & \\ 0 & 0 & & \sigma_{d,N}^2 \end{bmatrix}.
\tag{3.99}
$$

If the standard deviation of the model variables about m_0 is σ_m, then σ_m could be used to weight the terms in the second part of the objective function,

$$
S_2(m) = \frac{1}{2}\sum_i^M \left(\frac{m_i - m_{0,i}}{\sigma_{m,i}}\right)^2.
\tag{3.100}
$$

If all the data are equally accurate and all of the model variables are equally uncertain, then it is reasonable to let $\alpha = \sigma_d^2/\sigma_m^2$ in Eq. (3.97).

To investigate the behavior of this type of solution in greater depth, consider a slightly easier problem than considered previously. In this problem the data are direct measurements of the model variables in a one-dimensional model that is described by 40 model variables. Assume that the mean value of the variable field is approximately equal to 1.0, and the standard deviation of the model variables about the mean is 0.5. The true field is shown in Fig. 3.4. Note that the values vary between about 0.7 and 2, but that the average seems to be not too far from 1.0. Now, suppose that this is a reservoir and that wells have been drilled at $x = 6$, 19, and 32, so measurements of the model variables (permeability) are available at those locations. The manufacturer of the instrument

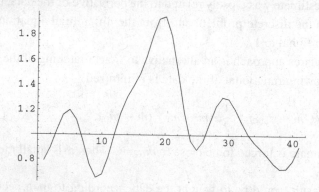

Figure 3.4. The true parameter field used to calculate data.

that measures permeability has estimated the accuracy of the instrument. It seems that the observed measurements are distributed normally about the true value and that the standard deviation of the observed values is $\sigma_d = 0.1$. Thus a reasonable estimate of the model variables can possibly be obtained by simply using $\alpha = 0.1^2/0.5^2 = 0.04$.

The solution (the value of m that minimizes Eq. (3.97)) can be written either as

$$m_{\text{est}} = m_0 + \left(\frac{\sigma_d^2}{\sigma_m^2}I_M + G^T G\right)^{-1} G^T(d_{\text{obs}} - Gm_0)$$

$$= m_0 + \frac{\sigma_m^2}{\sigma_d^2}\left(I_M + \frac{\sigma_m^2}{\sigma_d^2}G^T G\right)^{-1} G^T(d_{\text{obs}} - Gm_0) \qquad (3.101)$$

or as

$$m_{\text{est}} = m_0 + G^T \left(\frac{\sigma_d^2}{\sigma_m^2}I_{N_d} + GG^T\right)^{-1} (d_{\text{obs}} - Gm_0). \qquad (3.102)$$

Either formula can be used, but the second is more useful when the number of data is small, as it is in this problem.

The sensitivity matrix for this problem is

$$G = \begin{bmatrix} 0 & 0 & 0 & 0 & 0 & 1 & 0 \\ 0 & 0 & 0 & 0 & 0 & 0 & 0 \\ 0 & 0 & 0 & 0 & 0 & 0 & 0 \end{bmatrix} \cdots , \qquad (3.103)$$

where there is a 1 in the first row at the sixth column, in the second row at the 19th column, and in the third row at the 32nd column. GG^T is a 3×3 identity matrix so calculation of the solution in Eq. (3.102) is trivial. Figure 3.5 shows a weighted least-squares solution of this problem computed using $\alpha = 0.04$ which seems to be the proper value. The estimate is quite good at the locations of the measurements ($x = 6$, 19, and 32) but does not look at all like the truth. Perhaps it is not reasonable to expect to obtain a solution that looks like the truth from these three measurements, but the estimate from Eq. (3.102) is still unsatisfying because it is discontinuous. In fact, it

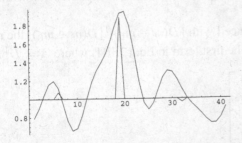
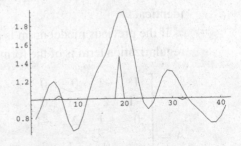

Figure 3.5. Two model estimates using only the ratio of variance as a measure of closeness to the prior. On the left, the estimate uses only the value of α predicted from the ratio of variances. On the right, the value of α is ten times larger.

takes the value m_0 everywhere except at the data locations. One explanation is that the estimate (or more accurately, the difference between the estimate and m_0) is a linear combination of the sensitivity functions, which happen to be delta functions.

It is not necessary to be limited to the simple least-squares weighting of Eq. (3.100). It is more general to use

$$S(m) = \frac{1}{2}\left[(m - m_0)^{\mathrm{T}}A(m - m_0) + (Gm - d_{\mathrm{obs}})^{\mathrm{T}}C_{\mathrm{D}}^{-1}(Gm - d_{\mathrm{obs}})\right], \tag{3.104}$$

where A is a symmetric, positive-definite matrix (i.e. one for which $x^{\mathrm{T}}Ax > 0$ for all nonzero vectors x) and $C_{\mathrm{D}} = \sigma_d^2 I_{N_d}$, in which case the value of m that minimizes 3.104 can be shown to be

$$m_{\mathrm{est}} = m_0 + \left(A + G^{\mathrm{T}}C_{\mathrm{D}}^{-1}G\right)^{-1}G^{\mathrm{T}}C_{\mathrm{D}}^{-1}(d_{\mathrm{obs}} - Gm_0). \tag{3.105}$$

To avoid the possibility of an estimate that is spatially discontinuous, we might try to minimize the magnitude of a derivative of the parameter field, or at least a finite-difference approximation to the derivative. To force the derivative of the estimate $f(x)$ to be small, we could minimize the integral of the square of the derivative, that is $\int f'(x)^2 \, dx$ while simultaneously trying to honor the data. For a discrete problem, the derivative of a parameter field can be approximated by applying the $(M - 1) \times M$ matrix D

$$D = \begin{bmatrix} -1 & 1 & 0 & 0 & \cdots & 0 \\ 0 & -1 & 1 & 0 & \cdots & 0 \\ 0 & 0 & -1 & 1 & & 0 \\ \vdots & \vdots & & \ddots & \ddots & \vdots \\ 0 & 0 & 0 & \cdots & -1 & 1 \end{bmatrix} \tag{3.106}$$

to the model. Letting $d_{i,j}$ denote the element in the ith row and jth column of D, Eq. (3.106) indicates that $d_{i,i} = -1$ and $d_{i,i+1} = 1$ for $i = 1, 2, \ldots, M - 1$, with all other entries of D equal to zero. Note that if $\|Dm\|^2 = 0$, then $Dm = 0$ which implies $-m_i + m_{i+1} = 0$ for $i = 1, 2, \ldots, M - 1$, i.e. all components of m must be identical.

If the previous model norm is replaced with $[D(m - m_0)]^{\mathrm{T}}[D(m - m_0)]$ the model regularization term is of the form of the first term in Eq. (3.104) where $A = D^{\mathrm{T}}D$:

$$A = \begin{bmatrix} 1 & -1 & 0 & 0 & \cdots & 0 \\ -1 & 2 & -1 & 0 & \cdots & 0 \\ 0 & -1 & 2 & -1 & \cdots & 0 \\ 0 & 0 & -1 & 2 & \ddots & \vdots \\ \vdots & \vdots & \vdots & \ddots & \ddots & -1 \\ 0 & 0 & 0 & \cdots & -1 & 1 \end{bmatrix}. \tag{3.107}$$

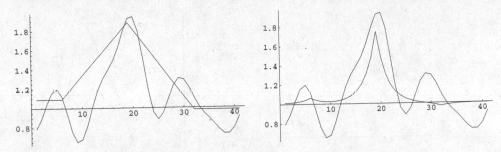

Figure 3.6. Two estimates using the size of the derivative as a measure of closeness to the prior. For the estimate on the left $A = D^T D$. On the right, $A = D^T D + 0.2I$.

This matrix is clearly singular because the columns are linearly dependent. (The sum of all the columns is the zero vector). It is not necessary to invert A in order to apply Eq. (3.105), however, and the sum $A + G^T C_D^{-1} G$ is invertible in this example. The least-squares solution that simultaneously minimizes the derivative and the data misfit is shown in Fig. 3.6. This estimate looks far different from the estimate that used only the variance estimates for the data and the model parameter. Note that the solution linearly interpolates between data locations and that the derivative is zero outside the data region. Also, note that the solution is not differentiable at the data locations. If a second regularization term of the type $\alpha \, (m - m_0)^T (m - m_0)$ is added, a quite different solution is obtained, as shown in the right-hand side of Fig. 3.6.

An estimate that is smooth even at data locations (i.e. differentiable everywhere) can be obtained by minimizing the second derivative of the model, or $[W(m - m_0)]^T W(m - m_0)$ where W is the following matrix approximation to the second derivative operator.

$$W = \begin{bmatrix} -1 & 2 & -1 & 0 & \cdots & 0 \\ 0 & -1 & 2 & -1 & \cdots & \\ \vdots & & \ddots & \ddots & \ddots & \\ 0 & & & -1 & 2 & -1 \end{bmatrix}. \tag{3.108}$$

The result with only the second derivative as a regularization term is shown on the left-hand side of Fig. 3.7. In this case, the estimate is very smooth, and approximately honors the data, but outside the range of the data, the estimate extrapolates as a straight line. When a second regularization term is added, which minimizes the size of the model, the result is quite different (right-hand side of Fig. 3.7). The estimate is still quite smooth, but reduces to the prior estimate far from the data.

3.3 Singular value decomposition

The singular value decomposition is sometimes useful for construction of approximate solutions to an inverse problem. Any $N \times M$ sensitivity matrix G can be decomposed

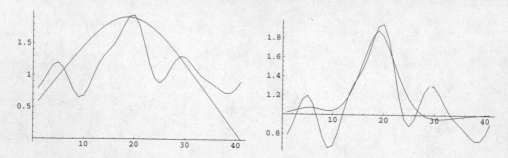

Figure 3.7. Two estimates using the size of the second derivative as a measure of closeness to the prior. For the estimate on the left the weighting matrix in the regularization term is $A = W^T W$. On the right, $A = W^T W + 0.2I$.

into a product of three matrices

$$G = U \Lambda V^T, \tag{3.109}$$

where U is $N \times N$ and orthogonal, i.e.

$$UU^T = U^T U = I_N. \tag{3.110}$$

Similarly, V is $M \times M$ and orthogonal, i.e.,

$$VV^T = V^T V = I_M. \tag{3.111}$$

The matrix Λ is $N \times M$ with nonzero elements only on the diagonal,

$$
\Lambda =
\begin{bmatrix}
\lambda_1 & 0 & \cdots & 0 & 0 & \cdots & 0 \\
0 & \lambda_2 & & 0 & 0 & \cdots & 0 \\
\vdots & & \ddots & \vdots & \vdots & & \vdots \\
0 & 0 & \cdots & \lambda_N & 0 & \cdots & 0
\end{bmatrix},
\tag{3.112}
$$

where for simplicity, we have assumed that $M > N$, which is the normal situation for inverse problems. The λ_i are called the *singular values* of the matrix G. The columns of U are called the *left singular vectors* and the columns of V are called the *right singular vectors* of G.

Consider the symmetric $N \times N$ matrix formed from the product of G with its transpose:

$$
\begin{aligned}
GG^T &= U \Lambda V^T (U \Lambda V^T)^T \\
&= U \Lambda V^T V \Lambda^T U^T \\
&= U \Lambda \Lambda^T U^T.
\end{aligned}
\tag{3.113}
$$

Note that $\Lambda \Lambda^T$ is an $N \times N$ diagonal matrix. Eq. (3.113) says that the columns of U are eigenvectors of GG^T and the singular values of G are the square roots of the eigenvalues of GG^T.

Reversing the order of the product of G with its transpose, we obtain a symmetric $M \times M$ matrix:

$$
\begin{aligned}
G^{\mathrm{T}}G &= (U\Lambda V^{\mathrm{T}})^{\mathrm{T}} U \Lambda V^{\mathrm{T}} \\
&= V\Lambda^{\mathrm{T}} U^{\mathrm{T}} U \Lambda V^{\mathrm{T}} \\
&= V\Lambda^{\mathrm{T}} \Lambda V^{\mathrm{T}},
\end{aligned}
\tag{3.114}
$$

where $\Lambda^{\mathrm{T}}\Lambda$ is an $M \times M$ diagonal matrix. Eq. (3.114) says that the columns of V are eigenvectors of $G^{\mathrm{T}}G$ and the singular values of G are the square roots of the eigenvalues of $G^{\mathrm{T}}G$.

In many cases of practical interest, the number, p, of nonzero singular values is less than N in which case the decomposition matrices can be further decomposed,

$$
G = U\Lambda V^{\mathrm{T}}
$$

$$
= \begin{bmatrix} U_p & \vdots & U_0 \end{bmatrix}
\begin{bmatrix} \Lambda_p & \vdots & 0 \\ \cdots & \vdots & \cdots \\ 0 & \vdots & 0 \end{bmatrix}
\begin{bmatrix} V_p^{\mathrm{T}} \\ \cdots \\ V_0^{\mathrm{T}} \end{bmatrix}
$$

$$
= U_p \Lambda_p V_p^{\mathrm{T}},
\tag{3.115}
$$

where U_p is $N \times p$ and $U_p^{\mathrm{T}}U_p = I_p$ but $U_p U_p^{\mathrm{T}} \neq I_N$; Λ_p is a $p \times p$ diagonal matrix whose elements are the square roots of the nonzero eigenvalues of GG^{T}. V_p is $M \times p$ and $V_p^{\mathrm{T}}V_p = I_p$ but $V_p V_p^{\mathrm{T}} \neq I_M$.

Consider the product of G with the matrix of columns of V_0. Recall that $G = U\Lambda V^{\mathrm{T}}$ so we can begin by examining the product of V^{T} with V_0:

$$
V^{\mathrm{T}}V_0 = \begin{bmatrix} V_p^{\mathrm{T}} \\ \cdots \\ V_0^{\mathrm{T}} \end{bmatrix} V_0 = \begin{bmatrix} V_p^{\mathrm{T}}V_0 \\ \cdots \\ V_0^{\mathrm{T}}V_0 \end{bmatrix} = \begin{bmatrix} 0 & 0 \\ \cdots & \cdots \\ I & 0 \end{bmatrix},
\tag{3.116}
$$

where all elements in the first p rows are 0 because of the orthogonality of V. Now we note that only the first p columns of Λ contain nonzero elements, so the product $\Lambda V^{\mathrm{T}}V_0$ is equal to zero. This is equivalent to the statement that

$$
GV_0 = 0
\tag{3.117}
$$

so the columns of V_0 are in the null space of G.

Example

Let us consider again the inverse problem of estimating inverse permeability from the acquisition of two independent measurements of pressure drop in steady linear flow. The sensitivity matrix for this problem that was discussed on page 28 was

$$G = \begin{bmatrix} 1 & 1 & 1 & 0 & 0 \\ 1 & 1 & 1 & 1 & 1 \end{bmatrix}. \tag{3.118}$$

Begin by computing the eigenvalues and eigenvectors of

$$GG^{T} = \begin{bmatrix} 3 & 3 \\ 3 & 5 \end{bmatrix}. \tag{3.119}$$

The eigenvalues of GG^{T} are approximately $7.162\,28$ and $0.877\,22$. The eigenvectors are the columns of

$$\begin{bmatrix} -0.584\,71 & -0.811\,242 \\ -0.811\,242 & 0.584\,71 \end{bmatrix}. \tag{3.120}$$

Similarly, we can compute the eigenvalues and eigenvectors of

$$G^{T}G = \begin{bmatrix} 2 & 2 & 2 & 1 & 1 \\ 2 & 2 & 2 & 1 & 1 \\ 2 & 2 & 2 & 1 & 1 \\ 1 & 1 & 1 & 1 & 1 \\ 1 & 1 & 1 & 1 & 1 \end{bmatrix}. \tag{3.121}$$

The nonzero eigenvalues of $G^{T}G$ are the same as the nonzero eigenvalues of GG^{T}. The eigenvectors of $G^{T}G$ are the columns of

$$\begin{bmatrix} -0.521\,609 & 0.247\,502 & 0.786\,332 & 0.055\,2388 & 0 \\ -0.521\,609 & 0.247\,502 & -0.583\,591 & 0.677\,867 & 0 \\ -0.521\,609 & 0.247\,502 & -0.202\,741 & -0.733\,106 & 0 \\ -0.303\,127 & -0.638\,838 & 0 & 0 & -0.707\,107 \\ -0.303\,127 & -0.638\,838 & 0 & 0 & 0.707\,107 \end{bmatrix}. \tag{3.122}$$

The singular value decomposition of G provides the square roots of the eigenvalues and a combination of eigenvectors from G^TG and from GG^T.

$$G = \begin{bmatrix} 0.5847 & -0.8112 \\ 0.8112 & 0.5847 \end{bmatrix} \begin{bmatrix} \sqrt{7.162} & 0 & 0 & 0 & 0 \\ 0 & \sqrt{0.8377} & 0 & 0 & 0 \end{bmatrix}$$

$$\times \begin{bmatrix} -0.5216 & -0.5216 & -0.5216 & -0.3031 & -0.3031 \\ 0.2475 & 0.2475 & 0.2475 & -0.6388 & -0.6388 \\ 0.7863 & -0.5836 & -0.2027 & 0 & 0 \\ 0.0552 & 0.6779 & -0.7331 & 0 & 0 \\ 0 & 0 & 0 & -0.7071 & 0.7071 \end{bmatrix} \qquad (3.123)$$

$$= \begin{bmatrix} 0.5847 & -0.8112 \\ 0.8112 & 0.5847 \end{bmatrix} \begin{bmatrix} 2.6762 & 0 \\ 0 & 0.9153 \end{bmatrix}$$

$$\times \begin{bmatrix} -0.5216 & -0.5216 & -0.5216 & -0.3031 & -0.3031 \\ 0.2475 & 0.2475 & 0.2475 & -0.6388 & -0.6388 \end{bmatrix}.$$

Note that the last three rows of V^T in the decomposition of G are clearly elements of the null space of G as they should be. Also, note that it was unnecessary to use those rows to reconstruct G.

Application of the singular value decomposition

If there are no zero singular values and Λ is square, then the solution to the linear inverse problem is

$$m = G^{-1}d_{obs} = V\Lambda^{-1}U^Td_{obs}. \qquad (3.124)$$

Of course, it is trivial to invert Λ in this case as it is a diagonal matrix. In general, for real problems we either find that some of the singular values are zero or that they are zero to the precision of the computer. In this case, we might consider the solution

$$m_{est} = V_p\Lambda_p^{-1}U_p^Td_{obs}. \qquad (3.125)$$

This is a particularly interesting solution as it has two useful properties. First, the estimate has no component from the null space of G. This can be shown by taking the product of the matrix of basis vectors for the null space (V_0^T) with m_{est}.

$$V_0^Tm_{est} = V_0^T(V_p\Lambda_p^{-1}U_p^Td_{obs}) = 0 \qquad (3.126)$$

because the columns of V_0 are orthogonal to the columns of V_p. Another way to see that the estimate has no component from the null space of G is to observe that m_{est} is constructed from a linear combination of the columns of V_p. This is obvious from Eq. (3.125).

The second useful property is that the residual prediction error from Eq. (3.125) can not be further reduced by any combination of model variables. In other words, no solution can do better at reducing the data mismatch. We verify this property by considering the product of the basis vectors for the data space, U_p^T, with the prediction error.

$$
\begin{aligned}
U_p^T e &= U_p^T[G m_{est} - d_{obs}] \\
&= U_p^T[(U_p \Lambda_p V_p^T)(V_p \Lambda_p^{-1} U_p^T d_{obs}) - d_{obs}] \\
&= U_p^T[(U_p U_p^T d_{obs}) - d_{obs}] \\
&= U_p^T d_{obs} - U_p^T d_{obs} = 0.
\end{aligned}
\tag{3.127}
$$

Example

On page 52 we computed the singular value decomposition of the sensitivity matrix for the problem of two independent measurements of pressure drop in steady linear flow. Let us use this decomposition in Eq. (3.125) to obtain an estimate for the distribution of inverse permeability. The data for this problem were provided on page 37.

$$
\begin{aligned}
m_{est} &= V_p \Lambda_p^{-1} U_p^T d_{obs} \\
&= \begin{bmatrix} -0.5216 & 0.2475 \\ -0.5216 & 0.2475 \\ -0.5216 & 0.2475 \\ -0.3031 & -0.6388 \\ -0.3031 & -0.6388 \end{bmatrix} \begin{bmatrix} 1/2.6762 & 0 \\ 0 & 1/0.9153 \end{bmatrix} \begin{bmatrix} 0.5847 & -0.8112 \\ 0.8112 & 0.5847 \end{bmatrix} \begin{bmatrix} 3.95 \\ 6.30 \end{bmatrix} \\
&= \begin{bmatrix} 1.317 \\ 1.317 \\ 1.317 \\ 1.175 \\ 1.175 \end{bmatrix},
\end{aligned}
\tag{3.128}
$$

which is the same solution that we obtained in Eq. (3.64) using the formula for purely underdetermined problems.

Small, nonzero singular values

To this point we have considered problems for which the partitioning of the decomposition of G into components that span the null space and components that are necessary for construction of an estimate was straightforward – we simply chose the singular values associated with nonzero singular values. Let us consider a decomposition of G such that the singular values are ordered from largest to smallest and then consider the effect of errors in the data on the model estimate provided by Eq. (3.125). Denote the errors in the observations by e, so

$$
d_{obs} = G m + e = U_p \Lambda_p V_p^T m + e.
\tag{3.129}
$$

There is no approximation in this statement as long as we include all nonzero singular values and associated singular vectors. Let

$$
\begin{aligned}
m_{\text{est}} &= V_p \Lambda_p^{-1} U_p^{\mathrm{T}} d_{\text{obs}} \\
&= V_p \Lambda_p^{-1} U_p^{\mathrm{T}} (U_p \Lambda_p V_p^{\mathrm{I}} m + e) \\
&= V_p V_p^{\mathrm{T}} m + V_p \Lambda_p^{-1} U_p^{\mathrm{T}} e \\
&= V_p V_p^{\mathrm{T}} m + \begin{bmatrix} & \\ V_p & \\ & \end{bmatrix} \begin{bmatrix} (e . u_{p,1})/\lambda_1 \\ \vdots \\ (e . u_{p,p})/\lambda_p \end{bmatrix}.
\end{aligned}
\tag{3.130}
$$

Note that the random errors are divided by the singular values. Small singular values can contribute large errors to the estimate. If the matrix G is square and $p = M$, then $V_p V_p^{\mathrm{T}} = I$ and the model resolution is as good as possible. Reducing the number of singular values and singular vectors reduces the ability to reproduce the true model since the columns of V_p will not span the model space. On the other hand, by including the singular vectors associated with small singular values, we run the risk of adding large error components to the estimate. It is often desirable to select a value of p that eliminates small but nonzero singular values for the construction of the estimate.

3.4   Backus and Gilbert method

Although almost all practical inverse problems, of any importance, must, in the end, be discretized so that the partial differential equations representing the system can be solved, or so that integrals can be evaluated, it is useful to begin by considering problems in which the unknown is a *function*. This helps prevent the common misconception that the number of variables should be kept relatively small in order to obtain a meaningful solution. In fact, the reverse is probably true. The number of variables should usually be made as large as computationally possible. It is, of course, always necessary to discretize the permeability field in a petroleum reservoir in order to solve the equations of fluid flow and transport.

The Backus–Gilbert approach [14] is useful for discussing the infinite-dimensional problem. Assume that some data d are related to a model m by a linear relationship:

$$
d_i = (G_i, m)_{\mathrm{M}} \qquad i = 1, N,
\tag{3.131}
$$

where the parentheses indicate an inner product and the subscript indicates that this is an inner product on the model space. Examples include well test data for small variation in permeability, mass of the Earth, and the inverse Laplace transform.

> **Remark 3.3.** In the density of the Earth problem (Section 2.1), the first datum, d_1, was the mass of the Earth; the model, m, was the density as a function of position; the first data kernel, $G_1 = 4\pi r^2$, and the inner product on the model space was an integral from the center of the Earth to the outer radius, i.e. $(f, h)_M = \int_0^a f(r)h(r)\,dr$.

Generalized linear prediction is the estimation of some property of the model that is related to the model as follows.

$$\psi = (\Psi, m)_M. \tag{3.132}$$

Examples might be the average value of the model over some region (such as a gridblock in a simulator model) or the value at a particular point. In this chapter, we will focus on the problem of estimation of the value of a function at a particular point. Consider the estimation of the value of m at a location r_0 where $m(r_0)$ is related to $m(r)$ by the inner product

$$m(r_0) = (\delta(r - r_0), m(r))_M. \tag{3.133}$$

Because the functional in (3.131) is linear, linear combinations of the data are related to linear combinations of the data kernels, G_i, as follows:

$$\sum_{i=1}^N a_i d_i = \sum_{i=1}^N a_i (G_i, m)_M = \left(\sum_{i=1}^N a_i G_i, m \right)_M. \tag{3.134}$$

If we can find a set of coefficients a_i such that $\sum_1^N a_i G_i = \hat{\delta}(r - r_0)$, an approximation of the delta function, then an estimate of the model, $m(r_0)$, can be constructed as follows,

$$
\begin{aligned}
\sum_{i=1}^N a_i d_i &= \left(\sum_{i=1}^N a_i G_i, m \right)_M \\
&= \left(\hat{\delta}(r - r_0), m \right)_M \\
&= \widehat{m}(r_0).
\end{aligned}
\tag{3.135}
$$

The goal, then, is to find a set of a_i such that $\hat{\delta}$ looks most like δ, in which case our estimate of m might be expected to be close to the true value. This assumption must be evaluated carefully because it impacts the interpretation of our estimates.

Consider the following geometrical analogy: given a vector v, find the vector $\hat{v}$ that looks most like v in a plane A. If our criterion for being close is the minimization of the length of the difference vector then the best approximation to v in the plane A is found by taking the projection of v onto A (see Fig. 3.8).

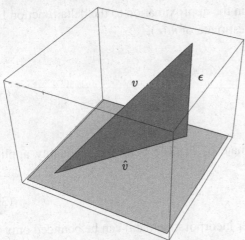

Figure 3.8. The projection of the true vector, v, onto the subspace spanned by the data kernels.

Note that the error ϵ is perpendicular to the plane A. For most inverse problems, the data are limited and hence so are the data kernels, G_i. Our estimates of m are restricted to lie within the subspace of M (the Hilbert space containing m) spanned by the data kernels. The difference between $\hat{m}$ and m is outside the data space, i.e. it is orthogonal to the data subspace just as ϵ is orthogonal to $\hat{v}$. The fundamental problem is to estimate the magnitude of ϵ after we have estimated $\hat{v}$. Unfortunately, without additional information there is no way to estimate ϵ; it is arbitrary.

For the inverse problem, the projection of m onto the data space is spanned by the G_i, $i = 1, N$, and the approximation to the delta function is a linear combination of the G_i. Denote the difference between the delta function and the approximation to the delta function constructed as a linear combination of data kernels as e, i.e.

$$\delta = \hat{\delta} + e = \sum_{n=1}^{N} a_n G_n + e. \tag{3.136}$$

The magnitude of e is just the norm of the difference between the delta function and its approximation:

$$\|e\| = \|\delta - \hat{\delta}\| = \left\| \delta - \sum_{n=1}^{N} a_n G_n \right\|. \tag{3.137}$$

The best approximation to δ is found by varying the a_i until the difference is minimized. Let Q be the minimum value of the difference, i.e.

$$Q \equiv \min \|e\|. \tag{3.138}$$

Eq. (3.138) formally gives the error in the approximation to the delta function but we really want to bound the error in the estimate of $m(r)$,

$$
\begin{aligned}
m &= (\delta, m) \\
&= (\hat{\delta}, m) + (\delta - \hat{\delta}, m) \\
&= (\hat{\delta}, m) + (e, m) \\
&= \hat{m} + (e, m).
\end{aligned}
\tag{3.139}
$$

Thus (e, m) is the error in the estimate of m. The Schwartz inequality applied to Eq. (3.139) gives

$$
|m - \hat{m}| \leq \|e\| \, \|m\| = Q \|m\|.
\tag{3.140}
$$

So the error in the estimate of m (after incorporating data) can be bounded only if we have a bound on the norm of m. Referring back to Fig. 3.8, the magnitude of the error in the estimate, ϵ, can only be bounded if the length of the vector v can be bounded.

3.4.1   Least-squares estimation for accurate data

What does it mean for $\sum_1^N a_i G_i$ to be close to $\delta(r - r_0)$? First we would have to choose a norm to measure the distance between two functions. It is convenient (but not always appropriate) to choose ℓ_2 as the norm. Backus and Gilbert [14] suggested several measures of "δ-ness." The most obvious measure is $\|\delta(r - r_0) - \hat{\delta}(r - r_0)\|$, in which case, the a_i can be determined by minimizing the misfit in the following functional

$$
Q^2 = \int_\Omega \left[\delta(r - r_0) - \sum_{i=1}^N a_i(r_0) G_i(r) \right]^2 dr
\tag{3.141}
$$

with respect to variation in the coefficients of the data kernels, a_i. Thus

$$
\begin{aligned}
\frac{\partial Q^2}{\partial a_i} &= \frac{\partial}{\partial a_i} \int_\Omega \left[\delta(r - r_0) - \sum_{i=1}^N a_i(r_0) G_i(r) \right]^2 dr \\
&= -2 G_i(r_0) + 2 \int_\Omega \left[G_i(r) \sum_{k=1}^N a_k(r_0) G_k(r) \right] dr
\end{aligned}
\tag{3.142}
$$

and at the minimum,

$$
\frac{\partial Q^2}{\partial a_i} = 0,
\tag{3.143}
$$

so we obtain

$$
G_i(r_0) = \sum_{k=1}^N a_k(r_0) \int_\Omega G_i(r) G_k(r) \, dr,
\tag{3.144}
$$

which is a system of N equations to be solved for the unknown a_i. In matrix notation, this can be written

$$
\begin{bmatrix} G_1(r_0) \\ \vdots \\ G_N(r_0) \end{bmatrix} = \begin{bmatrix} \int G_1 G_1 \, dr & \cdots & \int G_1 G_N \, dr \\ \vdots & \ddots & \vdots \\ \int G_1 G_N \, dr & \cdots & \int G_N G_N \, dr \end{bmatrix} \begin{bmatrix} a_1(r_0) \\ \vdots \\ a_N(r_0) \end{bmatrix}. \tag{3.145}
$$

Recall that these coefficients can be used in Eq. (3.135) to estimate the value of m at r_0, i.e.

$$
\hat{m}(r_0) = \sum_{i=1}^{N} a_i(r_0) d_i. \tag{3.146}
$$

Summary for Backus–Gilbert estimate

Our answer to the inverse problem can be interpreted in two ways: (1) $\langle m(r_0) \rangle$ is a localized average of the true m. If we filter the true m with $\hat{\delta}$ then we will obtain $\langle m(r_0) \rangle$. Of all possible filters the one we derived has the narrowest peak width. (2) $\langle m(r_0) \rangle$ is an estimate of the true value of m at r_0. In this interpretation, however, the bound on the error of the estimate is infinite.

Example: computation of the density of Earth

Reconsider the problem introduced in Section 2.1 – estimation of the density distribution inside the Earth from the mass and the rotational moment of inertia. The mass of the Earth is known to be approximately 5.976×10^{24} kg and the rotational moment of inertia is approximately 8.04449×10^{37} kg m^2, so from Eqs. (2.1) and (2.2),

$$
d_1 = \int_0^a (4\pi r^2) \rho(r) \, dr = 5.976 \times 10^{24} \text{ kg} \tag{3.147}
$$

and

$$
d_2 = \int_0^a \left(\frac{8\pi}{3} r^4 \right) \rho(r) \, dr
$$
$$
= 8.044\,49 \times 10^{37} \text{kg m}^2. \tag{3.148}
$$

The data kernels are

$$
G_1 = 4\pi r^2 \quad \text{and} \quad G_2 = \frac{8\pi}{3} r^4. \tag{3.149}
$$

The Backus–Gilbert solution to the estimation problem is given by Eq. (3.135) with the coefficients found by solving Eq. (3.144), which is repeated here in a slightly different form:

$$
a_k(r_0) = \sum_{i=1}^{N} G_i(r_0) \left[\int_\Omega G_i(r) G_k(r) \, dr \right]^{-1}. \tag{3.150}
$$

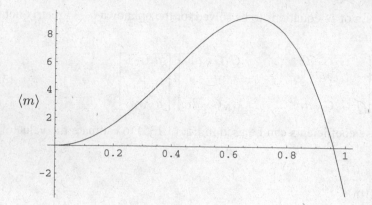

Figure 3.9. Estimated density of the Earth in g/cm^3 as a function of normalized radius, r/a.

The inner product of the data kernels can be calculated exactly for this problem:

$$\begin{bmatrix} (G_1, G_1) & (G_1, G_2) \\ (G_2, G_1) & (G_2, G_2) \end{bmatrix} = \begin{bmatrix} \int_0^a 16\pi^2 r^4 \, dr & \int_0^a (32\pi^2 r^6/3) \, dr \\ \int_0^a (32\pi^2 r^6/3) \, dr & \int_0^a (64\pi^2 r^8/9) \, dr \end{bmatrix}$$

$$= 16\pi^2 a^5 \begin{bmatrix} 1/5 & 2a^2/21 \\ 2a^2/21 & 4a^4/81 \end{bmatrix}. \tag{3.151}$$

Therefore, the coefficients of the data kernels for the estimation of density are

$$\begin{bmatrix} a_1(r) \\ a_2(r) \end{bmatrix} = \begin{bmatrix} (G_1, G_1) & (G_1, G_2) \\ (G_2, G_1) & (G_2, G_2) \end{bmatrix}^{-1} \begin{bmatrix} G_1(r) \\ G_2(r) \end{bmatrix}$$

$$= (64\pi^2 a^5)^{-1} \begin{bmatrix} 245 & -945/2a^2 \\ -945/2a^2 & 3969/4a^4 \end{bmatrix} \begin{bmatrix} 4\pi r^2 \\ 8\pi r^4/3 \end{bmatrix}, \tag{3.152}$$

where $a = 6.378 \times 10^6$ meters is the radius of the Earth.

The least-squares estimate of density based on mass and rotational inertia is

$$\langle \rho(r) \rangle = \begin{bmatrix} d_1 & d_2 \end{bmatrix} \begin{bmatrix} a_1(r) \\ a_2(r) \end{bmatrix}$$

$$= \begin{bmatrix} d_1 & d_2 \end{bmatrix} \begin{bmatrix} (G_1, G_1) & (G_1, G_2) \\ (G_2, G_1) & (G_2, G_2) \end{bmatrix}^{-1} \begin{bmatrix} G_1(r) \\ G_2(r) \end{bmatrix} \tag{3.153}$$

$$= 40\,620\,(r/a)^2 - 44\,040\,(r/a)^4$$

in units of kg/m^3. The estimate of density, $\langle \rho \rangle$, is plotted in Fig. 3.9 as a function of dimensionless radius.

If we interpret $\langle \rho \rangle$ as an estimate of the true density distribution, the result shown in Fig. 3.9 is disappointing. The density should not be negative at the surface of the Earth, nor should it be zero at the center. This "estimate" does, however, reproduce the data

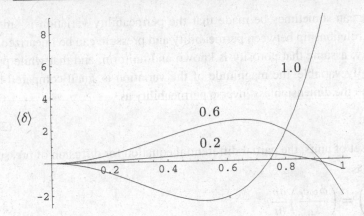

Figure 3.10. Resolution functions for the estimation of the density of the Earth at $r_0 = 0.2a, 0.6a,$ and $1.0a$.

exactly and in that sense it is a good estimate. Another way of viewing the solution is that it is a smoothed or filtered version of the true density distribution. To see what type of smoothing of the true distribution would result in $\langle \rho \rangle$ we must examine $\langle \delta(r - r_0) \rangle$, the approximation to the delta function at r_0.

Figure 3.10 shows resolution functions for the estimates of density at three different depths. The peak of the approximation to the delta function is close to the desired peak only for $r_0 = 0.6a$. Thus the estimates of density obtained are not averages of the density in the vicinity of the estimation location. Also, $\langle \delta \rangle$ is very wide in each case so the averaging or smoothing occurs over a large region of the Earth. Finally, note that $\langle \delta \rangle$ is approximately zero for the region near the center of the Earth. The estimates of density are insensitive to the true values in the deep interior (because the mass and the moment of inertia are insensitive to the density of that region).

It might seem that it would be easy to constrain the density to a particular value at the surface by constraining the density to equal the density of quartz at that location, i.e.

$$d_3 = \int_0^a \delta(r - a)\rho(r)\, dr \qquad (3.154)$$
$$= 2.65 \text{ gm/cm}^3,$$

but the data kernel for this type of measurement is not square integrable. The solution requires a different measure of the closeness to the delta function.

3.4.2   Example: estimation of permeability from well test data

The method of Backus and Gilbert has been applied to the problem of estimating a radial permeability distribution from well test data [15]. Although the problem is nonlinear, the assumption can sometimes be made that the permeability variation is small, in which case the relationship between permeability and pressure can be linearized.

For simplicity, assume that porosity is known and uniform, and that while permeability is spatially variable, the magnitude of the variation is small compared to the mean, k_0. Define the dimensionless inverse permeability as

$$f(r) = k_0/k(r).$$ (3.155)

In a consistent set of units, the partial differential equation for diffusion of pressure in 2D radial flow is

$$\frac{1}{r}\frac{\partial}{\partial r}\left(\frac{r}{f(r)}\frac{\partial p}{\partial r}\right) = \left(\frac{\phi \mu c_t}{k_0}\right)\frac{\partial p}{\partial t}.$$ (3.156)

The boundary condition at the well, for constant rate production is

$$\frac{\partial p}{\partial r} = -\left(\frac{qB\mu}{2\pi r_w k_0 h}\right)f(r_w),$$ (3.157)

where q is the production rate at surface conditions, B is the formation volume factor, μ is the fluid viscosity, r_w is the wellbore radius, and h is the thickness of the reservoir. We also assume that the pressure is initially uniform and equal to p_i. It is convenient to define the following dimensionless quantities:

$$p_D(r, t) = \frac{2\pi k_0 h(p_i - p(r, t))}{qB\mu}$$

$$t_D = \frac{k_0 t}{\phi \mu c_t r_w^2}$$

$$r_D = \frac{r}{r_w}.$$

Dropping the subscript D on dimensionless variables, Eq. (3.156) can then be written as

$$\frac{1}{r}\frac{\partial}{\partial r}\left(\frac{r}{f(r)}\frac{\partial p}{\partial r}\right) = \frac{\partial p}{\partial t}$$ (3.158)

and the boundary condition at the well as

$$\frac{\partial p}{\partial r} = f(1).$$ (3.159)

Because the variation in permeability is assumed to be small, it is convenient to separate the function $f(r)$ into a part that is constant and a part that describes the heterogeneity. Let the heterogeneous part be $\epsilon F(r)$ where $\|F(r)\| = 1$ and $\epsilon \ll 1$ so that,

$$\frac{1}{f(r)} = \frac{1}{1 - \epsilon F(r)} = 1 + \epsilon F(r) + O(\epsilon^2)$$ (3.160)

and assume a similar expansion for the pressure,

$$p(r, t) = p_0(r, t) + \epsilon p_1(r, t) + O(\epsilon^2).$$ (3.161)

Substitution of Eqs. (3.160) and (3.161) into Eq. (3.158) gives

$$\frac{1}{r}\frac{\partial}{\partial r}\left(r[1+\epsilon F + O(\epsilon^2)]\frac{\partial[p_0+\epsilon p_1+O(\epsilon^2)]}{\partial r}\right) = \frac{\partial[p_0+\epsilon p_1+O(\epsilon^2)]}{\partial t}. \quad (3.162)$$

Separating terms by order in ϵ, the following system of equations is obtained:

$$O(1): \quad \frac{1}{r}\frac{\partial}{\partial r}\left(r\frac{\partial p_0}{\partial r}\right) - \frac{\partial p_0}{\partial t} = 0 \quad (3.163)$$

$$O(\epsilon): \quad \frac{1}{r}\frac{\partial}{\partial r}\left(r\frac{\partial p_1}{\partial r}\right) - \frac{\partial p_1}{\partial t} = \frac{1}{r}\frac{\partial}{\partial r}\left(rF\frac{\partial p_0}{\partial r}\right). \quad (3.164)$$

The boundary condition for Eq. (3.163) is $\partial p_0/\partial r = 1$ at $r = 1$ and the boundary condition for Eq. (3.164) is $\partial p_1/\partial r = -F(1)$ at $r = 1$. The (approximate) solution of Eq. (3.163) is given by the exponential integral function:

$$p_0 = -\frac{1}{2}\,\mathrm{Ei}\left(-\frac{r^2}{4t}\right). \quad (3.165)$$

After solving for p_0, it is possible to solve Eq. (3.164) for $p_1(r, t)$. The solution of Eq. (3.164) at time t_i can be written in the form of an integral equation

$$p_{1,i} = \int_1^{\infty} G(r, t_i)F(r)\,dr$$
$$= \int_1^{\infty} G_i(r)F(r)\,dr. \quad (3.166)$$

The data kernels, $G_i(r)$, in Eq. (3.166) specify the influence of permeability on measured pressures. For a constant rate drawdown test, $G_i(r)$ is large near the well and decreases approximately as $1/r$ until the radius of investigation is reached (see Fig. 3.11). Beyond that distance $G_i(r)$ vanishes very rapidly. This behavior is reasonable since pressure drawdown depends strongly on skin (permeability near the well) and is insensitive to permeability beyond the radius of investigation.

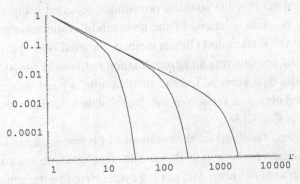

Figure 3.11. Well test data kernels for dimensionless time of 10^2, 10^4, and 10^6. (Adapted from Oliver [16], copyright SPE.)

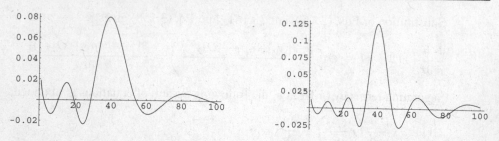

Figure 3.12. Resolving functions for estimation of permeability from well test data, using four data per log cycle (left) and eight data per log cycle (right).

The problem in well testing is to estimate the function $F(r)$ or, equivalently, the permeability given a finite set of pressures at times t_i for $i = 1, \ldots, N$. When only a finite data set is supplied, $F(r)$ can not be determined uniquely. Inaccuracy of the data adds further to the difficulty since it is possible that no function $F(r)$ will satisfy Eq. (3.166). Also, since the inverse problem is often unstable, the calculated solution may be highly oscillatory. The most common method of solving this problem (given the relationship shown in Eq. (3.166)) is simply to assume that $F(r)$ is piecewise constant, in which case we obtain a system of linear algebraic equations instead of an integral equation.

If only one value of permeability is to be estimated then, because the integral of $G_i(r)$ from $r = 1$ to $r = \infty$ is approximately $0.5(\ln t_i + c)$, the problem of estimation of permeability from well test data reduces to solving the following overdetermined system of equations for the average value of F.

$$p_{1,i} = 0.5 \bar{F}(\ln t_i + c).$$

In a real reservoir, the permeability almost certainly varies with location, and the problem of estimating that variation is of occasional importance. The method of Backus and Gilbert can be applied to this problem if the relationship between permeability at a point, and the pressure can be derived.

It is important to understand how the solution or estimate obtained using any procedure is related to other possible solutions of the permeability estimation problem. Recall that in this form of the Backus and Gilbert method, the goal is to optimize the resolution of the estimate by constructing an approximation to the delta function from a linear combination of the data kernels. Instead of obtaining a point estimate, the Backus and Gilbert method obtains a version of the "truth" which has been smoothed by a resolving function (see Eq. 3.134).

Figure 3.12 shows resolving functions for the estimation of permeability at a dimensionless radius of 40, based on pressure drawdown data. One example assumes four data per log cycle, the other assumes eight data per log cycle. Note that the approximation to a delta function is better (width is narrower) when eight data are used. Sidelobes are present in both cases.

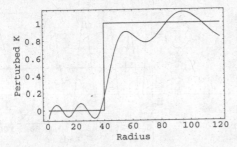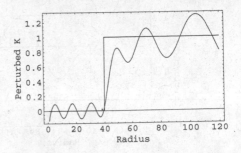

Figure 3.13. Estimation of permeability from well-test data, using four data per log cycle (left) and eight data per log cycle (right).

Variable permeability – accurate data

To illustrate the inversion method, a set of artificial pressure data were generated by solving Eq. (3.156) numerically for a composite radial reservoir. The permeability was assumed to be 1500 md for $1 < r < 40$ and 750 md for $r > 40$. All other properties were uniform. Since the sensitivities were computed assuming small perturbations to a uniform model, the pressure deviation from a uniform response was calculated and used for the estimation of permeability variation.

Eq. (3.145) was used to solve for the coefficients, a_i. The model estimates of inverse permeability were then computed using Eq. (3.146). The base model, from which the perturbation is calculated, is a constant 1500 md permeability throughout the reservoir. The "truth" is shown in Fig. 3.13 as a deviation from the 1500 md model. We see from Fig. 3.13 that the estimates of $F(r)$ are smoothed versions of the truth, and that the resolving functions appear to provide good estimators of the width of the region of smoothing.

Variable permeability – inaccurate data

In the previous example, the data was assumed to be completely accurate. Smoothing of the model occurred primarily because of the impossibility of constructing an exact delta function from a linear combination of the available data kernels, $G_i(r)$. Often, the lack of resolution of the pressure gauge imposes additional restrictions on the ability to invert the pressure data for estimation of the permeability distribution.

Assume that our measured pressure deviations, $p_{obs,i}$, differ from the true pressures, $p_{1,i}$, due to the presence of random, uncorrelated, zero-mean errors, ϵ_i, such that

$$p_{obs,i} = p_{1,i} + \epsilon_i. \tag{3.167}$$

Figure 3.14 shows the effect of noise in the data on the estimate. Clearly even small amounts of noise can have a large effect on the results.

Summary of well test example

In contrast to conventional well test analysis, the goal of which is to estimate a few parameters which characterize the reservoir, the Backus–Gilbert method applied here

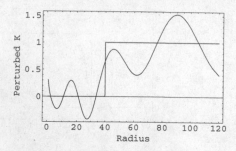
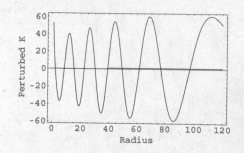

Figure 3.14. Estimation of permeability from noisy well test data, using four data per log cycle (left) and eight data per log cycle (right).

recognizes and accounts for the fact that the unknown permeability is a function of position. Of course, because the data are always limited in number, the problem of determining a map of permeability is inherently underdetermined. The "true" permeability distribution can not be determined from a finite set of data but it is possible to compute some universal properties which all solutions must share. In a crude sense, estimation of a permeability from the slope of a semilog plot accomplishes this; conventional analysis results in the calculation of an average permeability. The Backus and Gilbert approach outlined in this section provides a model which could be the truth (in the sense that it matches the data when the data is accurate) but, what is more important, it is known to be a smoothed version of the truth. In other words, filtering of any of the infinite number of permeability models which reproduce the pressure data with the resolving kernel would obtain the solutions shown in this section.

In general, it seems unlikely that permeability will vary in a radially symmetrical manner around a well, except perhaps in the near-wellbore region where the permeability has been altered by fines migration, filtrate invasion, or stimulation. It should be recognized, however, that unless other information is available, a radial permeability distribution is the best that can be obtained from a single well test. To increase the ability to map permeability requires the addition, into the inversion procedure, of pressure data from other wells. Information from interference tests can be fairly easily incorporated with the simple generalization that the perturbation to the pressure is assumed to be of the form

$$p(r, \theta, t) = \sum_{j=0}^{\infty} [\alpha_j(r, t) \cos j\theta + \beta_j(r, t) \sin j\theta].$$

The sensitivities of pressure in an interference test to general spatial variation in permeability and porosity were presented in Oliver [17]. For a single well test, only $\alpha_0(r, t)$ contributes to the estimate of permeability.

4 Probability and estimation

In this book we will consider several methods of generating solutions to the problem of estimating reservoir properties. Some solutions will honor the data exactly. Some solutions will be chosen because they possess special properties such as smoothness. Some solutions might be spatially discontinuous. We will find that some of the estimates of reservoir properties are not acceptable because they do not fit our preconceived notion of what the truth should be like. (This will be especially true when we use synthetic examples, in which case we know what the truth is supposed to look like.)

We require a way of representing our belief that some of the solutions are more likely to be correct than others, (1) either because they match the data better or (2) because they fit our notion of plausibility.

"Subjective probability" can be used to quantify the plausibility of models, and to represent our state knowledge. In Bayesian statistics, which is the philosophy we adopt, probability is used as a measure of uncertainty. It is legitimate, in this context, to assign probabilities to events that are not truly random. For example, we might assign a probability to the statement that Paris, France is further north than Montreal, Canada. We might discuss the probability that it will rain in Tulsa on December 13 of this year (which seems like it might be random), or we might discuss the probability that it rained on December 13, 1945. Although these seem reasonable (in a common sense way) it is difficult to imagine a frequency interpretation to this usage of probability.

In a similar vein, probability and randomness are used to describe the spatial distribution of rock properties in subsurface reservoirs, despite the fact that the laws that govern deposition, erosion, and diagenesis are deterministic. An appreciation of the limitations on predictability of the details of geology from physical laws can be gained from consideration of the evolution of stream meanders. The processes of erosion and deposition are highly nonlinear, so the evolution of meanders can be chaotic [18], with abrupt and unpredictable changes in the path. Detailed observations of meanders of the River Bollin in England over a three year period demonstrate the unpredictability of geologic processes (Fig. 4.1).

The goal for our application of probability to inverse problems is to be able to give a probabilistic answer to questions about future production. While it may not be meaningful to ask "What is the permeability at a particular location?" or "What will

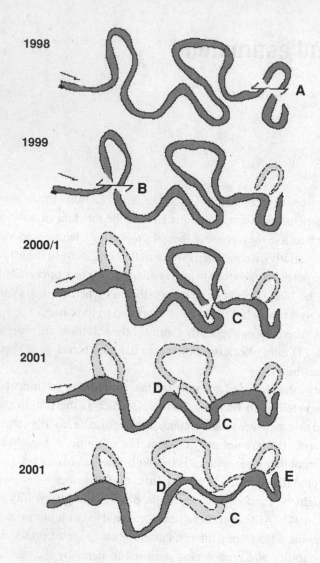

Figure 4.1. Evolution of a section of the River Bollin [18]. (Reprinted with permission of Blackwell Publishing who retain copyright.)

the production rate be for this well in 3 years?" it may be appropriate to ask "What is the probability that the permeability is between 50 and 100 md?" or "What is the probability that the production rate will exceed 150 m³/d in three years?"

The reservoir can be modeled as a chance outcome of complex geological processes. To characterize the reservoir for flow prediction, we need to know porosity, permeability, fluid properties, and saturations at every point in the reservoir. The language of random variables and geostatistics is useful for this purpose.

4.1   Random variables

Porosity and permeability are examples of reservoir properties that are sometimes modeled as continuous random variables. We say that a variable X is a continuous random variable if the probability of X taking the value x for any x is zero and we write this as

$$P(X = x) = 0, \qquad -\infty < x < \infty. \tag{4.1}$$

When this is the case, it is often more useful to speak of the probability density than the probability. A density is any nonnegative function f such that

$$\int_{-\infty}^{\infty} f(x)\, dx = 1. \tag{4.2}$$

If X is a continuous random variable having density f_X then the probability of X taking a value between a and b is

$$P(a \le X < b) = \int_{a}^{b} f_X(x)\, dx \tag{4.3}$$

and of course the probability of X taking a value in the interval $(-\infty, \infty)$ is

$$P(-\infty \le X \le \infty) = \int_{-\infty}^{\infty} f_X(x)\, dx = 1 \tag{4.4}$$

by Eq. (4.2). It is not difficult to extend this concept to several random variables such as porosity and permeability of a core plug, or the permeabilities of several core plugs from different depths. Let $X_1, \ldots, X_M$ be a set of random variables with density f_X such that the **joint probability** that a realization of the set lies within the region $a_1 \le X_1 < b_1, \ldots,$ and $a_M \le X_M < b_M$ is

$$P(a_1 \le X_1 < b_1 \text{ and} \ldots \text{and } a_M \le X_M < b_M)$$
$$= \int_{a_1}^{b_1} \cdots \int_{a_M}^{b_M} f_X(x_1, \ldots, x_M)\, dx_1 \ldots dx_M. \tag{4.5}$$

We will generally denote the set of random variables as a random vector, i.e. we typically let $X = [X_1, \ldots, X_M]^{\mathrm{T}}$ and write the joint probability that X is within the region A as

$$P(X \in A) = \int_{A} f_X(x)\, dx. \tag{4.6}$$

The **marginal probability** density for the random variable X_i is found by integrating the joint probability density over the remaining $M - 1$ variables. For example, the marginal density for the random variable X_1 is

$$f_{X_1}(x_1) = \int_{-\infty}^{\infty} \cdots \int_{-\infty}^{\infty} f_X(x_1, x_2, \ldots, x_M) \, dx_2 \, dx_3 \ldots dx_M. \tag{4.7}$$

Finally, we define the **conditional probability** density of X given $Y = y$ to be the probability density that X is equal to x when Y is equal to y. The conditional probability density is related to the joint and marginal densities:

$$f_{X|Y}(x|y) = \frac{f_{XY}(x, y)}{f_Y(y)}. \tag{4.8}$$

For reservoir characterization problems, we might be interested in the conditional probability density for the permeability at a particular location given a set of well-test data or we might be interested in estimating the porosity 50 feet from the location of a measurement. Although it may be reasonable to model the values of porosity as random, the values at nearby locations are not necessarily independent of each other.

From Eq. (4.8) we immediately see that

$$f_{XY}(x, y) = f_{X|Y}(x|y) f_Y(y) = f_{Y|X}(y|x) f_X(x). \tag{4.9}$$

When we are interested in the conditional probability of a variable given measurements of another variable, we rearrange the equations slightly to obtain:

$$f_{X|Y}(x|y) = \frac{f_{Y|X}(y|x) f_X(x)}{f_Y(y)}. \tag{4.10}$$

This result is called Bayes' rule. It provides the basis for computing the conditional probability of reservoir parameters from a combination of inaccurate measurements and a marginal probability density for reservoir parameters.

These relationships are also true when they are restricted by conditioning to other measurements. Eq. (4.8), for example, could be written

$$f_{X|YZ}(x|y, z) = \frac{f_{XY|Z}(x, y|z)}{f_{Y|Z}(y|z)}. \tag{4.11}$$

4.1.1   Example: permeability and porosity

Permeability–porosity trends based on core measurements from carbonates in the Wellington West field indicate several different trends depending on the facies.[1] The relationship between porosity and permeability for limestone and dolomite were found to be adequately described by the relationships

$$\ln K = \begin{cases} 18.9\phi - 0.29 & \text{(dolomite)} \\ 56.9\phi - 6.22 & \text{(limestone)}, \end{cases} \tag{4.12}$$

---

[1] The actual data for this field can be found in Bhattacharya *et al.* [19].

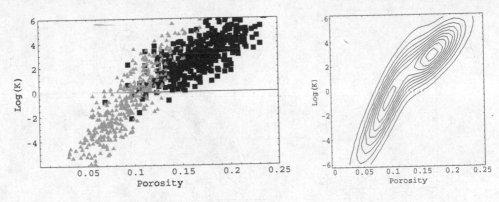

Figure 4.2. Core measurements of porosity and permeability from a reservoir with limestone (gray) and dolomite (black). An estimate of the joint probability density for porosity and permeability is on the right.

where K is the permeability in md and ϕ is the porosity. Even within a single rock type, these relationships are only approximate. Figure 4.2 shows a cross plot of porosity and permeability for a carbonate reservoir. The frequency of occurrence of values in various regions of the cross plot is an indication of the probability density for K and ϕ.

From this figure, it is apparent that the most probable combination of porosity and permeability is in the vicinity of $(0.17, \exp(3))$. There is another, smaller peak at $(0.09, \exp(-1))$. The first peak is due to the dolomite (which typically has high permeability) and the second peak to limestone (lower permeability). It is clear from the cross plot that there is a significant region for which plugs from limestone and dolomite have similar values of porosity and permeability.

The marginal probability densities for porosity and permeability can be computed using Eq. (4.7). The probability density for porosity $\Phi = \phi$ can be computed by simply integrating the joint density over all values of $\ln K$ evaluated at that value of ϕ. The result of this operation is shown in the lower right-hand side of Fig. 4.3. Alternatively, we could have estimated the marginal distribution for porosity from the original data base of core plug data, by simply counting the number of plugs whose porosity fall within certain ranges of values. In Fig. 4.3, we easily see the two peaks in the porosity distribution due to the differences in porosities of limestone and dolomite.

Figure 4.3 also shows the marginal probability density for log-permeability. If we were to select a core plug at random (regardless of porosity), this is how the values of log-permeability would be distributed. The computation of the marginal density for permeability is similar to the computation of the marginal density for porosity. We might ask if knowledge of the marginal densities is sufficient to estimate the joint density. Specifically, if the marginal probability densities for porosity and for log-permeability are known, can the joint distribution of porosity and log-permeability be

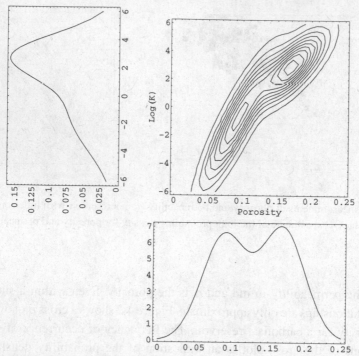

Figure 4.3. The joint density of porosity and log-permeability is shown on the upper right-hand. On the left is the marginal density for ln K, and below is the marginal density for ϕ.

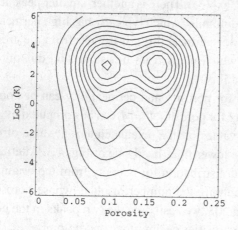

Figure 4.4. This joint probability density has the same marginal densities as the joint density shown in Fig. 4.2.

estimated for this example? The answer is no, as illustrated by Fig. 4.4, which shows a joint probability density for porosity and log-permeability that is much different from the correct joint probability density, but is entirely consistent with the marginal probability densities.

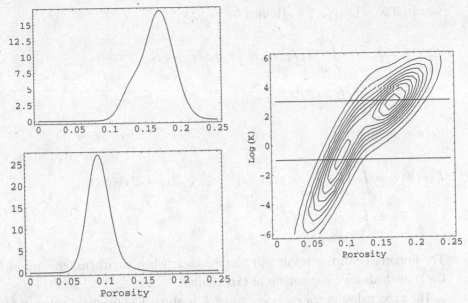

Figure 4.5. The conditional probability density for ϕ, given that $\ln K = 3$ (upper left), and for $\ln K = -1$ (lower left).

Suppose we selected a core plug at random and measured its log-permeability to be y_0. What is the probability density that this core plug's porosity is ϕ? This is a question about conditional probability. In this case we use Eq. (4.8) to compute the conditional probability density. It is not difficult – we simply normalize the joint density with $\ln K$ set to y_0. Figure 4.5 shows two examples: one for which the log-permeability was 3 and another for which the log-permeability was -1. Note that by providing a measurement of permeability, the uncertainty in the porosity is significantly less than it was from the marginal distribution (Fig. 4.3).

4.2   Expected values

It is frequently impractical to fully describe the probability density function for a random variable, so we instead compute summary statistics such as the mean, the median, and the variance to name a few. The expected value of a function g of a random variable X whose probability density is $f_X(x)$ is

$$E[g(X)] = \int_{-\infty}^{\infty} g(x) f_X(x) \, dx. \tag{4.13}$$

From a practical standpoint, it is useful to note that it is not necessary to know the joint distribution, in order to compute the expectation of one of the components. In other words, if $X = [X_1, \ldots, X_N]^T$, then

$$
\begin{aligned}
E[X_j] &= \int_{-\infty}^{\infty} \cdots \int_{-\infty}^{\infty} x_j f_X(x_1, \ldots, x_N) \, dx_1 \cdots dx_N \\
&= \int_{-\infty}^{\infty} x_j f_{X_j}(x_j) \, dx_j
\end{aligned}
\tag{4.14}
$$

or more generally,

$$
\begin{aligned}
E[g(X_j)] &= \int_{-\infty}^{\infty} \cdots \int_{-\infty}^{\infty} g(x_j) f_X(x_1, \ldots, x_N) \, dx_1 \cdots dx_N \\
&= \int_{-\infty}^{\infty} g(x_j) f_{X_j}(x_j) \, dx_j.
\end{aligned}
\tag{4.15}
$$

The importance of this result will become clear when we discuss the use of Monte Carlo methods for computations of expectations.

The expected value (or *expectation*) of X is also called the mean value and is often denoted μ_X. The expectation of the product of two random variables X and Y is

$$
E[XY] = \int_{-\infty}^{\infty} XY f_{XY}(x, y) \, dx \, dy.
\tag{4.16}
$$

Two random variables X and Y are said to be *independent* if

$$
f_{XY}(x, y) = f_X(x) f_Y(y)
\tag{4.17}
$$

so if X and Y are independent, then the expectation of the product is just the product of the expectations,

$$
E[XY] = E[X]E[Y].
\tag{4.18}
$$

Recall also from Eq. (4.10) that

$$
f_{XY}(x, y) = f_{X|Y}(x|y) f_Y(y) = f_{Y|X}(y|x) f_X(x)
\tag{4.19}
$$

so if X and Y are independent,

$$
f_{X|Y}(x|y) = f_X(x) \quad \text{and} \quad f_{Y|X}(y|x) = f_Y(y),
\tag{4.20}
$$

which means that the conditional probability density is the same as the prior (or marginal) probability density. A measurement of X does nothing to reduce the uncertainty in Y.

4.2.1   Variance, covariance, and correlation

The variance of a random variable X is defined in terms of the expectation:

$$
\begin{aligned}
\text{var}(X) &= E\big[(X - E[X])^2\big] \\
&= E\big[(X - \mu_X)^2\big] \\
&= E[X^2] - 2E[X\mu_X] + E[\mu_X^2] \\
&= E[X^2] - 2\mu_X E[X] + \mu_X^2 \\
&= E[X^2] - \mu_X^2 \\
&= \sigma_X^2
\end{aligned}
\tag{4.21}
$$

σ_X is the standard deviation of the random variable X.

Consider two random variables X and Y. The variance of the sum of X and Y can be computed as follows.

$$
\begin{aligned}
\text{var}(X + Y) &= E\big[((X + Y) - E[X + Y])^2\big] \\
&= E\big[((X + Y) - (\mu_X + \mu_Y))^2\big] \\
&= E\big[((X - \mu_X) + (Y - \mu_Y))^2\big] \\
&= E\big[(X - \mu_X)^2 + (Y - \mu_Y)^2 + 2(X - \mu_X)(Y - \mu_Y)\big] \\
&= \text{var}(X) + \text{var}(Y) + 2E\big[(X - \mu_X)(Y - \mu_Y)\big] \\
&= \text{var}(X) + \text{var}(Y) + 2\,\text{cov}(X, Y),
\end{aligned}
\tag{4.22}
$$

where the covariance of X and Y is defined to be

$$
\begin{aligned}
\text{cov}(X, Y) &= E\big[(X - \mu_X)(Y - \mu_Y)\big] \\
&= E[XY - X\mu_Y - Y\mu_X + \mu_X\mu_Y] \\
&= E[XY] - \mu_X\mu_Y - \mu_Y\mu_X + \mu_X\mu_Y \\
&= E[XY] - \mu_X\mu_Y
\end{aligned}
\tag{4.23}
$$

so $\text{cov}[X, Y] = 0$ whenever X and Y are independent. If the covariance of X and Y is equal to zero, then we also say that X and Y are uncorrelated.

The correlation of X and Y is defined as

$$
\rho_{X,Y} = \frac{\text{cov}(x, y)}{\sigma_X \sigma_Y},
\tag{4.24}
$$

where

$$
\sigma_X = \sqrt{\text{var}(X)}.
\tag{4.25}
$$

Random variables X and Y are uncorrelated if

$$
E[XY] = E[X]E[Y].
\tag{4.26}
$$

It is important to note that

- Independent random variables are uncorrelated.
- But uncorrelated random variables are not necessarily independent.

Example. Assume that the random variables (X, Y) take values $(1, 0)\,(0, 1)\,(-1, 0)$ $(0, -1)$ with probability equal to 1/4 for each, as shown in the following table.

	$y = -1$	$y = 0$	$y = 1$	$p(x)$
$x = -1$	0	1/4	0	1/4
$x = 0$	1/4	0	1/4	1/2
$x = 1$	0	1/4	0	1/4
$p(y)$	1/4	1/2	1/4	

It is easy to see that $E[X] = E[Y] = 0$. Note also that the product xy is equal to 0 for all combinations of x and y with nonzero probability so $E[XY] = 0$. The variables are therefore uncorrelated. But we also see that $P(X = 0) = 1/2$ and $P(Y = 0) = 1/2$ but $P(X = 0, Y = 0) = 0$ so $P(X = 0, Y = 0) \neq P(X = 0)P(Y = 0)$. Hence, while the variables are uncorrelated, they are not independent.

For this same example, we can easily compute the expectation and the variance of each variable from the marginal distributions.

$$\mu_y = \frac{1}{4} \cdot (-1) + \frac{1}{2} \cdot 0 + \frac{1}{4} \cdot (1) = 0 \tag{4.27}$$

$$\text{var}(y) = \sum_{i=1}^{3}(y_i - \mu_y)^2 P(y_i) = \frac{1}{2}. \tag{4.28}$$

4.2.2 Random vectors

Let X be a vector of random variables, i.e.

$$X = \begin{bmatrix} X_1 & X_2 & \cdots & X_M \end{bmatrix}^{\mathrm{T}} \tag{4.29}$$

and let μ_X be the expected value of X. The auto-covariance of the elements of the random vector X is defined as follows,

$$
\begin{aligned}
C_{XX} &= E\big[(X - \mu_X)(X - \mu_X)^{\mathrm{T}}\big] \\
&= E[XX^{\mathrm{T}} - \mu_X X^{\mathrm{T}} - X\mu_X^{\mathrm{T}} + \mu_X\mu_X^{\mathrm{T}}] \\
&= E[XX^{\mathrm{T}}] - E[\mu_X X^{\mathrm{T}}] - E[X\mu_X^{\mathrm{T}}] + E[\mu_X\mu_X^{\mathrm{T}}] \\
&= E[XX^{\mathrm{T}}] - \mu_X\mu_X^{\mathrm{T}} - \mu_X\mu_X^{\mathrm{T}} + \mu_X\mu_X^{\mathrm{T}} \\
&= E[XX^{\mathrm{T}}] - \mu_X\mu_X^{\mathrm{T}}.
\end{aligned} \tag{4.30}
$$

The covariance matrix for the vector X of random variables has the following entries

$$
C_X = \begin{bmatrix} E[X_1 X_1] - \mu_1 \mu_1 & E[X_1 X_2] - \mu_1 \mu_2 & \cdots & E[X_1 X_M] - \mu_1 \mu_M \\ E[X_2 X_1] - \mu_2 \mu_1 & E[X_2 X_2] - \mu_2 \mu_2 & \cdots & E[X_2 X_M] - \mu_2 \mu_M \\ \vdots & \vdots & \ddots & \vdots \\ E[X_M X_1] - \mu_M \mu_1 & E[X_M X_2] - \mu_M \mu_2 & \cdots & E[X_M X_M] - \mu_M \mu_M \end{bmatrix}.
$$

$$(4.31)$$

The components x_i and x_j are uncorrelated if

$$
E[X_i X_j] = \mu_i \mu_j
$$

$$(4.32)$$

in which case the covariance matrix is a diagonal matrix.

Consider a second random vector Y related to X by a linear transformation

$$
Y = AX.
$$

$$(4.33)$$

The expectation of Y is

$$
\begin{aligned}
\mu_y = E[Y] &= E[AX] = A E[X] \\
&= A\mu_x
\end{aligned}
$$

$$(4.34)$$

and the (auto-)covariance of Y is

$$
\begin{aligned}
C_{YY} &= E\big[(Y - E[Y])(Y - E[Y])^{\mathrm{T}}\big] \\
&= E\big[[AX - A\mu_X][AX - A\mu_X]^{\mathrm{T}}\big] \\
&= E[AXX^{\mathrm{T}}A^{\mathrm{T}} - AX\mu_X^{\mathrm{T}}A^{\mathrm{T}} - A\mu_X X^{\mathrm{T}}A^{\mathrm{T}} + A\mu_X \mu_X^{\mathrm{T}}A^{\mathrm{T}}] \\
&= A\big[E[XX^{\mathrm{T}}] - \mu_X \mu_X^{\mathrm{T}} - \mu_X \mu_X^{\mathrm{T}} + \mu_X \mu_X^{\mathrm{T}}\big]A^{\mathrm{T}} \\
&= A\big[E[XX^{\mathrm{T}}] - \mu_X \mu_X^{\mathrm{T}}\big]A^{\mathrm{T}} \\
&= A C_{XX} A^{\mathrm{T}}.
\end{aligned}
$$

$$(4.35)$$

This result is particularly interesting if X is a vector of independent identically distributed random deviates with variance equal to 1 because then C_{XX} is the identity matrix and $C_{YY} = AA^{\mathrm{T}}$.

Suppose we wanted to compute the variance in the average porosity for a composite core of M independent core plugs each drawn from the same distribution (identical means and variances). We can write the average porosity as the product of a vector

$$
A = \begin{bmatrix} 1 & 1 & \cdots & 1 \end{bmatrix}/M
$$

$$(4.36)$$

and the vector of random plug porosities

$$
\phi = \begin{bmatrix} \phi_1 \\ \vdots \\ \phi_M \end{bmatrix}.
$$

$$(4.37)$$

If the covariance of ϕ is denoted

$$C_{\phi\phi} = \begin{bmatrix} \sigma_\phi^2 & 0 & \cdots & 0 \\ 0 & \sigma_\phi^2 & \cdots & 0 \\ \vdots & \vdots & \ddots & \vdots \\ 0 & 0 & \cdots & \sigma_\phi^2 \end{bmatrix} \tag{4.38}$$

then the variance of the average porosity is found by applying Eq. (4.35)

$$
\begin{aligned}
\text{var}(\phi) &= A C_{\phi\phi} A^{\text{T}} \\
&= \frac{1}{M} \begin{bmatrix} 1 & \cdots & 1 \end{bmatrix} \begin{bmatrix} \sigma_\phi^2 & 0 & \cdots & 0 \\ 0 & \sigma_\phi^2 & \cdots & 0 \\ \vdots & \vdots & \ddots & \vdots \\ 0 & 0 & \cdots & \sigma_\phi^2 \end{bmatrix} \begin{bmatrix} 1 \\ \vdots \\ 1 \end{bmatrix} \frac{1}{M} \\
&= \frac{1}{M} \begin{bmatrix} 1 & 1 & \cdots & 1 \end{bmatrix} \begin{bmatrix} \sigma_\phi^2 \\ \sigma_\phi^2 \\ \vdots \\ \sigma_\phi^2 \end{bmatrix} \frac{1}{M} \\
&= \frac{\sigma_\phi^2}{M},
\end{aligned}
\tag{4.39}
$$

which is a well-known result.

4.3 Bayes' rule

Our primary use for Bayes' rule (4.10) will be to estimate the conditional probability density for model variables m given observations d_{obs}

$$f(m|d_{\text{obs}}) = \frac{f(d_{\text{obs}}|m)f(m)}{f(d_{\text{obs}})} = \frac{f(d_{\text{obs}}|m)f(m)}{\int_D f(d_{\text{obs}}|m)f(m)\,dm}. \tag{4.40}$$

The density $f(m)$ is called the prior probability density for m. It is the probability density for m before incorporating the observations. $f(d_{\text{obs}}|m)$ is the probability of d_{obs} given the model variables m. We also call the function $f(d_{\text{obs}}|m)$ in Eq. (4.40) the likelihood function for the model variables, given the data. Because of the importance of Bayes' rule, the remainder of this section discusses two non-Gaussian examples that demonstrate the reduction in uncertainty that results from inclusions of new data.

4.3.1 Example: total depth of a well

Suppose that you are sitting alone in your laboratory, and an assistant periodically brings core plugs to you for analysis. We will suppose also that you have been told that the well from which these plugs were obtained was cored from surface to total depth, and plugs were taken every foot. If the assistant chooses the plugs randomly, how would you estimate the depth of the well based on the depths of the plugs? And what would be the uncertainty in the estimate?

The plugs are known to be numbered from 1 to n, where n is the total depth of the well in feet. Because the variable to be estimated is discrete, we use probability instead of probability density for this problem. The objective is to calculate the conditional probability for n, given observations of depths of k randomly selected core plugs numbered $m_{obs,i}$ for $i = 1, \ldots, k$. Bayes' rule can be used to compute the conditional probability of n from the product of the prior probability of n and the probability of the observation given the number for total depth.

Because the plugs are randomly selected, the probability of selecting any depth less than the maximum is 1 divided by the number of possibilities. The probability of selecting a plug from a depth greater than the maximum is zero. In other words, the probability that we would observe a depth m, if the total depth is n is

$$p_{M|N}(m|n) = \begin{cases} 1/n & \text{for } m \leq n \\ 0 & \text{for } m > n. \end{cases} \tag{4.41}$$

If you initially believe that the maximum possible depth is n_0, then the prior probability for observing depth n is

$$p_N(n) = \begin{cases} 1/n_0 & \text{for } n \leq n_0 \\ 0 & \text{for } n > n_0. \end{cases} \tag{4.42}$$

This expresses the prior belief that any depth less than the upper bound is equally likely. (Of course this may not be a very good prior, so we will have to check the sensitivity of the results to the prior assumption.)

Suppose the depth of the first core plug is $m_{obs,1}$. The conditional probability distribution for the total depth n is given by

$$p(n|m_{obs,1}) = \frac{p(m_{obs,1}|n)p(n)}{\sum_1^\infty p(m_{obs,1}|n)p(n)}. \tag{4.43}$$

This distribution now reflects the uncertainty in the value of n (the total depth of the well), given the first observation. Assuming that the depth of the first core plug is less than the assumed upper bound, the conditional probability for n, after the first observation, is

$$p(n|m_{obs,1}) = \begin{cases} 0 & \text{for } n < m_{obs,1}, \\ c/n & \text{for } m_{obs,1} \leq n \leq n_0, \\ 0 & \text{for } n > n_0, \end{cases} \tag{4.44}$$

where

$$c = \left(\sum_{n=m_{\text{obs},1}}^{n_0} \frac{1}{n} \right)^{-1} \tag{4.45}$$

is a normalizing constant.

After observing another core plug, this time from depth $m_{\text{obs},2}$, you can compute the probability distribution for n, given the two observations, $m_{\text{obs},1}$ and $m_{\text{obs},2}$ using the same procedure that was used for one observation. The only difference is that we now take $p(n|m_{\text{obs},1})$ to be the prior probability distribution for n when we condition to the second observation. So,

$$p(n|m_{\text{obs},1}, m_{\text{obs},2}) \propto p(m_{\text{obs},2}|n) \, p(n|m_{\text{obs},1})$$
$$\propto p(m_{\text{obs},2}|n) \left[p(m_{\text{obs},1}|n) p(n) \right]$$
$$\propto \begin{cases} 0 & \text{for } n < \max[m_{\text{obs},1}, m_{\text{obs},2}] \\ 1/n^2 & \text{for } \max[m_{\text{obs},1}, m_{\text{obs},2}] \le n \le n_0 \\ 0 & \text{for } n > n_0 \end{cases} \tag{4.46}$$

and in general, after k observations,

$$p(n|m_{\text{obs},1}, \ldots, m_{\text{obs},k}) \propto p(m_{\text{obs},k}|n) \, p(n|m_{\text{obs},k-1})$$
$$\propto p(m_{\text{obs},k}|n) \cdots p(m_{\text{obs},1}|n) p(n)$$
$$\propto \begin{cases} 0 & \text{for } n < \max[m_{\text{obs},i}, i = 1, k], \\ 1/n^k & \text{for } \max[m_{\text{obs},i=1,k}] \le n \le n_0, \\ 0 & \text{for } n > n_0. \end{cases} \tag{4.47}$$

Note that even observations of depths that are less than the current maximum observed, add considerable information to the probability of n.

Numerical example

Suppose that we are confident that the well could not have been more than 20 000 feet deep, and the true total depth of the well is 9815 feet. The first 50 core plugs that are delivered to your laboratory (one at a time) are as follow.

$$\begin{aligned} m_{\text{obs}} = \ & [9043, 7118, 2869, 2211, 5979, 536, 7684, 6891, 8452, 5955, \\ & 2514, 9302, 4776, 4655, 7484, 8092, 2443, 8615, 212, 9740, 729, \\ & 6038, 7333, 6994, 1946, 7626, 3514, 1903, 8574, 6336, 7598, 4568, \\ & 4433, 4310, 5485, 1997, 5130, 1809, 1906, 2642, 323, 1747, 148, \\ & 1866, 9736, 7137, 4516, 4882, 5176, 4551]^{\text{T}}. \end{aligned} \tag{4.48}$$

Before any observations, the probability distribution for n is uniform on the interval $[1, 20\,000]$ as shown in the upper left-hand side of Fig. 4.6. After the first observation,

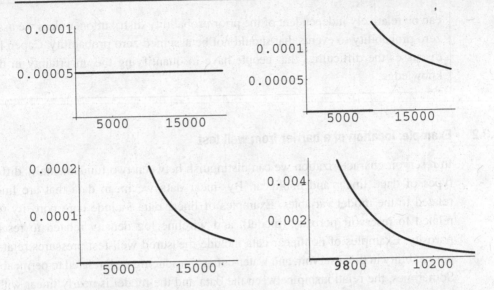

Figure 4.6. The prior probability for the total depth of the well is shown on the upper left-hand side. The distributions after one observation of depth, two observations of depths, and fifty observations are shown on the upper right-, the lower left-, and the lower right-hand side respectively. Note that the horizontal scale was changed for fifty observations.

the probability is zero for n less than 9043. After the second observation, the probability is still zero for n less than 9043 and decreases as $1/n^2$ for values of n between 9043 and 20 000. Finally, after 50 observations of depths, the range of values with significant probability is much narrower. In fact, the 95% confidence interval for total depth of the well is [9740, 10 354] for this example. The answer to the problem of estimating the total depth of the well after 50 observations is given by the probability distribution shown in the lower right-hand side of Fig 4.6. There are, however, a number of "best estimates" of the total depth that one could give. The expected value of total depth, computed using Eq. (4.13) is 9942. The value of depth with the highest probability is simply the largest observed value, 9740. The depth for which the probability is approximately 0.5 that the total depth is less than that value is 9879.

Remark 4.1. One objection that is sometimes raised to a Bayesian approach is the subjective nature of the prior probability. In the previous example, we assumed that the total depth of the well would not be more than 20 000 feet. If we had instead assumed that the upper bound was 30 000, the 95% confidence interval after the same 50 observations of core plug depths would still be [9740, 10 354]. If, however, we had been overconfident and had chosen an upper bound for the depth that was smaller than the actual depth of the well, the results could have been either very misleading or we could have arrived at an inconsistency such as observing a depth greater than the upper bound. The good news is that the results can be relatively independent of the prior probability distribution, unless we assign zero probability to events that should not be assigned zero probability. Capen [20] discusses the difficulties that people have in quantifying the uncertainty in their knowledge.

4.3.2   Example: location of a barrier from well test

In reservoir characterization we can distinguish between two fundamentally different types of data: linear and nonlinear. By linear data we mean data that are linearly related to the model variables. Examples of linear data include core porosity (data) related to reservoir porosity (model), and wireline log density related to reservoir porosity. Examples of nonlinear data include measured well-test pressures related to permeability in the reservoir, and water-cut at a producing well related to permeability. Sometimes, the relationship between the data and the model is nearly linear within a limited neighborhood, in which case a linear approximation may be valid.

The distinction between these two types of data is important when one wants to estimate or simulate the distribution of permeability in the face of insufficient information. When the probability distribution is multi-variate normal, it can be described completely by the mean and the covariance, and realizations can be generated using the standard methods for Gaussian co-simulation. When the data–model relationship is highly nonlinear the probability distribution is often multi-modal and the mean and covariance do not provide adequate measures of the distribution.

As an example, suppose that we have obtained well-test data from two wells that are located 500 m apart.[2] Analysis of one set of well-test data indicates the presence of a sealing fault at distance 220 ± 22 m from well 1 and analysis of a second well test indicates the presence of a fault at distance 100 ± 10 m from well 2. Where is the fault located? Or, more precisely, what is the probability distribution for the location of the fault given the interpretations from the well tests?

Let the location of the fault be parameterized by its distance, d, from well 1 and by the angle, θ, with respect to the line joining both wells. For this problem we can think of d_1^{obs} and d_2^{obs}, the measured distances of the fault from wells 1 and 2, as data while d and θ are the model variables. It is not difficult to see that the relationships between the model variables, d and θ, and the data are given by

$$d_1 = d \tag{4.49}$$

and

$$d_2 = d + 500 \, |\sin\theta| \tag{4.50}$$

---

[2] This example is from [21].

or by

$$d_2 = -d + 500\,|\sin\theta| \tag{4.51}$$

depending on whether the wells are on the same, or opposite, sides of the fault.

We can characterize our state of knowledge after the first well test by the joint probability distribution of the model parameters d and θ. The first test gives no information on the orientation of the fault but does specify the uncertainty in d, hence if we assume that the uncertainty is described adequately by a normal distribution we can write the probability density for d and θ after the first well test as

$$p(d, \theta|d_1) \propto \exp\left[-\frac{(d - d_1)^2}{2\sigma_1^2}\right]. \tag{4.52}$$

This probability density is the prior probability density for the second well test.

If the location (d and θ) of the fault was known, and a set of pressure data was generated from these known parameters, measurement errors were added, and the data analyzed, there would be some uncertainty in the estimation of the distance from the second well. The conditional probability of measuring the data d_2 in the second test given true values of the model variables is given by

$$p(d_2|d, \theta) = \frac{1}{\sigma_2\sqrt{2\pi}} \exp\left[-\frac{(g(d, \theta) - d_2)^2}{2\sigma_2^2}\right], \tag{4.53}$$

where $g(d, \theta)$ is the function of d and θ given by Eqs. (4.50) and (4.51). The a posteriori probability density of the model variables, given the observation d_2, can be found from Bayes' rule, for which we take the probability density in Eq. (4.52) to be the

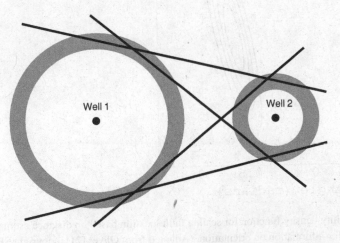

Figure 4.7. Locations of four faults corresponding to local maxima in the probability density function. (Adapted from Oliver [21], copyright SPE.)

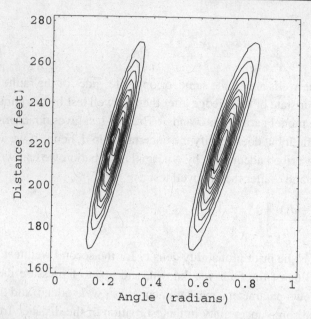

Figure 4.8. The probability density function for θ, the orientation of a sealing fault, and d_1, its distance from well 1 based on interpretation of two well tests. (Adapted from Oliver [21], copyright SPE.)

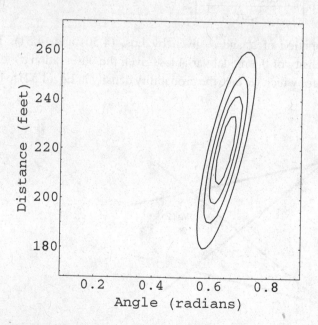

Figure 4.9. The probability density function for sealing fault location based on distance estimates from well tests and prior information on orientation. (Adapted from Oliver [21], copyright SPE.)

prior density.

$$p(d, \theta | d_1, d_2) \propto \exp\left[-\frac{(d_1 - d)^2}{2\sigma_1^2}\right] \exp\left[-\frac{(d_2 - g(d, \theta))^2}{2\sigma_2^2}\right]. \tag{4.54}$$

The fault locations that could have resulted in the observed pressure data are shown in Fig. 4.7. Although uncertainties in the data were assumed to be Gaussian, Fig. 4.8 shows that the a posteriori probability distribution for (d, θ) is clearly multi-modal and that for the assumed data there are four local maxima in the probability distribution. (The probability density function is symmetric about $\theta = 0$ so only positive values of θ are shown in Fig. 4.8.)

This simple two-parameter example illustrates the difficulty of always describing the probability distribution in terms of only the maximum likelihood estimate and the covariance. If, however, we were able to obtain additional information on fault direction from dipmeter or tectonic studies we might obtain a much different answer. Suppose, for example, that analysis of other data indicates that $\theta = 0.78 \pm 0.3$ radians ($\theta = 45° \pm 17°$), corresponding to a fault running NE to SW. The probability distribution that results from combining the well-test data with the new information has only one peak as shown in Fig. 4.9.

5 Descriptive geostatistics

5.1   Geologic constraints

It is not possible in general to constrain rock property distributions to honor the physical constraints that govern the erosion, transport, deposition, sorting, and diagenesis of sedimentary materials and the observations of properties at well locations. In many cases, however, the spatial covariance can be used to quantify the geological plausibility of a distribution of rock properties. Although there are many cases where this is not a good measure, it is important to understand what the covariance means in terms of the spatial relationship of the properties. We can begin by examining well-log traces from a well in Alaska.

Figure 5.1 shows two scaled well-logs from the same interval in the N. Kapilik 1 well (NK-1). One log is a spontaneous potential or SP log. The potentials are primarily created by diffusion of ions between the formation and the wellbore caused by differences in salinity in the fluids. Variation in the potential along the well path is due to variations in permeability as well as variations in salinity. Because clay particles have such a large effect on permeability, the SP log is often used as a shale indicator. On the other hand, the SP has relatively poor vertical resolution, and can not resolve small shales. The other log in Fig. 5.1 is a gamma ray (GR) log. The GR log measures the natural radiation of minerals in the formation. Shales and clays typically show higher levels of radioactivity than other rock types, such as sandstones, limestones, or dolomites. The GR log is very sensitive to the presence of clay and yields high resolution.

The logs have been normalized to remove long term trends due to changes in salinity. The response in both logs then is driven primarily by the presence of clay minerals. Both logs are used as clay indicators, yet anyone familiar with well-logs would recognize that the upper log must be the GR log and the lower log the SP. This is what we qualitatively mean by plausibility. The upper log is too rough to be an SP log.

5.2   Univariate distribution

As an introduction to geostatistics, we start with univariate distribution to review basic concepts and methods for statistical analysis, including a few common graphical

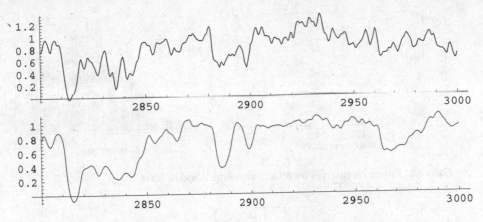

Figure 5.1. Two logs from same section. One is GR and the other is SP. Which is which?

methods for describing the frequency distribution of a set of experimental data in estimation of a single variable. Then we review a few key parameters representing important features of the graphical distributions. By necessity, the review in this chapter is fairly brief. A more comprehensive review of geostatistics as applied to inverse problems can be found in Kitanidis [22].

5.2.1   Histogram

A histogram is a useful way to approximate the probability density by counting the number of times that a sample from an experimental data set falls within each of the specified intervals or "bins." If a value is equal to the upper limit of the interval, it is included in the adjacent higher bin. The histogram plot is normally used to display data frequency distributions. The number of samples within an interval is proportional to the frequency with which the data fall in that interval. The important characteristics of the data distribution can often be identified from the shape of the histogram. It is, however, necessary to use enough bins to display the variability. Too many bins, on the other hand, reduces the number of data in each and amplifies the effects of random sampling error.

As an example, Fig. 5.2 shows proven oil reserves for most nations of the world. The histogram on the left-hand side shows the frequency of reserves on a linear scale in billions of barrels. The plot is useful for showing the skewness of the distribution but is of little value for estimating the distribution at the low end. The plot on the right-hand side in Fig. 5.2 shows the logarithm (base 10) of reserves in barrels. From this display, it is much easier to determine that the most frequent national reserves numbers are between 100 million and one billion barrels.

5.2.2   Cumulative distribution plot

P quantile refers to the value of the data point at which the cumulative probability distribution reaches the value P, where P is a number between 0 and 1. The cumulative

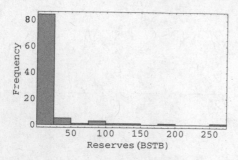
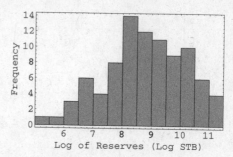

Figure 5.2. Proven oil reserves for 89 nations of the world in 2004.

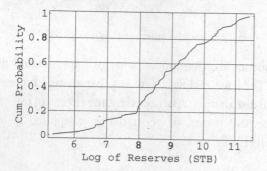

Figure 5.3. Ogive plot of national oil reserves.

probability can be computed as:

$$p_i = (i - 0.5)/n, \text{ for } i = 1, 2, \ldots, n, \tag{5.1}$$

where n is the total number of data samples, and i is the rank of the sorted datum. The cumulative distribution plot, also called the ogive, is used to present the relationship between the value of the data and the quantile. Figure 5.3 shows the ogive plot for the same reserves data used in Fig. 5.2. From this plot, it is straightforward to see that approximately 55% of the nations have reserves less than one billion barrels, and that the 0.5 quantile (or P50) appears to be approximately $10^{8.7}$ or 500 million barrels.

5.2.3   Box plot

The box plot conveniently summarizes key information about the frequency distributions for several sets of data. Each set of data is represented by a "box." Figure 5.4 shows a typical interpretive key for a typical box plot. The rectangular box shows the interval bounded by the 25th and 75th percentiles. The whiskers at each end of the polygon show the intervals between P1 and P10 and between P90 and P99. The P50 value is shown as a short line across the box and the mean is represented by a small cross. The details of box plots can vary with different applications. The maximum and the minimum data are often not displayed.

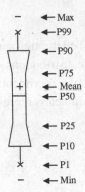

Figure 5.4. A key of box plot for display of representative values.

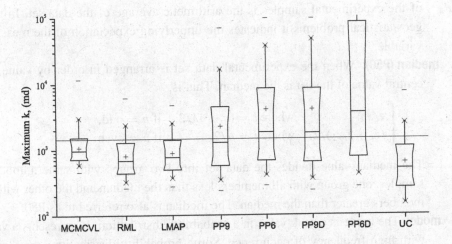

Figure 5.5. Distributions of conditional realizations of maximum permeability from the approximate sampling algorithms and from the very long MCMC (MCMCVL). The unconditional distribution (UC) and the true value are shown for comparison.

A box plot provides a simple way to summarize many important aspects of a data frequency distribution. A major application of the box plot is in qualitative comparison of data from two or more sets of data. Figure 5.5 shows an example of comparison of several data sets.[1] Each data set contains realizations from an algorithm as labeled below its corresponding box. The parameter estimated here is the maximum permeability of a one-dimensional reservoir model conditioned to transient pressure data. The group of estimations from the Very Long Markov chain Monte Carlo (MCMCVL) algorithm contains 320 million samples. The rest of the groups contain 5000 samples each. It is clear that the spread in the samples from the MCMCVL algorithm is the smallest among all the distributions and the spread is largest for the samples from the "PP6D" algorithm.

---

[1] See Section 10.8 for an explanation of the source of the samples in the box plots.

5.2.4   Representative values

The distribution of the experimental data is quantified and described through summary parameters. The most common summary parameters for univariate distributions include the mean, the median, the mode, and the variance. For a set of experimental data $z_1, z_2, \ldots, z_n$, the following formulas can be used to compute parameters representing the distribution of the data.

mean/arithmetic average:

$$\bar{z} = \frac{z_1 + z_2 + \cdots + z_n}{n} = \frac{1}{n} \sum_{i=1}^{n} z_i,$$

where n is the total number of samples, and z_i is the value of sample i. The mean of the experimental samples is the arithmetic average of the data set. In most geostatistical problems it indicates the underlying expectation of the measured variable.

median (P50): When the experimental data set is arranged in order by value, the central value of the list is the median. That is,

$$\bar{z} = \begin{cases} z_l, & \text{where } l = (n+1)/2, & \text{if } n = \text{odd}, \\ (z_l + z_{l+1})/2, & \text{where } l = n/2, & \text{if } n = \text{even}. \end{cases}$$

The median value divides the data set into two groups with same number of samples, one group with all members less than the median and the other with all members greater than the median. The median is also referred to as P50.

mode: The mode is a peak value in a probability distribution. It represents a value with high frequency of occurrence. Some probability distributions are too flat to have a mode, while others may have multiple modes. A typical distribution without a mode is the uniform distribution, in which the probability is the same everywhere within the domain. The Gaussian distribution has a unique mode at the mean of the distribution. The probability distribution for porosity in the field example (page 71) had two modes, corresponding to two different rock types. Examples with no mode, with single mode and with multiple modes are shown in Fig. 5.6.

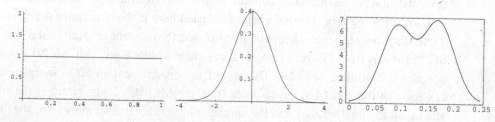

Figure 5.6. Probability distributions for the uniform distribution, the Gaussian distribution, and the multi-modal distribution, illustrating distributions with different numbers of modes.

The mean, the median and the mode coincide for symmetric probability distributions with unique mode.

Experimental variance: The experimental, or sample, variance quantifies the variability of the sample data. When the data are measurements, the variance reflects the measurement quality. The experimental variance is computed as:

$$\widehat{\text{var}} = \frac{(z_1 - \bar{z})^2 + (z_2 - \bar{z})^2 + \cdots + (z_n - \bar{z})^2}{n - 1}$$

$$= \frac{1}{n - 1} \sum_{i=1}^{n} (z_i - \bar{z})^2$$

$$= \frac{1}{n - 1} \sum_{i=1}^{n} z_i^2 - \frac{n}{n - 1} \bar{z}^2,$$

where $\bar{z}$ is the mean of the data set and n is the total number of data. The standard deviation is the square root of the variance: $s = \sqrt{\text{var}}$. It has the same unit as the measurement z_i. In a univariate Gaussian distribution, 95% of the experimental samples have the value within $2s$ of the mean/mode.

5.3 Multi-variate distribution

Close examination of the SP log in Fig. 5.1 suggests that if we knew the value of SP at a particular depth, we could probably do fairly well at predicting the value of the SP one foot away. On the other hand, it might be difficult to predict accurately the value of SP 100 feet away from a measurement. We can quantify the variability by examining the relationship of SP values that are some distance h apart. Figure 5.7 shows cross plots of values of the normalized SP at depth x and at depth $x + h$ for $h = 5, 10, 20$, and 40 feet. Note that for a distance of 5 feet the two values are highly correlated. The correlation progressively worsens as the distance between the points increases.

In the remainder of this section, we study the spatial relationships of multi-variate models. The main purpose is to model spatial distribution of a random field. The estimation of the auto-covariance models is of our prime interest.

5.3.1 Stationarity

The parameters of the geostatistical models for describing the spatial relationship within a random field are usually based on information in the sample data. In order for the geostatistical model to be adequate in describing the spatial relationship of the random variables, the sample data must represent the common features of the spatial random variables. In the cases where correlations should not be expected, such as geological

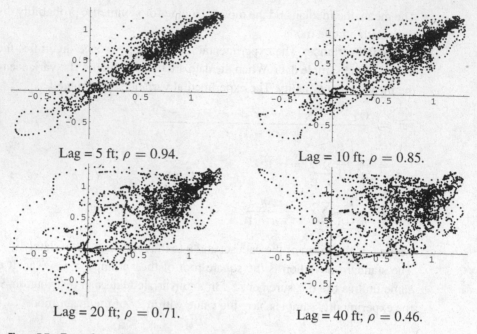

Lag = 5 ft; $\rho = 0.94$.

Lag = 10 ft; $\rho = 0.85$.

Lag = 20 ft; $\rho = 0.71$.

Lag = 40 ft; $\rho = 0.46$.

Figure 5.7. Crossplots of SP values at various lag distances.

unconformities, the domain of interest may need to be divided into regions for which the geostatistical model is assumed to be valid.

In linear estimation, it is adequate to use the mean and the covariance in modeling the spatial distribution of a random field. The mean of the multi-variate random field may be a function of the spatial location:

$$\bar{Z}(\mathbf{x}) = E[Z(\mathbf{x})], \tag{5.2}$$

where $\mathbf{x}$ specifies the spatial location of the variable. The spatial covariance between any two locations $\mathbf{x}$ and $\mathbf{x} + \mathbf{h}$ is defined as:

$$C[Z(\mathbf{x}), Z(\mathbf{x}+\mathbf{h})] = E\{[Z(\mathbf{x}) - E[Z(\mathbf{x})]][Z(\mathbf{x}+\mathbf{h}) - E[Z(\mathbf{x}+\mathbf{h})]]\}, \tag{5.3}$$

where $\mathbf{h}$ is a vector and $\mathbf{x} + \mathbf{h}$ can be any location within the interested region.

When the geostatistical model is constant within a region, the mean and the covariance models are no longer functions of spatial location. The first-order stationarity assumes the expectation value at any location in the specified region is constant:

$$\bar{Z} = E[Z(\mathbf{x})] = E[Z(\mathbf{x}+\mathbf{h})]. \tag{5.4}$$

A random process is second-order stationary if the covariance between the value of the random function at $\mathbf{x}$ and at $\mathbf{x}'$ is only a function of the distance between the two points:

$$C[Z(\mathbf{x}), Z(\mathbf{x}+\mathbf{h})] = C(h). \tag{5.5}$$

For $h = 0$, Eq. 5.5 becomes $\text{var}[Z(\mathbf{x})] = C(0)$, i.e. the variance is constant under the assumption of the second-order stationarity. A Gaussian random process is termed stationary if it satisfies the conditions for both first- and second-order stationarity.

Remark 5.1. In geostatistics, the variogram is commonly used instead of the covariance. The variogram is mathematically defined as half the variance between locations h distance apart. It is also called the semivariogram. For stationary models, it is expressed as:

$$\gamma(h) = \frac{1}{2} \text{var}[Z(x) - Z(x+h)]$$
$$= \frac{1}{2} E\{[Z(x) - Z(x+h)]^2\} - \frac{1}{2}\{E[Z(x) - Z(x+h)]\}^2.$$

When the distance $h = 0$:

$$\gamma(0) = \frac{1}{2} \text{var}[Z(x) - Z(x)] = 0.$$

With the assumption of first- and second-order stationarity, we obtain

$$\gamma(h) = C(0) - C(h) = C(0)\left[1 - \frac{C(h)}{C(0)}\right],$$

where $C(0)$ is the sill level for the variogram. The term $\frac{C(h)}{C(0)}$ is the correlation coefficient over the distance h and has the notation $\rho(h)$. The correlation coefficient $\rho(h)$ is often used to describe the spatial relationship. It equals to one for zero distance and typically decreases with increasing distance (see below).

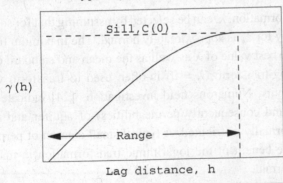

5.3.2 Transformation of variables

The Gaussian distribution is unimodal and symmetric. The probability density for a Gaussian variable is completely defined by the mean and the covariance.

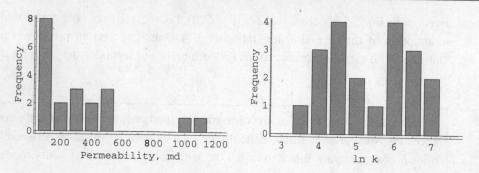

Figure 5.8. The log-transform of a permeability data set. Before (left) and after (right) transformation.

Characterization of skewed or multi-modal distributions can be much more complicated. Because many estimation and simulation methods rely on the assumption that the underlying PDF is approximately Gaussian, it is often useful to transform non-Gaussian variables to new variables that are approximately univariate Gaussian. There are many approaches to normalizing the distribution. Here we introduce three types of nonlinear transformation methods: log-transform, power transform, and normal score transform.

The Box–Cox transform [23] is often useful for transforming data to improve the linearity for regression. The log-transform is one example of the Box–Cox family of transformations:

$$z_\lambda = \begin{cases} \frac{z^\lambda - 1}{\lambda} & \text{for } \lambda \neq 0, \\ \ln z & \text{for } \lambda = 0. \end{cases}$$

The exponent for the transformation, λ, can be selected by assuming that for some value of λ, the probability density for z_λ is approximately normal. The maximum likelihood can be used to estimate the best value of λ as well as the mean and standard deviation of the distribution. The log-transform ($\lambda = 0$) is often used to transform values of permeability or transmissivity. Numerous field investigations [24] indicate that the hydraulic conductivities (and consequently permeabilities) of aquifer and reservoir rocks are frequently log-normally distributed. A typical small data set of permeability values (Fig. 5.8) shows the benefit of the logarithmic transformation in making the distribution more nearly normal.

The normal score transform utilizes a nonparameteric transformation to convert the frequency distribution to one that is univariate Gaussian. The steps in the method are illustrated in Fig. 5.9. Beginning with a non-Gaussian marginal PDF (lower right), the ogive is computed (upper right). A variable in the original distribution is then transformed to a variable in the normal distribution at the same value on the cumulative density function. In Fig. 5.9, the value 0.19 is transformed to a value of -0.1.

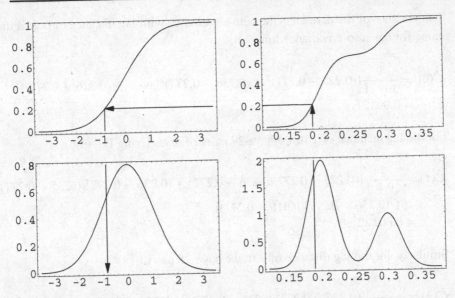

Figure 5.9. Using the normal score transform to convert a bimodal PDF into Gaussian PDF.

5.3.3   Experimental covariance

The experimental auto-covariance $\hat{C}(h)$ for a second-order stationary one-dimensional random field is calculated based on the sample data:

$$\hat{C}(h) = \frac{1}{n(h) - 1} \sum_{i=1}^{n(h)} [z(x_i) - \bar{z}][z(x_i + h) - \bar{z}], \tag{5.6}$$

where $\bar{z}$ is the mean of the data set: $\bar{z} = \frac{1}{n} \sum_{i=1}^{n} z_i$, h is the lag distance, $n(h)$ is the number of data pairs with a distance h apart.

The following simple example illustrates the computation of the experimental auto-covariance.

Example: calculate the auto-covariance of the porosity along a 30 feet core with 1 foot interval. The porosity data are provided below in the sequence of the core depth.

$\phi_{\mathrm{obs}} = \{0.225, 0.236, 0.235, 0.231, 0.225, 0.225, 0.230, 0.240, 0.252, 0.262,$
$\qquad 0.264, 0.260, 0.252, 0.243, 0.236, 0.237, 0.241, 0.254, 0.273, 0.296,$
$\qquad 0.313, 0.326, 0.338, 0.348, 0.353, 0.354, 0.353, 0.348, 0.336, 0.316\}.$

Solution: to compute the auto-covariance of the porosity distribution along the core, we need to first compute the average porosity:

$$\hat{m} = \frac{1}{20} \sum_{l=1}^{20} (0.225 + 0.236 + 0.235 + \cdots + 0.316)$$

$$= 0.277.$$

The variance of the data set is the auto-covariance with lag distance of 0. It is the peak value for the auto-covariance function.

$$\hat{C}(0) = \frac{1}{30-1}[(0.225-0.277)^2 + (0.236-0.277)^2 + \cdots + (0.316-0.277)^2]$$
$$= 0.002\,27.$$

For the lag distance of 1 ft, we have 29 pairs of data. Using Eq. (5.6),

$$\hat{C}(1) = \frac{1}{29-1}[(0.225-0.277)(0.236-0.277) + (0.236-0.277)(0.235-0.277) + \cdots$$
$$+ (0.336-0.277)(0.316-0.277)]$$
$$= 0.002\,22.$$

Similarly, for the lag distance of 2 ft, we have 28 pairs,

$$\hat{C}(2) = \frac{1}{28-1}[(0.225-0.277)(0.235-0.277) + (0.236-0.277)(0.231-0.277) + \cdots$$
$$+ (0.348-0.277)(0.316-0.277)]$$
$$= 0.002\,07$$

$$\vdots$$

$$\hat{C}(10) = \frac{1}{20-1}[[(0.225-0.277)(0.264-0.277) + (0.236-0.277)(0.260-0.277) + \cdots$$
$$+ (0.296-0.277)(0.316-0.277)]$$
$$= -0.000\,25.$$

The auto-covariance for the core data is plotted as a function of lag distance in Fig. 5.10. Note that the auto-covariance value becomes negative for lag distance of about 10 feet, implying that there may be some periodicity to the porosity.

The experimental auto-covariance for both the SP and the GR logs in the NK-1 well with separations between 0 and 60 feet are shown in Fig. 5.11. The auto-covariance

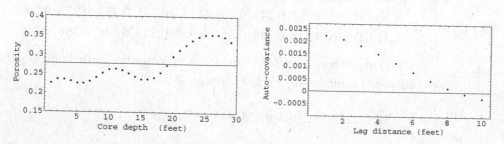

Figure 5.10. The plot on the left-hand side is the porosity along the core, the one on the right-hand side is the auto-covariance of the 1D porosity field.

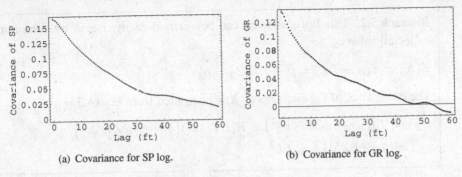

(a) Covariance for SP log.   (b) Covariance for GR log.

Figure 5.11. The experimental auto-covariance for the SP and GR logs.

curve for the GR log is nearly linear at the origin while the auto-covariance curve for the SP log appears to have a slope of 0 at the origin and more nearly a parabolic shape. This determines the roughness or differentiability of the random field. Both correlation curves show correlations that go to zero at distances on the order of 60 feet. We can use the correlation curve for a random field to characterize some aspects of the random process and to help determine whether an estimation or realization that we generate is plausible for that particular attribute.

5.4   Gaussian random variables

A random variable is said to be Gaussian with mean μ and variance σ^2 if its probability density is

$$f_x(x) = \frac{1}{\sigma\sqrt{2\pi}} \exp\left(-\frac{(x-\mu)^2}{2\sigma^2}\right). \tag{5.7}$$

Similarly, a random M-dimensional vector X is multi-variate Gaussian or multi-normal with mean μ and covariance C_X if the probability density for X is

$$f_X(x) = (2\pi)^{-M/2}\left|C_X^{-1}\right|^{1/2} \exp\left(-\frac{1}{2}(x-\mu)^T C_X^{-1}(x-\mu)\right). \tag{5.8}$$

Equivalently, if Z is a vector of independent identically distributed (iid) random variables with mean equal to 0 and variance equal to 1, then the random vector

$$X = \mu + LZ \tag{5.9}$$

is multi-variate Gaussian with mean μ and covariance $C_X = LL^T$.

Remark 5.2. This last statement can be verified easily. Begin by computing the expected value of X:

$$E[X] = E[\mu + LZ] = \mu + LE[Z] = \mu. \tag{5.10}$$

The covariance of the elements of X is computed from Eq. (4.35)

$$C_X = E[(X - \mu)(X - \mu)^\mathsf{T}] = E[LZ(LZ)^\mathsf{T}] = E[LZZ^\mathsf{T}L^\mathsf{T}]$$
$$= LE[ZZ^\mathsf{T}]L^\mathsf{T} = LL^\mathsf{T}. \tag{5.11}$$

5.4.1 Covariance models

We frequently make the assumption that the covariance between the random variable at h_i and the random variable at h_j is only a function of $|h_i - h_j|$. In this case the Gaussian random field is said to be stationary. There are many valid covariance functions but determining whether or not a general function can be a covariance function (and in which dimensions) is fairly complex. The subject is discussed more extensively in Section 5.4.3 or see the discussion in Christakos [25] who also provides a list of other valid covariance functions. The following covariance models are most commonly used for modeling random Earth processes: the spherical model for which

$$C(h) = \sigma^2 \begin{cases} 1 - \frac{3h}{2a} + \frac{h^3}{2a^3} & \text{for } 0 \leq h \leq a \\ 0 & \text{for } h > a \end{cases} \tag{5.12}$$

and the exponential family of covariance functions,

$$C(h) = \sigma^2 \exp(-3(|h|/a)^\nu). \tag{5.13}$$

For both formulas, h is the distance between two spatial locations and can be positive or negative. a is the correlation range. The exponential family of covariance functions is normally expressed as:

$$C(h) = \sigma^2 \exp(-(|h|/L)^\nu), \tag{5.14}$$

in which case the correlation range is $a = 3^{1/\nu}L$. It is common to refer to the covariance function in Eq. (5.14) as the exponential covariance function when $\nu = 1$ and as the Gaussian covariance when $\nu = 2$. Figure 5.12 shows several members of the exponential covariance family with different exponents and ranges.

In Fig. 5.13 we compare realizations of a Gaussian random field generated from the exponential covariance family with different parameters. On the left-hand side, the exponent is varied from $\nu \approx 0$ (white noise) to $\nu = 2$ (Gaussian covariance). Note that

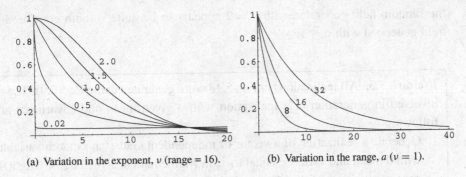

(a) Variation in the exponent, ν (range = 16).   (b) Variation in the range, a ($\nu = 1$).

Figure 5.12. Examples of exponential covariance functions for different values of ν and a.

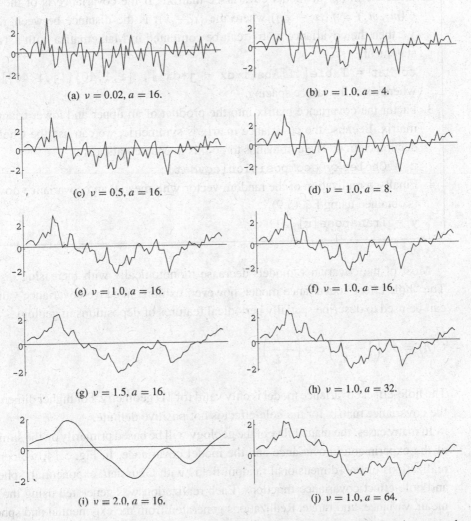

(a) $\nu = 0.02, a = 16$.   (b) $\nu = 1.0, a = 4$.

(c) $\nu = 0.5, a = 16$.   (d) $\nu = 1.0, a = 8$.

(e) $\nu = 1.0, a = 16$.   (f) $\nu = 1.0, a = 16$.

(g) $\nu = 1.5, a = 16$.   (h) $\nu = 1.0, a = 32$.

(i) $\nu = 2.0, a = 16$.   (j) $\nu = 1.0, a = 64$.

Figure 5.13. Random realizations of zero mean Gaussian random fields with covariance equal to $\exp(-3(|x - x'|/a)^\nu)$ for various values of ν and a. The total length of the field is 80.

the random field generated with $\nu = 2$ appears to be quite smooth compared to the field generated with $\nu \le 1$.

Remark 5.3. All realizations in Fig. 5.13 were generated using Eq. (5.9). The steps involved in generation of a realization with a given mean and covariance are as follows.

1. Generate a realization of a vector of independent Gaussian random variables Y with 0 mean and variance equal to 1. In Mathematica, a vector z of length 40 is generated with the following statement.
   ```
   z = RandomArray[NormalDistribution[0.,1.],40];
   ```
2. Generate the prior model covariance matrix. If the covariance is of the form $C(m_i, m_j) = f(|x_i - x_j|)$ where the $|x_i - x_j|$ is the distance between y_i and y_j, then the covariance matrix can be computed in Mathematica with a Table statement:
   ```
   covMat = Table[ f[Abs[i*dx - j*dx]], {i,1,40},{j,1,40}];
   ```
   where dx is the lattice spacing.
3. Factor the covariance matrix into the product of an upper and lower triangular matrix. Because the covariance matrix is symmetric, we can use the Cholesky decomposition. The commands to do this in Mathematica are
   ```
   u = CholeskyDecomposition[covMat];
   ```
4. Finally, a realization of the random vector with mean 0 and covariance covMat is obtained using Eq. (5.9).
   ```
   y = Transpose[u].z;
   ```

Most of the covariance models decrease monotonically with increasing distance. The "hole effect" covariance model, however, exhibits negative covariance values. It can be used to describe spatially periodical features of depositions in geology.

$$C(h) = \sigma^2 \left(1 - \frac{h}{L}\right) \exp\left(-\frac{h}{L}\right). \tag{5.15}$$

The hole-effect covariance model is only valid for 1D geometry. For higher dimensions, the covariance matrix for the hole-effect is not positive definite.

In many cases, the plausibility of the geology will be based primarily on the similarity of the experimental covariance and the model covariance. In Fig. 5.14, we compare realizations of one-dimensional random fields with Gaussian, exponential, spherical, and hole-effect covariance functions. Each realization was generated using the same mean, variance, and range. Realizations generated from the exponential and spherical covariance functions are nearly indistinguishable. The random fields are all continuous, but only the one with Gaussian covariance is differentiable.

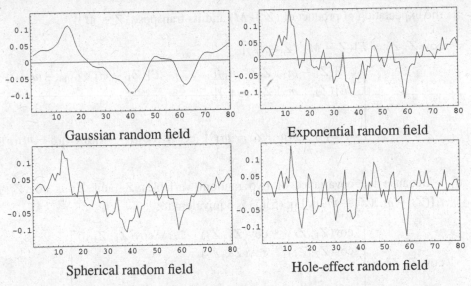

Gaussian random field   Exponential random field

Spherical random field   Hole-effect random field

Figure 5.14. Random multi-variate field with different covariance models and same mean and variance.

5.4.2   Covariance matrix

In cell-centered reservoir simulation models, the reservoir volume is discretized, and the cells are indexed two different ways: the i, j, k index refers to the row, column, and layer, while $m = i + N_i(j - 1) + N_i N_j(k - 1)$ indexes the grids with a single variable that takes values from 1 to N_{grid}. N_i is the number of gridblocks in a row, and N_j is the number of gridblocks in a column of the model. The following equation illustrates the process of vectorizing a two-dimensional 10×10 array into a one-dimensional 100 element vector.

$$Z = \begin{bmatrix} Z_{1,1} & Z_{1,2} & Z_{1,3} & \cdots & Z_{1,10} \\ Z_{2,1} & Z_{2,2} & Z_{2,3} & \cdots & Z_{2,10} \\ \vdots & & & & \vdots \\ Z_{10,1} & Z_{10,2} & Z_{10,3} & \cdots & Z_{10,10} \end{bmatrix}$$

$$\rightarrow \begin{bmatrix} Z_1 & Z_2 & Z_3 & \cdots & Z_{10} \\ Z_{11} & Z_{12} & Z_{13} & \cdots & Z_{20} \\ \vdots & & & & \vdots \\ Z_{91} & Z_{92} & Z_{93} & \cdots & Z_{100} \end{bmatrix}$$

$$\rightarrow [Z_1, Z_2, Z_3, \dots, Z_{100}].$$

(5.16)

Denote the expectation of the random property at each gridblock in this 100 cell model as $M = [m_1, m_2, m_3, \dots, m_{100}]$. The covariance of the properties on the grid is the expectation of product of $(Z - M)$ and its transpose $(Z - M)^T$:

$$\text{cov}(Z, Z) = E[(Z - M)^T(Z - M)]$$

$$= \begin{bmatrix} E[(Z_1 - m_1)(Z_1 - m_1)] & \cdots & E[(Z_1 - m_1)(Z_{100} - m_{100})] \\ E[(Z_2 - m_2)(Z_1 - m_1)] & & \\ \vdots & & \vdots \\ E[(Z_{100} - m_{100})(Z_1 - m_1)] & \cdots & E[(Z_{100} - m_{100})(Z_{100} - m_{100})] \end{bmatrix}.$$

$$(5.17)$$

Because the covariance of two random variables Z_i and Z_j is $\text{cov}(Z_i, Z_j) = E[(Z_i - m_i)(Z_j - m_j)]$, Eq. (5.17) is equivalent to:

$$\text{cov}(Z, Z) = \begin{bmatrix} \text{cov}(Z_1, Z_1) & \text{cov}(Z_1, Z_2) & \cdots & \text{cov}(Z_1, Z_{100}) \\ \text{cov}(Z_2, Z_1) & \text{cov}(Z_2, Z_2) & & \\ \vdots & & & \vdots \\ \text{cov}(Z_{100}, Z_1) & & \cdots & \text{cov}(Z_{100}, Z_{100}) \end{bmatrix}$$

$$= \begin{bmatrix} \text{cov}(Z_{1,1}, Z_{1,1}) & \text{cov}(Z_{1,1}, Z_{1,2}) & \cdots & \text{cov}(Z_{1,1}, Z_{10,10}) \\ \text{cov}(Z_{1,2}, Z_{1,1}) & \text{cov}(Z_{1,2}, Z_{1,2}) & & \\ \vdots & & & \vdots \\ \text{cov}(Z_{10,10}, Z_{1,1}) & & \cdots & \text{cov}(Z_{10,10}, Z_{10,10}) \end{bmatrix}.$$

$$(5.18)$$

The spatial covariance matrix is always symmetric. With the assumption of stationarity, the spatial covariance is only a function of lag distance. If each gridblock is 1 unit by 1 unit square and the values of the random variables are sampled at the gridblock centers, the covariance can be written in terms of the distance between grid centers:

$$\text{cov}(Z, Z) = \begin{bmatrix} C(0) & C(1) & C(2) & \cdots & C(9\sqrt{2}) \\ C(1) & C(0) & C(1) & \cdots & \\ \vdots & & & & \vdots \\ C(9\sqrt{2}) & & & \cdots & C(0) \end{bmatrix}.$$

$$(5.19)$$

5.4.3   Covariance model selection

In many practical situations, the engineer or geologist has general knowledge concerning the smoothness of the random property fields, and hence on the covariance. In the example of the GR and the SP logs, each log type had a different character. For a meandering stream, we may have a basic knowledge of the flow direction and the anisotropy of the deposition along the channel. After sampling from any permeability or porosity field, we can estimate the mean, the variability, and the spatial distribution. The prior information directs us in choosing a geostatistical model and estimating model parameters.

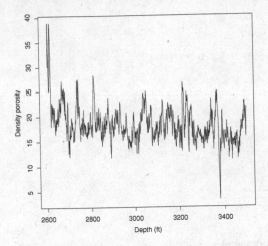

Figure 5.15. Raw data of the density porosity log from well NK-1.

Even though the covariance or variogram can be chosen based on information other than the observations, a structural analysis is generally based on data. The steps in structural analysis are as follows:

- examine the data;
  - plot raw data,
  - look for trends,
  - check the univariate distribution and the need for transformation,
  - check for obvious errors;
- transform the data if necessary to reduce outliers;
- compute and plot an experimental variogram;
- choose an appropriate variogram model and fit it to the experimental variogram.

As an example, we illustrate structural analysis of the porosity log from the NK-1 well. Figure 5.15 is the porosity log versus well depth from 2600 to 3500 feet. There is no obvious trend of porosity over this depth interval other than a slight decrease in porosity with depth. The peak at the depth of 2600 feet may be indication of gas. The univariate distribution of porosity is plotted in histogram as shown in Fig. 5.16. The porosity distribution is slightly skewed with a few outliers at high porosity between 0.35 and 0.40. Transformation is probably not necessary for this case.

The experimental variogram is the binned average of the squared differences between samples. It is calculated as:

$$\gamma(h) = \frac{1}{2 \times N(h)} \sum_{i=1}^{N(h)} (Z(s_i) - Z(s_j))^2,$$

where $Z(s_i)$ is the sample value at location s_i, and $N(h)$ is the number of distinct pairs (s_i, s_j) for which $s_i - s_j = h$. The form of the experimental variogram is very similar

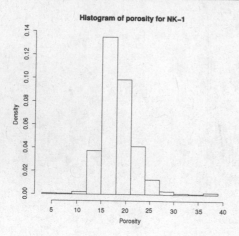

Figure 5.16. The histogram of the porosity samples.

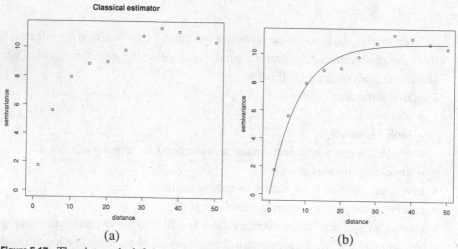

Figure 5.17. The plot on the left-hand side is the experimental variogram based on squared differences, and binning. The plot on the right-hand side is the least-squares fit to the experimental variogram from the full section of 1000 feet.

with that of the experimental covariance. The main difference between the two is that the experimental variogram does not require calculation of the mean. Figure 5.17 (a) shows the experimental variogram for the porosity log with lag distance from 1 foot to 50 feet. In Fig. 5.17 (b), we show a least-squares fit of an exponential variogram to the experimental variogram. The estimated variance is 10.7 and the practical range is 23 feet.

In cases when the amount of data is not adequate for providing accurate estimates of geostatistical parameters, we can end up with an experiential variogram far different from the truth. We illustrate the potential for this type of error by examining the results when data from the NK-1 porosity log in the depth interval from 2600 to 3800 feet is

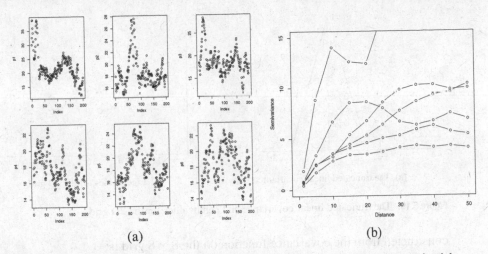

(a)   (b)

Figure 5.18. 200 foot depth intervals from the porosity log of well NK-1 and the experiential variograms computed for porosity in the depth intervals.

divided into six 200 feet sections. The raw data in each section are shown in Fig. 5.18(a). The plot on the upper left-hand corner contains the first 200 feet long section from 2600 to 2800 feet. The experiential variogram curves for each section of the raw data are very different, as shown in Fig. 5.18 (b). The bigger the lag distance, the greater the variation among the curves. The variation is caused either by the small number of data, or a violation of the assumption of stationarity. In practice, adequate numbers of samples can not always be available and it is meaningless to attempt to fit every wiggle in the covariance or variogram accurately. However, the geostatistical model should reflect the major features of the experimental variogram, such as the linearity at short lag distance, and the scale of the correlation range.

Validity of Covariance Functions

Not all functions are valid for use as covariance functions in all dimensions but both the exponential and the Gaussian covariance models can be used for all dimensions. The spherical covariance model can be used for up to three dimensions, while the hole-effect and the truncated linear covariance models can only be used for one dimension. The truncated linear covariance model is of the form:

$$C(h) = \begin{cases} 1 - h/a & \text{for } h \leq a, \\ 0 & \text{for } h > a. \end{cases}$$

Although the truncated linear covariance is not common in practice, it is sometimes used in synthetic one-dimensional models because of its simplicity. Here we demonstrate the lack of validity of the truncated linear covariance model in two dimensions.

Begin by assuming that Y is a 2D random variable on an 8×8 grid with truncated linear covariance as shown in Fig. 5.19 (a). The covariance matrix for Y that is

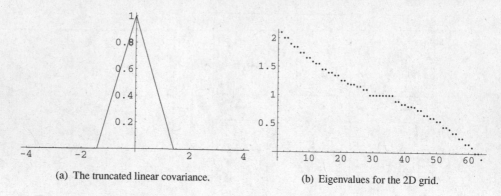

(a) The truncated linear covariance.   (b) Eigenvalues for the 2D grid.

Figure 5.19. The truncated linear covariance and eigenvalues for the 2D grid.

constructed from the covariance function on the 8×8 grid is:

$$
C_Y = \begin{bmatrix}
1 & 0.2928 & 0 & \cdots & 0 & 0.2928 & \cdots \\
0.2928 & 1 & 0.2928 & 0 & \cdots & 0 & \ddots \\
0 & 0.2928 & 1 & 0.2928 & 0 & \cdots & \ddots \\
& \ddots & & \ddots & & \ddots & \ddots
\end{bmatrix}
$$

A valid covariance function must be positive semidefinite (recall Eq. 5.12), in which case the eigenvalues must be nonnegative. The covariance matrix computed using the truncated Gaussian function is not positive definite in 2D, however, since one of the eigenvalues of C_Y is negative, as shown in Fig. 5.19(b).

A second indication of the fact that the truncated linear function can not be used in two dimensions is that the estimated variance can be negative in some situations. Consider a possible random function Y with a truncated linear function for a covariance on an 8×8 grid (Fig. 5.20). Compute the variance of the sum of Y on the white squares minus the Y on the black squares, i.e. $\text{var}(\sum(Y_w - Y_b))$ and define the "checkerboard average," $a = \{-1, 1, -1, 1, -1, 1, -1, 1, 1, -1, 1, -1, 1, -1, 1, -1, -1, 1, -1, 1, -1, 1, -1, 1, 1, -1, 1, -1, 1, -1, 1, -1, -1, 1, -1, 1, -1, 1, -1, 1, 1, -1, 1, -1, 1, -1, 1, -1, -1, -1, 1, -1, 1, -1, 1, -1, 1, 1, -1, 1, -1, 1, -1, 1, -1, 1, -1, 1\}^T.$

Write Y as a 64 elements column vector and let μ_Y be the mean of the random field Y at each gridblock. The variance $\text{var}(\sum(Y_w - Y_b))$ is equivalent to $\text{var}(a^T Y)$.

$$
\begin{aligned}
\text{var}(a^T Y) &= E\left[(a^T Y - a^T \mu_Y)(a^T Y - a^T \mu_Y)^T\right] \\
&= E\left[a^T Y Y^T a - a^T Y \mu_Y^T a - a^T \mu_Y Y^T a + a^T \mu_Y \mu_Y^T a\right] \\
&= a^T E\left[Y Y^T - Y \mu_Y^T - \mu_Y Y^T + \mu_Y \mu_Y^T\right] a \\
&= a^T \left(E[Y Y^T] - \mu_Y \mu_Y^T\right) a \\
&= a^T C_Y a \\
&= -1.584\,16,
\end{aligned}
\tag{5.20}
$$

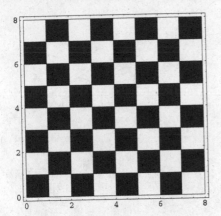

Figure 5.20. Checkerboard averaging operator.

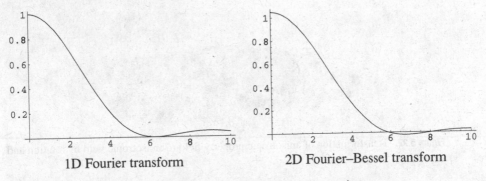

1D Fourier transform   2D Fourier–Bessel transform

Figure 5.21. 1D and 2D Fourier transform of the truncated linear covariance.

which is clearly impossible, hence the truncated linear function is not a valid covariance in two dimensions.

There are two standard methods to check covariance functions for validity. One is to compute the eigenvalues of the covariance *matrix* to ensure that all are nonnegative. The other method is to compute the Fourier transform of the covariance *function*:

$$F(k) = (2\pi)^{n/2} k^{1-n/2} \int_0^\infty r^{n/2} f(r) J_{n/2-1}(rk)\, dr.$$

The Fourier transforms of valid covariance functions are nonnegative. The 1D Fourier transform of the truncated linear covariance is

$$F(k) = \frac{1}{k^2} - \frac{\cos k}{k^2},$$

which is greater than or equal to zero for any k. The Fourier transforms of the truncated linear covariances in 2D, however, are negative for values of k between 6 and 8 (Fig. 5.21), therefore the 2D truncated linear covariance is not valid. The same approach can be used to show that the hole-effect covariance function is invalid in 2D.

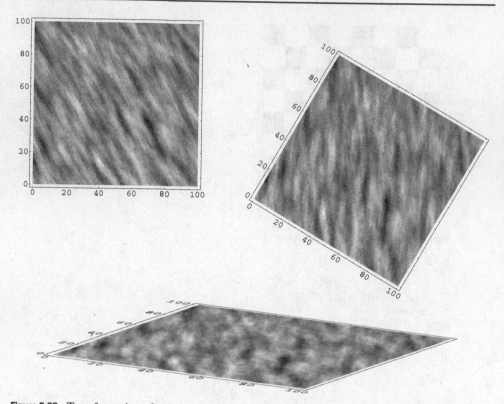

Figure 5.22. Transformation of anisotropic property field to an isotropic field by rotation and stretching.

5.4.4   Anisotropic covariance

To this point we assumed that the covariance of value of a variable at two locations is only a function of the distance between the locations, i.e.

$$\text{cov}(x, x') = f(\|x - x'\|)$$

but in many cases, the correlation depends on the direction, also. The most obvious geologic example is the anisotropy induced by channel deposits in which case the correlation length is much longer in the along-channel direction than in the cross-channel direction. A property field is called geoanisotropic if the covariance of the property field can be made isotropic by *rotating* the coordinate system, then *stretching* one of the coordinates. Figure 5.22 shows an example of a property field that clearly has a longer correlation length in the NNW–SSE direction than in the ENE–WSW direction. Rotation by 30°, followed by a factor of 4 stretching in the horizontal direction results in a field that appears to have the same correlation length in all directions. The rotation from the $x-y$ coordinate system to the $x'-y'$ system is accomplished by

$$x' = x \cos \theta + y \sin \theta,$$
$$y' = -x \sin \theta + y \cos \theta,$$

while the subsequent stretching is given by

$$x'' = x'$$
$$y'' = \alpha y'.$$

These transformations are easily written in matrix form:

$$\begin{bmatrix} x'' \\ y'' \end{bmatrix} = \begin{bmatrix} 1 & 0 \\ 0 & \alpha \end{bmatrix} \begin{bmatrix} \cos \theta & \sin \theta \\ -\sin \theta & \cos \theta \end{bmatrix} \begin{bmatrix} x \\ y \end{bmatrix}$$
$$= MT \begin{bmatrix} x \\ y \end{bmatrix}$$

in which case the distance measure is provided by

$$x''^2 + y''^2 = [x \quad y] T^T M^T M T \begin{bmatrix} x \\ y \end{bmatrix}$$
$$= [x \quad y] H \begin{bmatrix} x \\ y \end{bmatrix}$$

and where H is determined by multiplying $(MT)^T MP$,

$$H = \begin{bmatrix} \cos^2 \theta + \alpha^2 \sin^2 \theta & \cos \theta \sin \theta - \alpha^2 \cos \theta \sin \theta \\ \cos \theta \sin \theta - \alpha^2 \cos \theta \sin \theta & \alpha^2 \cos^2 \theta + \sin^2 \theta \end{bmatrix}.$$

In the x'', y'' coordinate system, everything is isotropic so we can use regular distance.

Figure 5.23 shows experimental variograms from an anisotropic two-dimensional random field. The direction of shortest correlation range appears to be 30°, and the direction of longest correlation range appears to be 120°. Once the principal directions have been determined, the correlation lengths in the two principal directions are estimated by fitting variogram models to the data as shown in Fig. 5.24.

Least-squares estimation provides nearly identical estimates of the variance parameter in the short and long directions, 50.6 and 49.8, respectively. The estimated range parameters in the two principal directions are 4.8 and 14.7. For the spherical variogram model, the range parameter is the same as the practical correlation length.

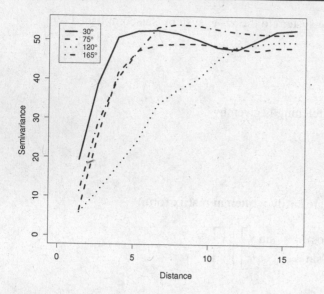

Figure 5.23. Experimental variograms computed in four directions (30°, 75°, 120°, 165°).

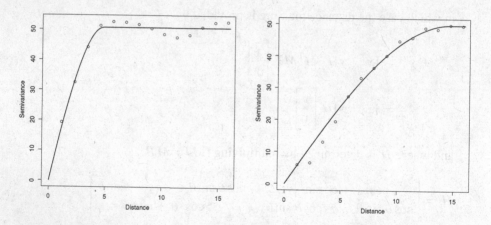

Figure 5.24. Fit the variogram in the direction of shortest correlation length and in the direction of longest correlation length.

5.5   Random processes in function spaces

Denote a random function by $x, y \rightarrow \Phi(x, y)$, and a *realization* of Φ by $x, y \rightarrow \phi(x, y)$. For given x, y, $\Phi(x, y)$ is an ordinary random variable. We will denote its probability density function by $f(\phi; x, y)$. Knowing $f(\phi; x, y)$ we can calculate the mean value of the random function and the standard deviation.

$$\overline{\Phi}(x, y) = \int \phi \, f(\phi; x, y) \, d\phi \tag{5.21}$$

and

$$\sigma_\Phi(x, y)^2 = \int [\phi - \overline{\Phi}(x, y)]^2 f(\phi; x, y) d\phi. \tag{5.22}$$

Now consider the relationship between the random variables $\Phi(x_1, y_1)$ and $\Phi(x_2, y_2)$. Clearly, the univariate probability density function does not give any information on the correlations between the random variables. This information is contained in the 2D joint probability density function $f(\phi_1, \phi_2; x_1, y_1, x_2, y_2)$. The covariance function (sometimes called the auto-covariance function) can be calculated using the following formula.

$$C_\Phi(x, y, x', y') = \int d\phi \int d\phi' [\phi - \Phi(x, y)] [\phi' - \Phi(x', y')] f(\phi, \phi'; x, y, x', y')$$

$$\tag{5.23}$$

A Gaussian random function $\Phi(x, y)$ is completely described by the mean value $\overline{\Phi}(x, y)$ and the covariance. Realizations of a stationary random function are n times differentiable if the correlation function $C(z, 0)$ is $2n$ times differentiable at $z = 0$ [see 4, pg. 423]. Thus a Gaussian random function with an exponential covariance is not differentiable, while a Gaussian random function with a Gaussian covariance is infinitely differentiable.

6 Data

To get an explicit solution of a given boundary value problem is in this age
of large electronic computers no longer a basic question. The problem can
be coded for the machine and the numerical answer obtained. But of what
value is the numerical answer if the scientist does not understand the peculiar
analytical properties and idiosyncrasies of the given operator? [26]

The main purpose of this chapter is to develop an understanding of the spatial dependence of the sensitivity of measurements to reservoir variables, particularly porosity and permeability. The measurements provide information that improve the quality of predictions of reservoir performance. Different types of measurements are sensitive to model variables in different volumes of reservoir, and have much different complexity. Because the focus in this chapter is on qualitative understanding, for each type of data we present a plot of the sensitivity to values of reservoir properties at various locations, without equations. A straightforward, but inefficient, approach to estimating sensitivities would be to make a small change to the value of permeability or porosity in a region, then compute the change in the theoretical measurement. Vela and McKinley [27] used this approach to estimate sensitivity of pulse test data (a type of interference test between wells) to permeability and porosity.

Even without a formal inversion methodology, sensitivities can be useful for parameter estimation. When the predicted data disagree with the observations, the sensitivity information tells which regions of the reservoir could be changed to obtain a better match.

6.1 Production data

Production data refers to measurements of flow rate, pressure, or ratios of flow rates, made in producing or injecting wells. Production observations are made at well locations, which are usually quite limited in number. On the other hand, the measurements are typically repeated frequently (sometimes as often as every few seconds), so the amount of production data for a field can be quite large. These measurements are often closely related to variables of interest for economic evaluation, for example the amount

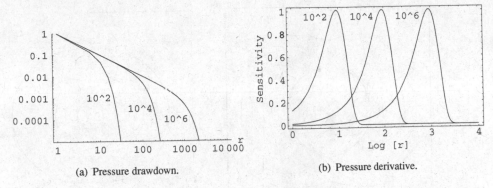

(a) Pressure drawdown.   (b) Pressure derivative.

Figure 6.1. Sensitivity functions for well-test data as functions of distance from the well. (Adapted from Oliver [16], copyright SPE.)

of water that will be produced during the next year. Because the relationships between a measurement such as bottom hole flowing pressure or producing water cut and the spatial distributions of permeability and porosity in the reservoir are typically quite complex, it is difficult to develop intuition for the dependence. The following sections present the dependencies graphically, for a reservoir that is nearly homogeneous.

6.1.1   Drawdown tests

A drawdown test is one in which the pressure drop in the wellbore is measured as a function of time while the well is produced. In general, the pressure drop depends not only on the properties of the reservoir, but also on conditions in the wellbore. For the development of understanding, however, we focus on an ideal situation in which the measurement device is isolated from the wellbore volume by a packer, the near-well region can be neglected, and the fluid is slightly compressible, in which case the response will be primarily controlled by the permeability and porosity of the reservoir. It has of course long been known that the pressure measurement at the end of a producing time t_p is only sensitive to the reservoir properties within a region of investigation whose radius is approximately $r_D = 1.79\sqrt{t_D}$ [28]. This approach says nothing, however, about the relative magnitude of sensitivity of the permeability near the well, versus the permeability near the radius of influence.

Figure 6.1(a) shows that the actual sensitivity is more complex than can be described by the radius of investigation. In fact, the drawdown pressure is highly sensitive to the permeability close to the wellbore and the sensitivity decreases approximately as $1/r$ out to a distance near the radius of investigation at which point the decrease becomes more rapid [16]. The permeability of a reservoir beyond the dimensionless distance $r_D = 2.34\sqrt{t_D}$ can not be determined from the pressure data of length t_D because the measurements are insensitive to reservoir properties beyond that distance. In general, the permeability of the formation is estimated either from the pressure derivative with

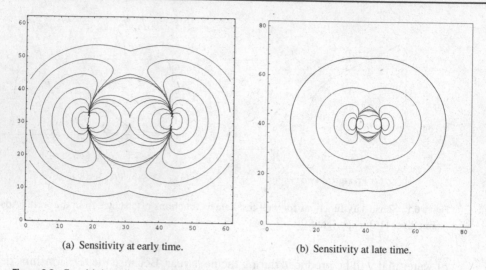

(a) Sensitivity at early time.   (b) Sensitivity at late time.

Figure 6.2. Sensitivity of pressure at the observation well to permeability in the reservoir as functions of location. (Reproduced/modified from Oliver [17] by permission of American Geophysical Union.)

respect to the logarithm of time [29] or from the slope of a plot of pressure drop versus the logarithm of time [30]. As a result, it is probably more important to understand the sensitivity of the pressure derivative to the permeability and porosity fields than to understand the sensitivity of the pressure itself. The pressure derivative with respect to time (Fig. 6.1b) is insensitive to reservoir properties from the well out to a distance of $r_D = 0.12\sqrt{t_D}$, and the maximum sensitivity occurs at a distance $r_D = 0.92\sqrt{t_D}$. The fact that the pressure derivative $(\partial p_w/\partial t)$ appears to be insensitive to the properties of the near-well region is consistent with standard analysis techniques for well test.

6.1.2   Interference tests

In a simple interference test, pressure is measured in a well that is shut in, while another well is produced at a constant rate. The resulting drop in pressure at the observation well is often thought to depend only on the properties between the wells [31, 32]. The actual sensitivity of pressure at the observation well to general two-dimensional nonuniform permeability is, however, relatively complex and can depend strongly on the permeability outside of that region. Figure 6.2 shows that the influence of permeability on pressure drawdown at the observation well is a function of time as well as location. At early times (but after the drawdown is detected) the region of sensitivity is relatively small and somewhat elliptical (Fig. 6.2a).

The sensitivity at a later time (Fig. 6.2b) extends further into the reservoir and becomes more nearly circular. The pressure at the observation well is highly sensitive to the permeability near both wells, even at late times. The sensitivity of the near-well

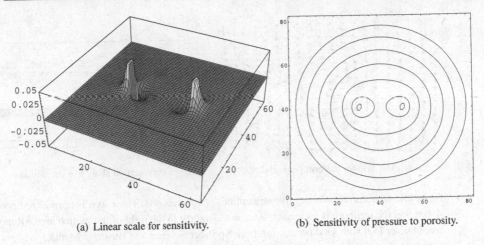

(a) Linear scale for sensitivity.   (b) Sensitivity of pressure to porosity.

Figure 6.3. Sensitivity functions for permeability and porosity as functions of location. (Reproduced/modified from Oliver [17] by permission of American Geophysical Union.)

region is even more obvious in Fig. 6.3(a). Note, however, that because of the mix of positive and negative contributions, if a small-scale nonuniformity in the transmissivity is radially symmetric about the well, the negative and positive contributions to the drawdown will exactly cancel each other.

Although the sensitivity to permeability is quite complex, we can see from Figs. 6.2(a) and 6.2(b) that the boundary of the area of nonzero influence is at least approximately elliptical with foci at the active and observation well locations. Oliver [17] showed that the outer edge of the area of influence approximately satisfies the relation $(r_1 + r_2) = 2r_{inv}$, where r_1 is the distance from the pumping well to a point on the edge of the influence area, r_2 is the distance from the observation well to a point on the edge of the influence area, and r_{inv} is the radius of investigation of a single well drawdown test of identical duration.

Porosity is also likely to vary within the region of influence of an interference test. If the variations in permeability and porosity are both fairly small, the combined effect of porosity and permeability variations can be obtained by summing the influence of variation in each property. The sensitivity of drawdown pressure with respect to variation in porosity at a time corresponding to Fig. 6.2(b) is shown in Fig. 6.3(b). The most obvious difference in the two sensitivities is that the sensitivity to porosity is radially symmetric in the near-wellbore region.

6.1.3   Tracer tests

In a tracer test, an easily identifiable substance is injected into one well in a reservoir and the concentration of the tracer at producing wells is monitored repeatedly. The presence of the tracer in the produced fluid is often used as a qualitative indicator of

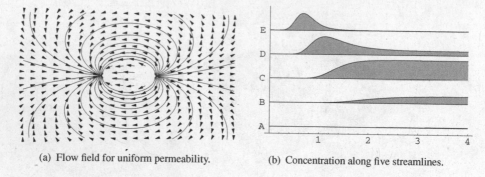

(a) Flow field for uniform permeability.   (b) Concentration along five streamlines.

Figure 6.4. Concentration along five streamlines at 10^6 seconds. The x axis is normalized potential, Φ/m, and the wellbore is located at $\Phi/m = 6$, slightly to the right of the plotted area. (Reproduced from Oliver [39] with kind permission from Springer Science and Business Media.)

connectivity between the injector and the producer. The concentration history can be used quantitatively to estimate porosity and permeability [33–38] but these estimates are typically made under the assumption that the porosity and permeability fields are uniform. When the reservoir is not uniform, the interpretation of data is not so simple. For a nearly homogeneous reservoir, the sensitivity of concentration at the producer depends on the locations of the producers and injectors, but understanding the sensitivity can be developed from consideration of one injector/producer pair.

Imagine that a pulse of tracer is injected in one well, and the concentration of tracer is measured at a second, nearby well. If the field is nearly homogeneous, the flow pattern would be as shown in Fig. 6.4(a). Some fluid travels in a nearly straight path from the injector to the producer at a relatively high velocity. Other fluid flows from the injector to the producer along longer paths at a lower average velocity. Tracer is carried along by the fluid so, neglecting transverse dispersion, the computation of concentration on a streamline is a one-dimensional problem. The tracer is advected and dispersed with time. As the pulse of tracer travels along the flowpath, the concentration slowly disperses as shown in Fig. 6.4(b). The produced concentration reflects the contribution of tracer from all streamlines that intersect the well. In order to compute the producing concentration we need to compute the concentration at a particular time on each streamline, then sum over the streamlines. After several days or weeks, the tracer begins to appear at the producer. The concentration increases for some time, then slowly decays back to the initial level.

The speed of advance of tracer along the flowpath is a function of the average properties along the path, and the speed of advance determines whether or not a particular flowpath contributes tracer to production at any time. The sensitivity of the produced concentration to porosity changes with time as the flowpaths that contribute tracer change with time. Figure 6.5 shows the sensitivities at three different times. Note that only at the earliest times is the concentration sensitive to the region between the wells.

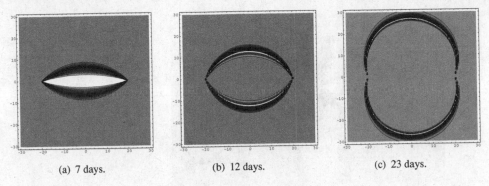

(a) 7 days.   (b) 12 days.   (c) 23 days.

Figure 6.5. Sensitivity of produced tracer concentration to porosity at three different times after start of injection of tracer. (Reproduced from Oliver [39] with kind permission from Springer Science and Business Media.)

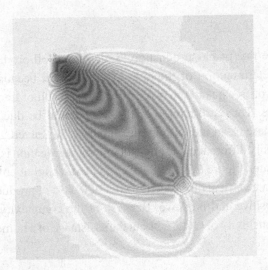

Figure 6.6. Sensitivity of producing water–oil ratio with respect to porosity in a five spot pattern at 1800 days. (Reproduced from Wu *et al.* [41], copyright SPE.)

6.1.4   Water–oil ratio

For a water-flood reservoir, determination of the ratio of the rate of water production to the rate of oil production is in many respects similar to the determination of concentration in a tracer test. Water can be thought of as a tracer, whose concentration is measured at the production well. It is not surprising then that the sensitivity of water–oil ratio to porosity is similar to the sensitivity of tracer concentration to porosity (compare Fig. 6.6 to Fig. 6.5b) and streamline methods are sometimes used to compute sensitivities [40]. The differences are primarily due to the long transition from irreducible water saturation to residual oil saturation, and the mobility variation in a water flood.

Figure 6.7. Sensitivity of producing gas–oil ratio with respect to vertical permeability in vertical section at 400 days. Figure from Li *et al.* [42] is being used with permission from the Petroleum Society of CIM, who retains the copyright.

6.1.5   Gas–oil ratio

Determination of the ratio of the rate of gas production to the rate of oil production (GOR) is more complex than determination of the ratio of water to oil because gas production is a result of the flow of free gas in the reservoir, and the result of transport of dissolved gas in the oil phase. Thus an increase in the GOR could be due to an increase in the flow of the gas phase or an increase in the amount of dissolved gas. In addition, because of the large density difference, the effect of gravity segregation is more pronounced with gas than with water. Figure 6.7 shows the computed sensitivity of gas–oil ratio to vertical permeability in a vertical cross section after 400 days of production. Values of sensitivity are both positive and negative. Because of the complexity, it is nearly impossible to predict the sensitivity of GOR without the assistance of a numerical simulator.

6.1.6   Sources of errors in production data

The data in both interference and drawdown tests are obtained from pressure gauges. Although modern gauges often have resolution of ± 0.01 psi, the accuracy is typically not that good [43]. Reasons include transient temperature effects and drift in the electronics. Of course the reason for quantifying the magnitude of the errors is that it is pointless to force model to match data to greater accuracy than required by the measurement accuracy. The problem of building a model and matching the data by adjusting parameters depends to a large degree on the accuracy of the assumptions and the approximations in the model. It would be pointless, for example, to attempt to match pressures in the early times of a drawdown or buildup test if the simulator did not correctly model wellbore storage. Similarly, the magnitude of the tidal effects in some interference tests can be as large as the pressure variation induced by the production

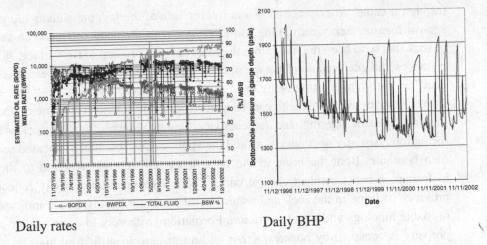

Daily rates   Daily BHP

Figure 6.8. Variability of production data.

rate variation. Although the tidal effects are not actually part of the accuracy of the measurement, the neglect of tidal effects in the modeling would make the data very difficult to match to the accuracy of the gauge.

Rate measurements may be obtained from a separator whose accuracy could probably be estimated reasonably well. In may cases, however, a well is actually tested at most once per month, and intermediate rates are estimated from the well-head pressure. It makes little sense to use the accuracy of the instantaneous measurement as the accuracy of the monthly rate, because it is actually the modeling error that is more important than the measurement error in this case.

Figure 6.8 shows daily measurements of rates and daily measurements of bottom-hole pressure for a well in a large fractured carbonate reservoir. For this reservoir, the rates are typically measured weekly, and daily rates are interpolated from the measured rates. The pressure variability is large, but very little is due to actual measurement error; most is a result of operational fluctuations. Unfortunately, it is not uncommon to use actual pressure measurements (obtained only once per month) with average monthly rates in history matching. The two are not compatible and the result can be classified as modeling error.

6.2 Logs and core data

Two types of commonly acquired data are sensitive only to the region of the reservoir in the immediate vicinity of the wellbore. Well-logging is carried out by lowering a probe equipped with sensors into a well. The sensors make measurements that can be used to estimate rock and fluid properties in the near-well region, and to monitor the well construction. The signals are transmitted from the sensors to the surface through a cable. After processing and interpretation, the logging signals are plotted against measurement depth in log formats. Core measurements are made on samples of rock that have been removed from the reservoir and brought to the surface. In many instances, the interpretations that are obtained from cores and logs are treated as "hard data," an unfortunate terminology that implies absolute accuracy.

Well-logging measurements can be placed into three major categories: electrical logging, acoustic logging, and radioactivity logging. Electric logging measures the formation resistivity or conductivity and the spontaneous potential in uncased sections of a borehole. Both the brine in the pore space and the water bound to the clay in formation rocks conduct electric current. The measured conductivity reflects the presence of brine in the rock and water bound to clay minerals, hence indicates the probable lithology and water saturation. Formations with very low porosity or whose porosity is occupied by nonconductive oil and gas are identified by high resistivity segments in electric logs.

Acoustic logging tools carry an acoustic source in one end and receivers in the other end. The acoustic waves emitted from the source travel through the near-well formation and are recorded by the receivers. The traveltime of the acoustic waves is recorded against the measurement depth. Continuous measurement of acoustic wave traveltime along the well path provides information about porosity, lithology, mechanical rock properties, and the existence of fractures.

Several types of radioactivity logs measure the natural radioactivity of the formation, while others emit radioactive particles and measure the response from the formation. Natural gamma logging recognizes lithologies and soil contents by measuring the natural radiation emitted from the rock. In general, potassium, uranium, and thorium are the primary contributors to natural radioactivity in the sedimentary formations. These elements primarily occur in clay minerals. The natural gamma ray log simply provides a measurement of the total contribution, while the spectral gamma ray log provides an estimate of the contribution from various isotopes.

Two typical logging tools with artificial radioactive sources are the neutron porosity log and the density log. The density log emits gamma rays and records the intensity of scattered gamma rays at the detector. The log provides an estimate of the density of electrons in the formation. The mass density can be estimated fairly accurately from knowledge of the density of electrons. (A correction must be made for hydrogen.) If the densities of the matrix grains and the fluid in the pore space are known, the porosity of the formation can be estimated. The neutron porosity log releases neutrons from its radioactive source, and records the intensity of the returned gamma ray emitted from hydrogen nuclei after neutrons have been captured. The intensity of the signal is proportional to the amount of hydrogen present. The neutron porosity log tends to respond to zones with high porosity (if the fluid is water or oil) and to zones with high clay content because of the bound water. Radioactivity logs can usually be used in both open and cased hole because the gamma ray has sufficient penetration power to the steel casing and the formation.

The only means for direct measurement of formation rock and fluid properties is through representative core samples that have been brought to the surface for analysis. The set of measurements normally are carried out on core plugs approximately 1 or 1.5 inches in diameter instead of on the whole core, which is typically between 2 and 5 inches in diameter. The basic rock and fluid properties to be measured include a lithologic description, porosity, permeability, grain density, fluid saturation, relative permeabilities, capillary pressure relations, and fluid PVT. Measurements are often made at room temperature and at either atmospheric confining pressure or formation confining pressure, or both.

6.2.1   Errors in log and core data

Both the log data and the core data are the measurement of a property of the formation in a near-wellbore region that is typically much smaller than the normal size of a reservoir simulation gridblock. The depths of investigation for well-logging vary considerably in heterogeneous conditions, and with different rock properties. Errors in logging data result from many factors, including randomness in the phenomenon measured (e.g. the count rate for radioactive particles), lack of calibration, and incorrect choice of parameter such as the grain density. In fact, however, it is the interpretation, not the actual "data" from logs that are used by petroleum engineers for reservoir characterization. That is, we use "porosity" from logs that may actually measure neutron density and electron density. Or we may use water saturation as if it were data, instead of the capture cross section for thermal neutrons. The difficulty with this approach is that the calibration of errors propagated to the final result is not easy, so the estimation of the magnitude of errors is usually based on a comparison between multiple estimates.

Although core measurement data are also contaminated by random noise and interpretation error, they are often regarded as accurate data for reservoir characterization. Bias or errors in sampling and handling are probably more important than actual measurement errors for many core properties. Unless great care is taken, fractured or unconsolidated cores may fall apart during coring or transport to the lab, and hence may not be available for measurements. Clays or shales may dry or crack, or wettability might be altered from core cleaning. Even the PVT sample may not be representative of virgin reservoir if it is obtained in the wellbore or at the surface.

6.3   Seismic data

Seismic data are measurements of wave signals that have traveled from a source at the surface, through the formation, are reflected from a subsurface layer, and are recorded by receivers at the surface. The waves carry information about the properties of the formation through which they travel, and are used to map the formation structure and

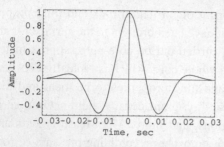

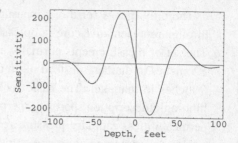

(a) Typical wavelet with most energy at 50 Hz.   (b) Sensitivity of seismic amplitude to porosity.

Figure 6.9. Sensitivity of amplitude to porosity for formation with P-wave velocity of 7000 feet per second.

property fields. To date, a seismic survey is the most widely used geophysical method in subsurface mapping. It has been applied in every stage of the oilfield development. In the early exploration stage, when few wells have been drilled, a seismic survey with large areal coverage is helpful in understanding sedimentary series, lithology, overall formation thickness, and other large-scale properties of sedimentation. In the stage of oilfield development, a detailed seismic mapping provides knowledge of the amount of hydrocarbons in place, which affects well placement and the surface facilities. For mature oil fields, an accurate seismic survey reveals the distribution of by-passed hydrocarbon, which assists on infill-well placement for improving recovery.

The interpretation of seismic data for determination of geologic structure from recorded wave signals has been studied for decades [44–48]. Seismic interpretation can be categorized based on the resolution: seismic stratigraphic interpretation (low resolution) and seismic lithological interpretation (high resolution). Seismic stratigraphic interpretation uses major seismic features to determine sedimentary series, lithology, and other large-scale properties of sedimentation. Structures that are favorable for oil and gas reserves can be recognized by this procedure. Seismic lithological interpretation focusses on a single or a small group of reflection horizons. From inversion of the amplitude, attenuation, velocity, and frequency information of the reflected seismic wave, lithotypes, thickness, porosity, and fluid properties of the target layers can be estimated.

The seismic amplitude is a convolution of the wavelet (Fig. 6.9a) with the derivative of the porosity [see 49] with respect to depth. The sensitivity of seismic reflection amplitude (Fig. 6.9b) is limited to the neighborhood of the measurement point. The range is determined both by the frequency of the seismic signal and the velocity of the wave through the medium.

The seismic wave reflection and refraction amplitudes are determined by four independent acoustic rock properties: the compression wave velocity V_p, the shear wave velocity V_s, the rock density, and the wave attenuation Q. All other seismic attributes depend on the above four properties and their spatial distribution. Therefore, the key task for seismic inversion of rock properties is to determine and model a robust relationship between the reservoir properties (including porosity, lithology, and saturation of fluids) and the seismic amplitudes, which are carried by these four independent attributes. Seismic amplitude is highly sensitive to porosity change, followed by the sensitivity to the lithology and the fluid properties. The pore pressure effects and the saturation variation have the least significant impact on the seismic response, although the order varies in some cases (gas saturated rock, for example).

6.3.1 Seismic data acquisition

Methods of seismic data acquisition are characterized by the type of the seismic source, the dimensions of the receiver array, and the recording tools. When the amplitude of a received seismic wave signal is plotted against the arrival time, it forms a trace. Multiple traces from various shot points are normally recorded at each receiver for later processing. Among the artificial seismic sources, the most common are the explosives, vibrators, air guns, gas exploders, electric sparkers, and electro-acoustic transducers. Geophones are used for receivers on land; they convert the mechanical motion of the Earth into electric signals to be recorded. Hydrophones are used offshore to convert water pressure changes into electric signals. In general, analog amplifiers, filters, and other modern instruments are built into the receivers to preprocess the signals before they are recorded.

The simplest seismic data acquisition is made in 1D. The source and receiver are at nearly the same location on the surface (zero offset), and the two-way time (TWT, time taken for the signals to travel to the interfaces of the formations and reflect back) is measured from the amplitude of the reflection wave. The TWT provides information for estimation of the reflection surface depth. When the formation is dipped, the first arrival reflection wave does not indicate the depth of the formation, but the distance from the source to the formation.

In 2D seismic measurement, a series of sources and receivers are positioned along a straight line at fixed intervals. Seismic signals reflected from a location on the subsurface are normally recorded by more than one receiver. Traces from several source–receiver pairs are stacked in data processing to reduce noise level. Before the 3D seismic survey became popular, an areal geologic map was obtained by shooting a number of parallel 2D seismic lines and interpolating between them.

A 3D seismic data cube is formed after shooting a large number of closely spaced 2D parallel lines. For marine 3D seismic, where the placement of the sources and the receivers is not limited by the surface conditions, shooting 2D parallel lines is affordable and efficient. A typical 3D seismic land survey is more expensive than the marine seismic due to all kinds of restrictions on land, such as inaccessibility due to buildings, forests, mountains, desserts, and so on. Moreover, shallow wells are normally necessary for placement of the seismic source in the base rock instead of the weathered layer. Swath shooting is a common pattern, in which the receivers are laid out in close parallel lines that are perpendicular to the line of sources.

The fourth dimension for the 4D seismic survey (time lapse seismic) is time. By repeating the 3D seismic survey on the same field, changes in hydrocarbon distribution, pore pressure, and rock properties are reflected by a change in the acoustic impedance, which is the product of the compression wave velocity and the media density. This information can be used to improve oil recovery and for evidence of sealing fault and local compartmentation.

6.3.2   Seismic data processing

It is important to understand the basics of seismic processing because it is common to use processed traces for interpretation and inversion. The objective of seismic data processing is to translate recorded seismic traces into visible geological cross sections. The processed data form a trace amplitude plot with the coordinates as distances in horizontal direction(s) and TWT in vertical direction. When the vertical travel velocity profile is known, the time coordinate can be converted into depth.

Seismic data processing is a fairly complicated technique. Yilmaz [45] provides a detailed explanation of procedures and techniques for seismic processing. The three primary stages in seismic data processing are: deconvolution, stacking, and migration. Although the three procedures are generally carried out in that order, the sequence can be altered depending on the requirement of certain data sets.

Deconvolution The process of deconvolution improves the resolution of a trace by compressing the basic seismic wavelet. The impedance contrast among the formation layers causes the seismic wave reflections. A recorded trace can be regarded as the result of the convolution of reflectivity contrasts with a wavelet. Ideally, deconvolution filters out the wavelet components and only keeps the formation reflectivity spikes in a trace. However, in reality, some of the high-frequency noises are also boosted, and the data require extra filtering after the deconvolution.

CMP stacking Common midpoint stacking is a data compression and noise reduction technique. After corrections on time, amplitude, phase, and the normal move-out (NMO), the traces with the same reflection point on the subsurface are added together (stacked) to reduce the random noise carried with an individual trace. CMP stacking also attenuates some coherent noises which have different stacking velocities with the reflected signal. The stacked traces with a common midpoint approximate a zero offset trace with reduced noise level.

Migration When the formation is dipped, it is necessary to migrate and relocate the trace from its CMP position in flat formation to its true subsurface location. The quality of a migrated section is dependent on the previous processing work, the signal to noise (S/N) ratio, and the velocity profile used in migration. If a velocity model that is significantly different from the true velocity is used, then the migrated section can be misleading.

6.3.3   Sources of errors in seismic data

Seismic data interpretation and inversion require an estimate of the level of noise in seismic data. Noise and errors are introduced in the acquisition of the data and in the complicated data processing procedures. Because of the complexity, quantification of the error level from a variety of sources can be very challenging. Here we list a few sources of uncertainties with the processed signal.

1. There are various sources of ambient noise in the recorded traces, such as a poorly planted geophone, the wind motion, the wave motion in the water that vibrates the receivers and cables, the static noise from the recording instruments, and the production operations in an adjacent oil field. The noise filtering process works well when the signal to noise ratio is high. The ambient noise becomes dominate in the deep portion of a stacked section where the signal attenuates. The repeatability is important for 4D seismic survey; unfortunately, the ambient noise varies from one survey to another and is nonrepeatable.

2. The weathered near-surface layer varies in the surface topography, the thickness, and the brine content. It is not practical to place each source and receiver in a hole that reaches under the weathered layer, so corrections for trace arrival time must be made with the assumption of known true velocity profile and the thickness of the weathered layer everywhere in the survey region. As these assumptions are unrealistic, the corrections to the arrival time are only approximate, and the error varies across the survey region.

3. Procedures in the seismic data processing are often interpretive, and the decisions on parameter settings may be based on geological intuition and personal preference. The same set of data processed by different organizations will yield different results. The uncertainty in the data processing is hard to calibrate. In a 4D seismic study, the recorded traces obtained years ago need to be reprocessed along with the data from the recent seismic survey to reduce the errors caused by differences in data processing.

6.3.4   Seismic data inversion

Seismic inversion is a process to integrate seismic data, well-logs, velocity information, and geology to form a robust 3D reservoir model. Based on the input of the techniques, there are two popular methods, the AVO inversion and the acoustic impedance (AI) inversion. Inversion problems are mostly underdetermined, so an infinite number of models could match the seismic data well, and also be distinct with the true lithology.

Although the results of seismic inversion are actually highly processed estimates, they are often used as "data" for a history-matching problem. One example is the use of impedance change data from 4D seismic surveys for history matching of reservoir rock properties [50, 51]. The impedance is closely correlated with the porosity and lithofacies. Seismic amplitudes and the offset carried by the prestack gathering contains information of the lithology contrast and the fluid property contrast, which cause amplitude anomalies.

Unlike other types of data discussed in this chapter, the seismic data for history matching are very dense in comparison with the size of the grids for reservoir dynamic flow simulation. They are also spatially correlated. The history-matching problems require quantification of the error variance and the assumption of data error to be in normal distribution. There are three major problems with error quantification because of the unique features of seismic data. First, the seismic data exist in every gridblock and the errors in data can be expected to be correlated. It is computationally expensive to match such large amount of data, and the data covariance matrix has to be computed and stored. The spatial nonstationarity of the error requires extra work for the conventional history-matching framework. Second, as introduced in the last section, the errors in seismic data for history matching are from various sources, and some noise in the traces has been rescaled and reshaped with the complicated data processing. It is difficult to quantify the residual noise level, and the error in data can be non-Gaussian. Third, the data processing is nonrepeatable because of the assumptions and intuitive parameter selection. The uncertainty with data processing can be large and typically heterogeneous in the cross section.

7 The maximum a posteriori estimate

Although the full solution of an inverse problem is the a posteriori probability density function for the model variables, in petroleum reservoir characterization, the complete characterization of the PDF is generally impractical. One exception is the linear inverse problem with Gaussian prior for model variables and Gaussian noise in the measurements. In this case, the posterior density is also Gaussian, and is completely characterized by the mean and covariance.

7.1 Conditional probability for linear problems

A linear inverse problem is one for which the theoretical data are related to the model variables by

$$d = Gm, \tag{7.1}$$

where d is an N vector of theoretical data, m is an M vector of model variables, G is an $N \times M$ matrix that describes the linear theoretical relationship between the variables and the data. If the observation (measurement) errors in the data, ϵ, can be quantified, and there is no modeling error, then the relationship between observations and model variables is

$$d_{\text{obs}} = Gm + \epsilon. \tag{7.2}$$

The observation errors are often assumed to be normally distributed with mean 0 and covariance matrix C_D.

For this relationship, the probability of measuring data d_{obs}, given the model variables, is simply the probability of ϵ, i.e.

$$p(d_{\text{obs}}|m) = p(\epsilon = d_{\text{obs}} - Gm)$$
$$\propto \exp\left[-\frac{1}{2}(d_{\text{obs}} - Gm)^{\text{T}} C_{\text{D}}^{-1}(d_{\text{obs}} - Gm)\right]. \tag{7.3}$$

If the values of the model variables are initially uncertain and the uncertainty can be represented by a Gaussian probability density function, then the prior uncertainty on model variables m is described by the following:

$$p(m) \propto \exp\left[-\frac{1}{2}(m - m_{pr})^{T}C_{M}^{-1}(m - m_{pr})\right], \tag{7.4}$$

where m_{pr} is the prior estimate of the variables and C_M is the prior model variable covariance.

Bayes' rule allows us to infer the probability distribution of the model parameters conditional to the data, from the product of the likelihood function (Eq. 7.3) and the prior probability (Eq. 7.4),

$$p(m|d_{obs}) \propto p(d_{obs}|m)p(m)$$

$$\propto \exp\left[-\frac{1}{2}(d_{obs} - Gm)^{T}C_{D}^{-1}(d_{obs} - Gm)\right]\exp\left[-\frac{1}{2}(m - m_{pr})^{T}C_{M}^{-1}(m - m_{pr})\right]$$

$$\propto \exp\left[-\frac{1}{2}(d_{obs} - Gm)^{T}C_{D}^{-1}(d_{obs} - Gm) - \frac{1}{2}(m - m_{pr})^{T}C_{M}^{-1}(m - m_{pr})\right] \tag{7.5}$$

or

$$p(m|d_{obs}) \propto \exp[-S(m)], \tag{7.6}$$

where

$$S(m) = \frac{1}{2}(d_{obs} - Gm)^{T}C_{D}^{-1}(d_{obs} - Gm) + \frac{1}{2}(m - m_{pr})^{T}C_{M}^{-1}(m - m_{pr}). \tag{7.7}$$

7.1.1 Posteriori mean

If the terms in the argument of the exponential in Eq. (7.5), can be rearranged in the following quadratic form

$$(x - A^{-1}B)^{T}A(x - A^{-1}B) = x^{T}Ax - x^{T}B - B^{T}x + B^{T}A^{-1}B \tag{7.8}$$

then the probability density is Gaussian with covariance matrix given by A^{-1} and mean equal to $A^{-1}B$.

Expanding the terms in Eq. (7.7) we obtain:

$$\begin{aligned}
2S(m) &= m^{T}C_{M}^{-1}m - m_{pr}^{T}C_{M}^{-1}m - m^{T}C_{M}^{-1}m_{pr} + m_{pr}^{T}C_{M}^{-1}m_{pr} \\
&\quad + m^{T}G^{T}C_{D}^{-1}Gm - d_{obs}^{T}C_{D}^{-1}Gm - m^{T}G^{T}C_{D}^{-1}d_{obs} + d_{obs}^{T}C_{D}^{-1}d_{obs} \\
&= m^{T}(G^{T}C_{D}^{-1}G + C_{M}^{-1})m - (m_{pr}^{T}C_{M}^{-1} + d_{obs}^{T}C_{D}^{-1}G)m \\
&\quad - m^{T}(C_{M}^{-1}m_{pr} + G^{T}C_{D}^{-1}d_{obs}) + m_{pr}^{T}C_{M}^{-1}m_{pr} + d_{obs}^{T}C_{D}^{-1}d_{obs}.
\end{aligned} \tag{7.9}$$

Comparison of terms in Eqs. (7.8) and (7.9) shows that the posteriori covariance matrix is

$$C_{M'} = (C_{M}^{-1} + G^{T}C_{D}^{-1}G)^{-1} \tag{7.10}$$

and the mean of the Gaussian PDF is

$$\begin{aligned}
\langle m \rangle &= (C_M^{-1} + G^T C_D^{-1} G)^{-1}(C_M^{-1} m_{pr} + G^T C_D^{-1} d_{obs}) \\
&= C_{M'}(C_M^{-1} m_{pr} + G^T C_D^{-1} d_{obs}) \\
&= C_{M'} C_M^{-1} m_{pr} + C_{M'} G^T C_D^{-1} d_{obs} \\
&= C_{M'}(G^T C_D^{-1} G + C_M^{-1} - G^T C_D^{-1} G) m_{pr} + C_{M'} G^T C_D^{-1} d_{obs} \\
&= C_{M'}(C_{M'}^{-1} - G^T C_D^{-1} G) m_{pr} + C_{M'} G^T C_D^{-1} d_{obs} \\
&= m_{pr} + (C_M^{-1} + G^T C_D^{-1} G)^{-1} G^T C_D^{-1} (d_{obs} - G m_{pr}).
\end{aligned} \tag{7.11}$$

These equations for the mean and the covariance are particularly useful when the number of model parameters is fairly small. When the number of data is considerably less than the number of model parameters, it is more useful to use another form of the solution:

$$\langle m \rangle = m_{pr} + C_M G^T (G C_M G^T + C_D)^{-1} (d_{obs} - G m_{pr}). \tag{7.12}$$

Although the preceding two equations are equivalent, in practice, computing $\langle m \rangle$ with Eq. (7.11) requires the numerical solution of the $M \times M$ matrix problem

$$(C_M^{-1} + G^T C_D^{-1} G) x = G^1 C_D^{-1} (d_{obs} - G m_{pr}) \tag{7.13}$$

for x, whereas, application of Eq. (7.12) requires the solution of the $N_d \times N_d$ matrix problem

$$(C_D + G C_M G^T) y = (d_{obs} - G m_{pr}) \tag{7.14}$$

for y. Thus, the choice of formula appears to be largely dictated by which of the two preceding matrix problems can be solved more efficiently. Since computation efficiency is directly related to the size of the matrix problem, Eq. (7.11) seems preferable if $M \ll N$. However, Eq. (7.11) involves C_M^{-1}. If this inverse must be computed from a given C_M, then it is necessary to solve an $M \times M$ matrix problem M times in order to compute C_M^{-1} and if M is large, this significantly degrades the computation efficiency of Eq. (7.11). For problems of practical interest, M may be on the order of 10^4 to 10^6, and thus direct application of Eq. (7.11) is usually not recommended unless the model can be reparameterized to reduce M. If measured data can be reasonably thinned to a relatively small number N where $N \ll M$, then Eq. (7.12) is usually quite computationally efficient even if M is large. Reparameterization of the model and computational efficiency issues will be discussed in much greater detail when nonlinear inverse problems are considered.

7.1.2   Maximum of the posterior PDF

The model variables with the highest probability density can be estimated by maximizing the logarithm of the posterior probability density function (Eq. 7.5), or minimizing

$S(m)$ in Eq. (7.7). The expression to be minimized is often termed the *objective function*. For a linear relationship between the model variables and the data, the objective function is quadratic in m. Although the determination of the minimum of the objective function can be accomplished easily for a Gaussian using the method of the preceding section, it is worthwhile to repeat the exercise by determining the value for which the gradient of the objective function vanishes. The objective function (Eq. 7.7) is repeated here as

$$S(m) = \frac{1}{2}(m - m_{\mathrm{pr}})^{\mathrm{T}} C_{\mathrm{M}}^{-1}(m - m_{\mathrm{pr}}) + \frac{1}{2}(Gm - d_{\mathrm{obs}})^{\mathrm{T}} C_{\mathrm{D}}^{-1}(Gm - d_{\mathrm{obs}}). \tag{7.15}$$

Using the operational results on gradients presented in Section 3.2.1, it is straightforward to show that

$$\nabla S = C_{\mathrm{M}}^{-1}(m - m_{\mathrm{pr}}) + G^{\mathrm{T}} C_{\mathrm{D}}^{-1}(Gm - d_{\mathrm{obs}}), \tag{7.16}$$

with the associated Hessian matrix given by

$$H = \nabla[(\nabla S)^{\mathrm{T}}] = C_{\mathrm{M}}^{-1} + G^{\mathrm{T}} C_{\mathrm{D}}^{-1} G. \tag{7.17}$$

H can be shown to be positive definite and hence $S(m)$ has a unique global minimum which can be found by setting $\nabla S = 0$ in Eq. (7.16) and solving for m. The resulting variables with maximum a posteriori probability density are denoted by m_{map} and given by

$$m_{\mathrm{map}} = m_{\mathrm{pr}} + (C_{\mathrm{M}}^{-1} + G^{\mathrm{T}} C_{\mathrm{D}}^{-1} G)^{-1} G^{\mathrm{T}} C_{\mathrm{D}}^{-1}(d_{\mathrm{obs}} - Gm_{\mathrm{pr}}). \tag{7.18}$$

By applying the result of Eq. (7.44), an equivalent expression for m_{map} is obtained:

$$m_{\mathrm{map}} = m_{\mathrm{pr}} + C_{\mathrm{M}} G^{\mathrm{T}}(C_{\mathrm{D}} + G C_{\mathrm{M}} G^{\mathrm{T}})^{-1}(d_{\mathrm{obs}} - Gm_{\mathrm{pr}}). \tag{7.19}$$

7.1.3   The posteriori covariance and variance

For the Gaussian distribution, the covariance describes the joint uncertainty in all variables. When the problem contains many variables, it is impractical to compute the entire covariance matrix and the form of the a posteriori covariance derived by completing the square

$$C_{\mathrm{M}'} = \left(C_{\mathrm{M}}^{-1} + G^{\mathrm{T}} C_{\mathrm{D}}^{-1} G\right)^{-1} \tag{7.20}$$

is not well suited for large models. Instead, when the number of data is small compared to the number of model variables, it is better to use an alternative formulation:

$$C_{\mathrm{M}'} = C_{\mathrm{M}} - C_{\mathrm{M}} G^{\mathrm{T}}\left(G C_{\mathrm{M}} G^{\mathrm{T}} + C_{\mathrm{D}}\right)^{-1} G C_{\mathrm{M}}. \tag{7.21}$$

(See page 142 for a demonstration of the equivalence of Eqs. (7.20) and (7.21).) For large models, even this form results in an impractically large computation, with results that are not easy to visualize. In these cases, it is often preferable to compute only the variance of each model variable. If $C_{M,i}$ denotes the ith column of C_M then the variance of the ith model variable is

$$(C_{M'})_{i,i} = (C_M)_{i,i} - C_{M,i}^T G^T (G C_M G^T + C_D)^{-1} G C_{M,i}. \tag{7.22}$$

7.2   Model resolution

Neither seismic data, nor production data are conducive to the estimation of reservoir properties at a single point in space, and while it is possible to estimate a value, it must be recognized that the estimate is actually an average of the true properties over a region. If there were no errors in the data then the observed data would be related to the true model parameters by the equation

$$d_{obs} = G m_{true}.$$

Upon substitution into Eq. (7.12), we see that

$$\langle m \rangle = m_{pr} + C_M G^T (G C_M G^T + C_D)^{-1} (G m_{true} - G m_{pr}) \tag{7.23}$$
$$= m_{pr} + C_M G^T (G C_M G^T + C_D)^{-1} G (m_{true} - m_{pr}).$$

In other words, the difference between the maximum a posteriori estimate of the model parameters and the prior estimate is the product of a matrix and the difference between the true model parameters and the prior estimate, i.e.

$$\langle m \rangle - m_{pr} = R(m_{true} - m_{pr}), \tag{7.24}$$

where the model resolution, R, is given by

$$R = C_M G^T (C_D + G C_M G^T)^{-1} G. \tag{7.25}$$

How do we interpret the model resolution? Consider the individual terms in the definition of resolution.

$$
\begin{bmatrix}
\langle m_1 \rangle - m_{pr,1} \\
\langle m_2 \rangle - m_{pr,2} \\
\vdots \\
\langle m_M \rangle - m_{pr,M}
\end{bmatrix}
=
\begin{bmatrix}
R_{11} & R_{12} & \cdots & R_{1M} \\
R_{21} & R_{22} & & \vdots \\
\vdots & & \ddots & \\
R_{M1} & \cdots & & R_{MM}
\end{bmatrix}
\begin{bmatrix}
m_{true,1} - m_{pr,1} \\
m_{true,2} - m_{pr,2} \\
\vdots \\
m_{true,M} - m_{pr,M}
\end{bmatrix}. \tag{7.26}
$$

Clearly, then, if we want to know how much "averaging" went into our estimate of m_j, we should look at the jth row of R. An easy way to obtain this, without computing the entire resolution matrix, is to compute the jth column of R^T, where

$$R^T = G^T (G C_M G^T + C_D)^{-1} G C_M. \tag{7.27}$$

When the number of data is small compared to the number of model variables, the jth column of R^T is computed efficiently from

$$(R^T)_j = G^T(GC_M G^T + C_D)^{-1} G C_{M,j}, \tag{7.28}$$

where $C_{M,j}$ is the jth column of the model covariance matrix C_M.

7.2.1 Spread of the resolution

The model resolution, for the estimated value of a variable at a particular location, can be displayed as a map showing the influence of values at one part of the reservoir on the estimate at a particular location. But, because it is a map, it is difficult to tell if changing the type of data results in better or worse estimates of the model parameters. For this type of decision, a single number to measure the "goodness of the resolution" is useful. Since the model resolution of a perfectly resolved model is the identity transform, a measure of the closeness of the model resolution to the identity matrix is an appropriate choice. The Dirichlet spread function is such a measure [2]:

$$S(R_j) = \sum_{i=1}^{M} (r_{ij} - \delta_{ij})^2. \tag{7.29}$$

Note that $S(R_j) = 0$ if $r_{ij} = \delta_{ij}$ and $S(R_j) > 0$ for any other resolution, so a small spread means that the model is well resolved.

7.2.2 1D example – well-logs

By placing sensors at different spacing, induction logs often record two curves corresponding to a medium and deep volume of investigation. In addition, a microresistivity log or a shallow laterlog (a logging tool in which the current is forced to flow laterally away from the borehole) might measure the resistivity close to the wellbore. For simplicity, we will assume that the actual sensitivities of the three measurements are as shown in Fig. 7.1. In each case, the sensitivity of the tool is greatest near the wellbore but the deep measurement "sees" further into the reservoir.

During drilling operations, it is common for some of the mud filtrate to invade the formation. The mud filtrate often has a different resistivity than the formation fluid it displaces. If the resistivity of the uninvaded zone, and the resistivity of the fully invaded zone can both be estimated, it is possible to estimate the resistivity of the in situ water. If there were no mudcake and the depth of invasion can be estimated, then it is also possible to estimate formation permeability.

Consider the case in which, prior to any measurement, the best estimate of formation resistivity is 100 ohm-m, with an uncertainty described by a Gaussian covariance with

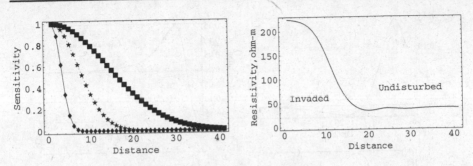

Figure 7.1. The sensitivities of shallow, medium, and deep resistivity logs to formation resistivity (left), and the true resistivity profile (right).

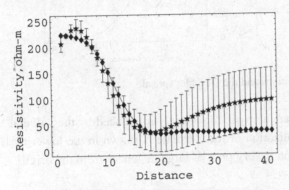

Figure 7.2. Comparison of the truth, with the MAP estimate. Error bars are $\pm 2\sigma$ from the estimate.

practical range equal to about 2.9 units and a variance of 900 (ohm-m)^2, that is

$$C_M = 900 \exp\left(-\frac{x^2}{5^2}\right).$$

Three data, corresponding to the sensitivities and synthetic true resistivity of Fig. 7.1, were generated. The measurements were assumed to be independent and quite accurate ($\sigma_D = 0.01$ ohm-m).

$$d_{obs} = [679.29, 1607.59, 2359.5]^T.$$

Because the number of data (3) is small compared to the number of model variables (40), it is appropriate to use Eq. (7.19) to compute the maximum a posteriori model estimate, and to use Eq. (7.22) to compute the posterior variance. The MAP estimate is compared to the truth in Fig. 7.2. The agreement is quite good, except for the region beyond $x = 25$ where the true values are approximately 30 ohm-m and the MAP estimate has returned to the prior value of 100 ohm-m because none of the data are sensitive to that region.

Figure 7.3 shows the model resolution at three different locations, $x = 5$, 10, and 15. The estimate at $x = 5$ is seen to be an average of values of resistivity in the range

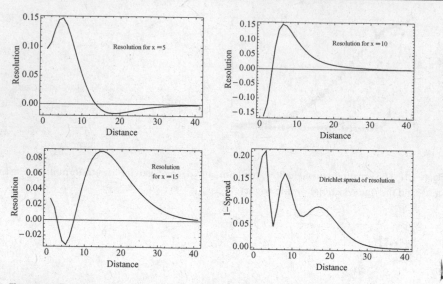

Figure 7.3. Resolution at three locations and the Dirichlet spread.

$0 < x < 10$, while the estimate at $x = 10$ is largely influenced by the value at $x = 7$. A scalar measure of the localization of the estimate is shown in the lower right-hand side of Fig. 7.3. The resolution is very poor at large distances from the origin.

7.2.3   Well-testing example

To illustrate the usefulness of the resolution concept in optimizing data acquisition, consider the problem of estimation of the average gridblock permeability at a location between several wells (Fig. 7.4). The data to be used for the estimates are assumed to come from well tests at locations W1, W2, and W3 and from an interference test between W1 and W2. Assuming that well tests have already been run at W1 and W2, is it better to add information from a third well test at W3, or to run an interference test between W1 and W2? Of course, if porosity is also unknown the best test to run for the estimation of permeability might be different from the best test to run for the estimation of porosity.

The calculation of the resolution requires calculation of the sensitivity of pressure to variation in each of the model parameters. Normally, one would calculate G numerically for arbitrary permeability and porosity distributions using a reservoir simulator [53, 54]. For this example, we have instead used analytical expressions from Oliver [16], Jacquard [55], and Chu and Reynolds [56] for the sensitivity to well pressure observations and derivatives of pressure with respect to time, and expressions from Oliver [17] for sensitivity to observation-well pressures. The analytic expressions for sensitivity are based on small variations of model parameters from a uniform distribution.

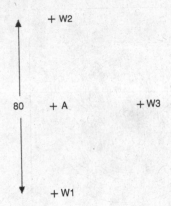

Figure 7.4. The location of the estimation points, A, in relation to the surrounding wells (W1, W2, and W3) from which pressure data can be obtained. Distances are given in meters [52].

In well tests, the data are measurements of pressure at discrete times. Although pressure derivatives are not actually data, pressure derivatives are used for this example as if they were data because the shapes of the sensitivity functions for pressure derivative are more conducive to estimation of properties, and the estimation of the pressure derivatives from pressure data is straightforward.

The dimensionless sensitivity of the pressure derivative from well test data to variation in the average of the logarithm of permeability over a gridblock of area A_B is approximately [56]

$$G_Y^{wt} = -\frac{A_B r^2}{8\pi t^2} \exp\left(-\frac{r^2}{2t}\right)\left[K_1\left(\frac{r^2}{2t}\right) + K_0\left(\frac{r^2}{2t}\right)\right], \tag{7.30}$$

where r is the dimensionless distance from the center of the gridblock to the well location, t is dimensionless time, and K_0 and K_1 are modified Bessel functions of the second kind. Slightly better approximations could be obtained by integrating the point sensitivity values over the gridblock area instead of simply using the midpoint value but, because we used gridblocks that are relatively small compared to the variation in the sensitivity function, the correction for this approximation would be very small everywhere except in the vicinity of the well.

If interference-test pressures, caused by drawdown at location $x = -a$, $y = 0$, are observed at location $x = a$, $y = 0$, the sensitivity of observed pressures to variation in log of permeability at grid (x, y) is given by [17]

$$G_Y^{int} = -\frac{A_B(x^2 - a^2 + y^2)}{8\pi r_1^2} \int_0^t \frac{1}{(t - t')^2} \exp\left[-\frac{r_1^2}{4t'} - \frac{r_2^2}{4(t - t')}\right] dt', \tag{7.31}$$

where $r_1 = \left[(x + a)^2 + y^2\right]^{1/2}$ is the dimensionless distance from the drawdown well and r_2 is the dimensionless distance from the observation point.

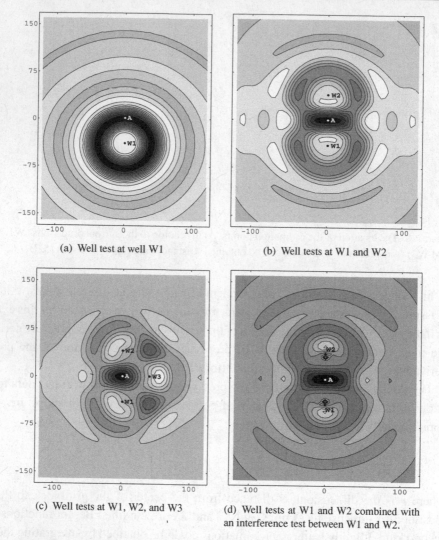

(a) Well test at well W1

(b) Well tests at W1 and W2

(c) Well tests at W1, W2, and W3

(d) Well tests at W1 and W2 combined with an interference test between W1 and W2.

Figure 7.5. Maps of the resolution for estimation of permeability at location A from observations of pressure derivative. Darker shades indicate a greater influence on the estimate [52].

7.2.4   The resolution of permeability

Recall from Eq. (7.24) that $m^{\text{est}} - m^{\text{pr}} = R(m^{\text{true}} - m^{\text{pr}})$, so the jth component of m^{est} is a weighted average of the true model parameters. The weighting is given by the jth row of R, that is,

$$m_j^{\text{est}} - m_j^{\text{pr}} = \sum_{k=1}^{M} r_{jk}(m_k^{\text{true}} - m_k^{\text{pr}}), \tag{7.32}$$

so a plot of the value of r_{jk} at the location of gridblock k shows the weighting given to the permeability within that gridblock.

Figure 7.5(a), which shows the resolution of the permeability at the estimation point, A, confirms that a single well test is not very useful for estimating the permeability at a point in the interwell region because it contains no directional information. The estimate of permeability at point A, based on pressure derivative data from well W1 only, is an average of the permeabilities in an annulus around the well. Although it is not shown, the estimate of permeability at A is increased by higher porosity near well W1 and decreased by higher porosity in the annular region surrounding W1 out to the radius of location A. The influence of variation in porosity, however, is smaller than the influence of variation in permeability.

Figure 7.5(b) shows significant improvement in the resolution of the estimate due to the addition of well-test data from the well at W2. Although the estimate of permeability at A is influenced by permeabilities in the annular regions surrounding both wells, the largest influence is from the neighborhood of the estimation point.

Adding an interference test to the available data does little in this case to improve the spread of the resolution of permeability. Indeed, the shape of the resolution, as seen in Fig. 7.5(d), is remarkably similar to the shape of the resolution obtained from only the two well tests. Adding data from a third well test is a more effective way to estimate the permeability at the midpoint location than running the interference test. Figure 7.5(c) shows that the primary effect of adding the well test at W3 is to reduce the "east–west" width of the peak in the resolution at A.

7.3 Doubly stochastic Gaussian random field

In standard geostatistical practice, the covariance of permeability or porosity is typically estimated from point measurements of these properties. This covariance estimate is subsequently used for *estimation* of property fields at locations where measurements were not available and also for *simulation* of values at those locations. Uncertainty in the variogram or mean is seldom dealt with systematically although it has been discussed a number of times in the geostatistical literature [57–59].

Let us consider a simplification of the "doubly stochastic model" of Tjelmeland *et al.* [57]. In their model, both the mean and the variance were allowed to be unknowns to be determined by the data. Here, we consider only the simpler case in which the prior mean is uncertain, but the prior variance in the local roughness is known. The reservoir characteristics, X, will be written as the sum of an uncertain uniform mean, Θ, with prior mean, μ_θ, and variance, C_θ, and a Gaussian random field, U_m, with mean, μ_m, and spatial covariance given by C_U.

$$[X|\Theta = \theta] = E\theta + U_m, \tag{7.33}$$

where θ is a low-order vector of real variables that are added to the log-permeability and porosity fields to adjust the entire fields up or down in value. The random field U_m is potentially infinite dimensional although, in practice, U_m is only evaluated on a finite-dimensional lattice. E is a matrix whose ith column has value 1 for every lattice point corresponding to θ_i and 0 elsewhere, thus $E\theta$ is the matrix of vectors that take the value θ_i everywhere on the lattice.

> **Example:** suppose that X is the random vector of gridblock porosities and log-permeabilities,
>
> $$
> X = \begin{bmatrix} \phi_1 \\ \vdots \\ \phi_{N_g} \\ \ln k_1 \\ \vdots \\ \ln k_{N_g} \end{bmatrix}, \quad
> E = \begin{bmatrix} 1 & 0 \\ \vdots & \vdots \\ 1 & 0 \\ 0 & 1 \\ \vdots & \vdots \\ 0 & 1 \end{bmatrix}, \quad \text{and} \quad
> \theta = \begin{bmatrix} \theta_\phi \\ \theta_{\ln k} \end{bmatrix}, \quad \text{so} \quad
> E\theta = \begin{bmatrix} \theta_\phi \\ \vdots \\ \theta_\phi \\ \theta_{\ln k} \\ \vdots \\ \theta_{\ln k} \end{bmatrix}.
> $$

If we now define

$$
m = \begin{bmatrix} U_m \\ \theta \end{bmatrix}, \qquad m_{\text{prior}} = \begin{bmatrix} \mu_m \\ \mu_\theta \end{bmatrix}, \tag{7.34}
$$

$$
C_M = \begin{bmatrix} C_U & 0 \\ 0 & C_\theta \end{bmatrix}, \tag{7.35}
$$

we can write,

$$
x = \begin{bmatrix} I & E \end{bmatrix} m. \tag{7.36}
$$

Under these assumptions, the prior probability density for M is

$$
P_M(m) = C_1 \exp\left[-\frac{1}{2}(m - m_{\text{prior}})^{\mathrm{T}} C_M^{-1}(m - m_{\text{prior}}) \right]. \tag{7.37}
$$

It is useful to note that this can also be written as

$$
P(U_m, \theta) = C_1 \exp\left[-\frac{1}{2}(U_m - \mu_m)^{\mathrm{T}} C_U^{-1}(U_m - \mu_m) - \frac{1}{2}(\theta - \mu_\theta)^{\mathrm{T}} C_\theta^{-1}(\theta - \mu_\theta) \right].
$$

The conditional probability density for the reservoir characteristics X given the mean θ is

$$
P(x|\theta) = C_2 \exp\left[-\frac{1}{2}(x - (\mu_m + E\theta))^{\mathrm{T}} C_U^{-1}(x - (\mu_m + E\theta)) \right],
$$

therefore, if the mean is known, the PDF for X is Gaussian with covariance given by C_U.

For problems in which the reservoir characteristics are related to the observed data by

$$
d_{\text{obs}} = G_x x + U_d,
$$

the likelihood of the reservoir characteristics, given the data, is

$$P(d_{\text{obs}}|M = m) = C_2 \exp\left[-\frac{1}{2}(G_x x - d_{\text{obs}})^{\text{T}} C_{\text{D}}^{-1}(G_x x - d_{\text{obs}})\right], \tag{7.38}$$

where C_{D} is the covariance of the measurement errors (and includes the modeling error if that is necessary).

The a posteriori probability for m is the product of the probability densities in Eqs. (7.37) and (7.38), except for a constant factor which makes the probability density integrate to 1. From Eq. (7.36) we note that $G_x x = G_x \begin{bmatrix} I & E \end{bmatrix} m = G_m m$ so

$$G_m = G_x \begin{bmatrix} I & E \end{bmatrix}. \tag{7.39}$$

and the maximum of the a posteriori probability density function occurs when

$$\begin{aligned} S(m) &= \frac{1}{2}(m - m_{\text{prior}})^{\text{T}} C_{\text{M}}^{-1}(m - m_{\text{prior}}) + \frac{1}{2}(G_x x - d_{\text{obs}})^{\text{T}} C_{\text{D}}^{-1}(G_x x - d_{\text{obs}}) \\ &= \frac{1}{2}(m - m_{\text{prior}})^{\text{T}} C_{\text{M}}^{-1}(m - m_{\text{prior}}) + \frac{1}{2}(G_m m - d_{\text{obs}})^{\text{T}} C_{\text{D}}^{-1}(G_m m - d_{\text{obs}}) \end{aligned} \tag{7.40}$$

is a minimum. To find the maximum a posteriori model we must solve

$$\begin{aligned} \nabla_m S(m) &= C_{\text{M}}^{-1}(m - m_{\text{prior}}) + G_m^{\text{T}} C_{\text{D}}^{-1}(G_m m - d_{\text{obs}}) \\ &= C_{\text{M}}^{-1}(m - m_{\text{prior}}) + G_m^{\text{T}} C_{\text{D}}^{-1}(G_m m - G_m m_{\text{prior}} + G_m m_{\text{prior}} - d_{\text{obs}}) \\ &= (C_{\text{M}}^{-1} + G_m^{\text{T}} C_{\text{D}}^{-1} G_m)(m - m_{\text{prior}}) + G_m^{\text{T}} C_{\text{D}}^{-1}(G_m m_{\text{prior}} - d_{\text{obs}}) \\ &= 0 \end{aligned} \tag{7.41}$$

for m. We obtain

$$\begin{aligned} m_{\text{map}} &= m_{\text{prior}} + (C_{\text{M}}^{-1} + G_m^{\text{T}} C_{\text{D}}^{-1} G_m)^{-1} G_m^{\text{T}} C_{\text{D}}^{-1}(d_{\text{obs}} - G_m m_{\text{prior}}) \\ &= m_{\text{prior}} + C_{\text{M}} G_m^{\text{T}}[G_m C_{\text{M}} G_m^{\text{T}} + C_{\text{D}}]^{-1}[d_{\text{obs}} - G_m m_{\text{prior}}]. \end{aligned} \tag{7.42}$$

The a posteriori covariance for the model parameters is

$$\begin{aligned} C_{\text{M}'} &= (C_{\text{M}}^{-1} + G_m^{\text{T}} C_{\text{D}}^{-1} G_m)^{-1} \\ &= C_{\text{M}} - C_{\text{M}} G_m^{\text{T}}(C_{\text{D}} + G_m C_{\text{M}} G_m^{\text{T}})^{-1} G_m C_{\text{M}}. \end{aligned} \tag{7.43}$$

While these equations appear to be identical to those obtained for the standard linear inverse problem with a Gaussian prior, we need to recognize that G_m should be replaced by $G_x \begin{bmatrix} I & E \end{bmatrix}$ everywhere it occurs in Eqs. (7.42) and (7.43).

7.3.1   Doubly stochastic model – example

In this simple one-dimensional model, we illustrate the procedure for MAP estimation when the mean is uncertain. The unknown random function is single valued at each lattice point. We wish to estimate x_i for $i = 1, \ldots, 60$ from measurements of x at

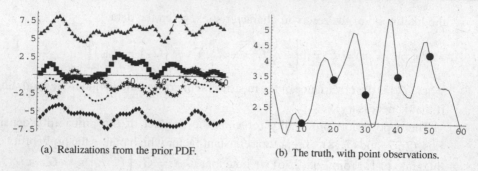

(a) Realizations from the prior PDF.   (b) The truth, with point observations.

Figure 7.6. Doubly stochastic model with uncertain mean.

locations $i = 10, 20, 30, 40, 50$, and from a diffuse measurement of a weighted average of x corresponding to a sensitivity that decays away from the origin:

$$G = [1, 0.928, 0.861, 0.800, 0.741, 0.687, \ldots, 0.012].$$

Prior to assimilation of the measurements, knowledge of the random field is represented by an uncertainty in the uniform mean, which is assumed to be 0 with a variance of 16, and an uncertainty in the heterogeneity which is represented by a Gaussian covariance with a practical range equal to 6 and a variance equal to one. Figure 7.6(a) shows a collection of realizations from the prior PDF, and Fig. 7.6(b) shows the one true realization with the five-point measurements.

Eqs. (7.42) and (7.43) are used to compute the MAP estimate and the a posteriori covariance. As noted, the only difference between this estimate and the estimate without a variable mean is the need to define $G_m = G_x \begin{bmatrix} I & E \end{bmatrix}$ where

$$X = \begin{bmatrix} x_1 \\ \vdots \\ x_{60} \end{bmatrix}, \quad E = \begin{bmatrix} 1 \\ \vdots \\ 1 \end{bmatrix}, \quad \text{and} \quad \theta = \begin{bmatrix} \theta_x \end{bmatrix}, \quad \text{so} \quad E\theta = \begin{bmatrix} \theta_x \\ \vdots \\ \theta_x \end{bmatrix}.$$

As a result of the data assimilation, the estimate of the mean is much improved:

$$\mu_x^{\text{prior}} = 0 \qquad \mu_x^{\text{post}} = 3.08 \qquad \mu_x^{\text{true}} = 3.27.$$

The variance of the estimate of the mean is also substantially reduced by the incorporation of data

$$\sigma_\mu^{\text{prior}} = 4 \qquad \sigma_\mu^{\text{post}} = 0.37.$$

Figure 7.7 compares the maximum a posteriori estimate of X with the truth. The standard deviation of the estimate is also shown. The MAP estimate is less variable than the truth, but the estimate is consistent with the truth, as are the error bars.

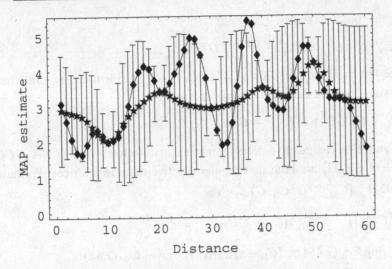

Figure 7.7. Comparison of the truth to the MAP estimate with 2σ error bars.

7.4 Matrix inversion identities

In general, there are multiple equivalent formulations for the solutions of the inverse problem; the choice of the best to use is dictated by considerations of computational efficiency. The equivalence of two useful forms is established in this section. Throughout this section, C_M represents an $M \times M$ positive-definite matrix, C_D an $N_d \times N_d$ positive-definite matrix, G is an $N_d \times M$ matrix, and for any positive integer j, I_j denotes the $j \times j$ identity matrix.

In order to show the equivalence of the two formulas (Eqs. 7.11 and 7.12) for computing the maximum a posteriori estimate, it is sufficient to show that

$$\left(C_M^{-1} + G^T C_D^{-1} G\right)^{-1} G^T C_D^{-1} = C_M G^T \left(C_D + G C_M G^T\right)^{-1}. \tag{7.44}$$

First note that

$$G^T + \left(G^T C_D^{-1}\right)\left(G C_M G^T\right) = G^T C_D^{-1} C_D + \left(G^T C_D^{-1}\right)\left(G C_M G^T\right)$$
$$= G^T C_D^{-1}\left(C_D + G C_M G^T\right),$$

or deleting the middle equality,

$$G^T + G^T C_D^{-1} G C_M G^T = G^T C_D^{-1}\left(C_D + G C_M G^T\right). \tag{7.45}$$

Similarly,

$$G^T + G^T C_D^{-1} G(C_M G^T) = C_M^{-1}(C_M G^T) + G^T C_D^{-1} G(C_M G^T)$$
$$= \left(C_M^{-1} + G^T C_D^{-1} G\right) C_M G^T,$$

or deleting the middle equality,

$$G^T + G^T C_D^{-1} G C_M G^T = \left(C_M^{-1} + G^T C_D^{-1} G\right) C_M G^T. \tag{7.46}$$

Since the left-hand sides of Eqs. (7.45) and (7.46) are identical, so must be their right-hand sides. Setting the right-hand sides equal gives

$$G^T C_D^{-1} \left(C_D + G C_M G^T\right) = \left(C_M^{-1} + G^T C_D^{-1} G\right) C_M G^T. \tag{7.47}$$

Since C_D and C_M are positive-definite matrices, it follows that $\left(C_D + G C_M G^T\right)$ and $\left(C_M^{-1} + G^T C_D^{-1} G\right)$ are both nonsingular positive-definite matrices. Premultiplying Eq. (7.47) by $\left(C_M^{-1} + G^T C_D^{-1} G\right)^{-1}$ gives

$$\left(C_M^{-1} + G^T C_D^{-1} G\right)^{-1} G^T C_D^{-1} \left(C_D + G C_M G^T\right) = C_M G^T. \tag{7.48}$$

Postmultiplying Eq. (7.48) by $\left(C_D + G C_M G^T\right)^{-1}$ gives Eq. (7.44).

The second important identity demonstrates the equivalence of two forms of the maximum a posteriori covariance estimate:

$$\left(C_M^{-1} + G^T C_D^{-1} G\right)^{-1} = C_M - C_M G^T \left(C_D + G C_M G^T\right)^{-1} G C_M. \tag{7.49}$$

Note that

$$\begin{aligned} I_M &= \left(C_M^{-1} + G^T C_D^{-1} G\right)^{-1} \left(C_M^{-1} + G^T C_D^{-1} G\right) \\ &= \left(C_M^{-1} + G^T C_D^{-1} G\right)^{-1} C_M^{-1} + \left(C_M^{-1} + G^T C_D^{-1} G\right)^{-1} \left(G^T C_D^{-1}\right) G. \end{aligned} \tag{7.50}$$

Using Eq. (7.44) to rewrite Eq. (7.50) gives

$$I_M = \left(C_M^{-1} + G^T C_D^{-1} G\right)^{-1} C_M^{-1} + C_M G^T \left(C_D + G C_M G^T\right)^{-1} G. \tag{7.51}$$

Postmultiplying Eq. (7.51) by C_M and rearranging terms gives Eq. (7.49).

8 Optimization for nonlinear problems using sensitivities

8.1 Shape of the objective function

We let $f(m|d_{\text{obs}})$ denote the posterior conditional PDF for the model m given the observed N_d-dimensional vector of data d_{obs}. Assuming that the prior PDF for the model m is Gaussian and that measurement and modeling errors are Gaussian, the posterior PDF is given by

$$f(m|d_{\text{obs}}) = a \exp\left[-\frac{1}{2}(m - m_{\text{prior}})^{\text{T}} C_M^{-1}(m - m_{\text{prior}})\right.$$
$$\left. -\frac{1}{2}(g(m) - d_{\text{obs}})^{\text{T}} C_D^{-1}(g(m) - d_{\text{obs}})\right], \tag{8.1}$$

where m is an N_m-dimensional column vector, a is the normalizing constant, C_M is the $N_m \times N_m$ prior covariance matrix for the random vector m, C_D is the $N_d \times N_d$ covariance matrix for data measurement errors and modeling errors, $g(m)$ is the assumed theoretical model for predicting data for a given m, d_{obs} is the N_d-dimensional column vector containing measured conditioning data, and m_{prior} is the prior mean of the N_m-dimensional column vector m. The precise assumptions involved in deriving Eq. (8.1) are discussed later.

The maximum a posteriori (MAP) estimate of the model is denoted by m_{MAP} and is the model that maximizes $f(m)$ or equivalently minimizes

$$O(m) = \frac{1}{2}(m - m_{\text{prior}})^{\text{T}} C_M^{-1}(m - m_{\text{prior}}) + \frac{1}{2}(g(m) - d_{\text{obs}})^{\text{T}} C_D^{-1}(g(m) - d_{\text{obs}}), \tag{8.2}$$

i.e.

$$m_{\text{MAP}} = \underset{m}{\text{argmin}}\, O(m). \tag{8.3}$$

Because $g(m)$ is nonlinear, the PDF may have many modes, each corresponding to a different local minimum of $O(m)$. In fact, $O(m)$ may have multiple global minima (multiple MAP estimates) as well as multiple local minima. If there exist local minima, a gradient based algorithm that always goes "down hill" will not necessarily converge to a MAP estimate. In such a situation, the minimum obtained by a gradient based algorithm will depend on the initial guess.

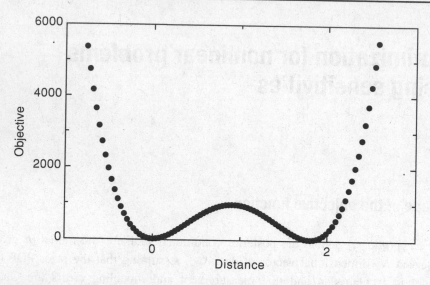

Figure 8.1. Values of the objective function, evaluated along a line in the model space that connects two minima of the function.

Zhang *et al.* [60] considered the problem of conditioning a stochastic channel to pressure data. As expected, they showed that the resulting objective function could have two minima because even though pressure data may be able to resolve the distance to the nearest channel boundary, it can not uniquely determine the direction from the nearest channel boundary to the fault. Multiple minima also occur when trying to determine the location of a fault from pressure data (page 85).

Here, we provide another example of apparent multiple minima. The problem investigated has four producers and one injector in a two-dimensional 11×11 reservoir simulation grid. We created an unconditional Gaussian random field of permeability values from a spherical variogram with a range of approximately 3.5 gridblocks. Using this model as the true reservoir, simulated synthetic pressure data at the wells were created and then used as observed data. We minimized Eq. (8.2) starting from five different initial models and using the Gauss–Newton method to obtain five conditioned models. Four of these results were essentially identical but the other was distinctly different, i.e. two distinct minima were obtained. We call these two minima $m_{\min,1}$ and $m_{\min,2}$. To visualize the shape of the objective function, we calculate $O(m)$ for models on a line between $m_{\min,1}$ and $m_{\min,2}$, parameterized by a scalar α:

$$m = m_{\min,1} + \alpha \frac{(m_{\min,2} - m_{\min,1})}{(m_{\min,2} - m_{\min,1})^{\mathrm{T}} C_M^{-1} (m_{\min,2} - m_{\min,1})}. \tag{8.4}$$

Figure 8.1 shows $O(m)$ as a function of α in the neighborhood of both minima. From the figure, the objective function appears to be fairly simple and differentiable. It, however, appears to be at least bimodal, and may have other local minima that we did not find. Note both mimimizing models give very small values of the objective function and thus give values of $O(m)$ that are close to its global minimum. Thus, both models give excellent matches and the results of Fig. 8.1 indicate that both models are highly probable. Thus, if one simply wishes to make predictions or decisions based on a model which has a high probability of being correct, then either model will do. There are however at least two difficulties that could arise for a multimodal posterior PDF.

1. If one intends to characterize uncertainty in m by approximating the posteriori PDF as a Gaussian centered on a single minimum obtained, then uncertainty may be significantly underestimated if the PDF is multimodal.

2. If the objective function has multiple local minima which are at markedly different levels, a gradient based algorithm may converge to a model which corresponds to a high value of the objective function (low value of the posterior PDF). In this case, this model can not be expected to provide a good approximation to the MAP estimate and predictions and decisions based on this model can not be expected to be reasonable. The question is then how to decide whether a mimimizing model gives a reasonable MAP estimate. As discussed below, if the model gives a good approximation of m_{MAP}, then the value of $O(m)$ should be close to $N_d/2$. If it is not, there is no assurance that we have obtained a global minimum.

Further examination of the preceding example uncovered an interesting result. By evaluating the objective function along a set of lines, we discovered that the two minima of Fig. 8.1 are connected by a long fairly flat horseshoe-shaped valley. The value of the objective function along this curve is not significantly higher than the value of the objective function at the two minima.

8.1.1   Evaluation of quality of estimate

Regarding the comment about the expected magnitude of the objective function at convergence, it can be shown [4] that if the assumed theoretical model is linear so $d = g(m) = Gm$, where G is the $N_d \times N_M$ sensitivity matrix, then the minimum of $2O(m)$ has a χ^2 distribution with N_d degrees of freedom (the mean is equal to N_d and the variance is equal to $2N_d$). It is reasonable to expect that this result also approximately applies in the nonlinear case, and computations that we have done with synthetic history-matching problems are consistent with this supposition. Thus, we assume that the following equation is approximately correct:

$$E[O(m_{MAP})] = \frac{N_d}{2}. \tag{8.5}$$

Assuming that any legitimate realization of this χ^2 distribution should be within five standard deviations of the mean, implies that the following inequality should hold

$$N_d - 5\sqrt{2N_d} \le 2O(m_{MAP}) \le N_d + 5\sqrt{2N_d}. \tag{8.6}$$

If a gradient based estimation converges to a model m_{MAP} which does not satisfy Eq. (8.6), one should question whether a reasonable MAP estimate has been obtained. Failure to satisfy Eq. (8.6) may indicate that we have converged to a local minimum; this possibility can be investigated by starting with a different initial model. If regardless of the starting model, the gradient algorithm converges to a model m_{MAP} which does not satisfy Eq. (8.6), it is likely that one has either used a poor implementation of the gradient algorithm or an incorrect assumption has made. For example, we may have significantly under estimated the variance of the measurement and or modeling errors, or our prior model may be inappropriate.

8.2   Minimization problems

We assume that the prior PDF for m is Gaussian with mean m_{prior} and covariance C_M so that the prior PDF is given by

$$ f_M(m) = \frac{1}{(2\pi)^{N_m/2}\sqrt{\det C_M}} \exp\left[-\frac{1}{2}(m - m_{\text{prior}})^{\text{T}} C_M^{-1}(m - m_{\text{prior}}) \right], \tag{8.7} $$

where m is a column vector of dimension N_m. The theoretical relationship between a model m and an N_d-dimensional column vector of predicted data d is given by

$$ d = g(m). \tag{8.8} $$

In the context of discrete inverse theory, a linear problem refers to one in which $g(m) = Gm + a$ where G is an $N_d \times N_m$ matrix and a is a constant N_d-dimensional column vector.

For the practical solution of inverse problems, Eq. (8.8) allows us to calculate the data vector corresponding to a given model m. In Eq. (8.8), the predicted data vector d is not corrupted by measurement error. For problems of interest to us, Eq. (8.8) is based on physical relationships which may be only approximate. For example, if the data vector d represents production data that could be measured at two wells producing an oil reservoir, then given a particular reservoir model m, we typically assume that we could calculate d by solving the partial differential equations which describe fluid flow in a porous media together with the appropriate boundary conditions at the wells and reservoir boundaries, i.e. by solving an appropriate initial boundary-value problem (IBVP). But in practice, the IBVP is solved by a numerical method (e.g. a finite-difference method) in which the rigorous boundary conditions at the wells are replaced by approximate relations, i.e. we use a reservoir simulator (possibly linked to a model for flow in the wellbore) to calculate the production data, d. In this case, Eq. (8.8) represents running the simulator with a given reservoir model m to calculate d. Because the numerical model only approximates the true physics, Eq. (8.8) contains modeling error.

Under the strong assumption that the modeling error can be represented by a random vector η which is independent of m, the relation between exact theoretical data d (from the true theoretical model) and the approximate theoretical model $g(m)$ is given by

$$d = g(m) + \eta. \tag{8.9}$$

If each component of d can be measured and d_{obs} denotes the vector of data that can be measured, then

$$d_{\text{obs}} = d + \epsilon = g(m) + \eta + \epsilon, \tag{8.10}$$

where ϵ is a random vector which represents measurement error. Note in Eq. (8.10), d_{obs} is itself a random vector which depends on the model m as well as the random vectors η and ϵ. Assuming the random vectors η and ϵ are independent Gaussian with covariance matrices given respectively by C_{D1} and C_{D2} and with means given by the zero vector, then from basic statistical theory [61], the random vector

$$e = \eta + \epsilon \tag{8.11}$$

is Gaussian with mean given by the N_d-dimensional zero vector and covariance given by

$$C_D = C_{D1} + C_{D2}. \tag{8.12}$$

For a given model m, it follows that d_{obs} is a random vector with mean given by $g(m)$ and covariance given by C_D so the conditional PDF for d_{obs} given m can be written as

$$f_{D_{\text{obs}}|M}(d_{\text{obs}}|m) = \frac{1}{(2\pi)^{N_d/2}\sqrt{\det C_D}} \exp\left[-\frac{1}{2}(d_{\text{obs}} - g(m))^{\mathrm{T}} C_D^{-1}(d_{\text{obs}} - g(m))\right]. \tag{8.13}$$

From this point on, we will use condensed notation and abandon the subscripts on the expressions for PDFs, e.g. we write the left-hand sides of Eqs. (8.7) and (8.13), respectively, as $f(m)$ and $f(d_{\text{obs}}|m)$. C_D will be referred to as the data covariance matrix or sometimes as the covariance for measurement errors, but one should bear in mind that both measurement and modeling errors are involved, and it is quite possible that modeling errors may dominate measurement errors [62].

8.2.1   Maximum likelihood estimate

It is customary to also denote an actual observation (measurement) of data as d_{obs} but this d_{obs} is actually a realization of the random vector $d_{\text{obs}}(m_{\text{true}})$ where m_{true} is the true model. Given the measured data, d_{obs}, the likelihood function is defined by

$$L(m|d_{\text{obs}}) = \frac{1}{(2\pi)^{N_d/2}\sqrt{\det C_D}} \exp\left[-\frac{1}{2}(d_{\text{obs}} - g(m))^{\mathrm{T}} C_D^{-1}(d_{\text{obs}} - g(m))\right]. \tag{8.14}$$

Note the right-hand sides of Eqs. (8.13) and (8.14) are notationally identical but in Eq. (8.13), m is fixed and d_{obs} is a random vector, whereas, in Eq. (8.14), d_{obs} is fixed and m is a random vector.

The maximum likelihood estimate of m is defined to be the model which maximizes the likelihood function or equivalently, minimizes the objective function $O_d = O_d(m)$ defined by

$$O_d(m) = \frac{1}{2}(d_{obs} - g(m))^{T} C_D^{-1}(d_{obs} - g(m)).$$

(8.15)

Denoting the maximum likelihood estimate by m_{ML}, it follows that

$$m_{ML} = \underset{m}{\operatorname{argmin}}\, O_d(m).$$

(8.16)

If one assumes that the components of the data error vector e of Eq. (8.11) are independent Gaussian random variables with all means equal to zero and the variance of the ith component of e given by $\sigma_{d,i}^2$, then C_D is a diagonal matrix with the i diagonal entry given by $\sigma_{d,i}^2$ and Eq. (8.15) reduces to

$$O_d(m) = \frac{1}{2} \sum_{k=1}^{N_d} \frac{(d_{obs,k} - g_k(m))^2}{\sigma_{d,k}^2},$$

(8.17)

where $d_{obs,k}$ denotes the kth component of the vector d_{obs}. Minimizing the preceding objective function gives the model which is referred to as the (weighted) least-squares estimate. Thus, under the assumptions that data measurement/modeling errors are independent Gaussians, the least-squares estimate is equivalent to the maximum likelihood estimate. If the data errors can be modeled as independent identically distributed Gaussian random variables with mean zero and variance σ_d then

$$C_D = \sigma_d^2 I$$

(8.18)

and minimizing the objective function of Eq. (8.15) is equivalent to minimizing the ordinary least-squares objective function

$$O_{ols} = \frac{1}{2} \sum_{k=1}^{N_d} (d_{obs,k} - g_k(m))^2.$$

(8.19)

8.2.2   MAP estimate

The maximum likelihood estimate ignores prior information for m. For cases where a prior model exists, Bayesian probability provides a convenient framework for integrating prior information with measured data. From Bayes' theorem, it follows that

$$f(m|d_{obs}) = \frac{f(d_{obs}|m)f(m)}{f(d_{obs})},$$

(8.20)

where $f(d_{obs})$ is the PDF for d_{obs}. From the viewpoint of Bayesian estimation, explicit knowledge of $f(d_{obs})$ is unnecessary. Given d_{obs}, Eq. (8.20) can be written as

$$f(m|d_{obs}) = af(d_{obs}|m)f(m) = aL(m|d_{obs})f(m), \tag{8.21}$$

where a is the normalizing constant which is defined by the requirement that the integral of $f(m|d_{obs})$ over the space of all allowable models must equal unity. The conditional PDF, $f(m|d_{obs})$ is referred to as the posterior or a posteriori PDF. With the prior PDF for m given by Eq. (8.7), the posterior PDF is given by Eq. (8.1). From the Bayesian viewpoint, the best estimate of m is the model which maximizes the a posteriori PDF. This estimate is referred to as the maximum a posteriori (MAP) estimate (Eq. 8.3) and is obtained by minimizing the objective function of Eq. (8.2).

8.2.3   Objective function for sampling with RML

For the purpose of evaluating uncertainty, we wish to generate an approximate sampling of the posterior PDF of Eq. (8.1). As discussed in Chapter 10, to generate a single sample (realization) from the conditional PDF using the randomized maximum likelihood method (RML), we add a realization of the measurement error to the observed data to obtain d_{uc}. We also generate an unconditional model realization m_{uc} by sampling the prior PDF for m. A sample of the posterior PDF is then given by

$$m_c = \operatorname*{argmin}_m O_r(m), \tag{8.22}$$

where the objective function $O_r(m)$ is defined by

$$O_r(m) = \frac{1}{2}(m - m_{uc})^{\mathrm{T}} C_M^{-1}(m - m_{uc}) + \frac{1}{2}(g(m) - d_{uc})^{\mathrm{T}} C_D^{-1}(g(m) - d_{uc}). \tag{8.23}$$

This minimization process generates a single sample of the posterior PDF. To generate additional samples, we repeat the process with different sets of (d_{uc}, m_{uc}).

Similar to Eq. (8.6), we expect that at convergence, the objective function of Eq. (8.23) will satisfy

$$N_d - 5\sqrt{2N_d} \leq O(m_c) \leq N_d + 5\sqrt{2N_d}. \tag{8.24}$$

8.3   Newton-like methods

In this section, we consider the use of Newton's method for minimization of an objective function, $O(m)$, which we assume has continuous second derivatives. Newton's method is considered first and is then modified to the Gauss–Newton method.

8.3.1   Newton's method for minimization

Two derivations of Newton's method are provided. The one presented here is based on the application of Newton–Raphson iteration to solve the system of equations represented by setting the gradient of the objective function equal to zero. The second is obtained by approximating the objective function we wish to minimize by its second-order Taylor series; see subsection 8.3.4.

At a minimum of the objective function $O(m)$, the gradient with respect to m is equal to the N_m-dimensional zero vector, i.e.

$$\nabla_m O = 0. \tag{8.25}$$

Eq. (8.25) represents a system of N_m nonlinear equations in N_m unknowns with the ith equation given by

$$\frac{\partial O(m)}{\partial m_i} = 0, \tag{8.26}$$

for $i = 1, 2, \ldots N_m$. Applying the Newton–Raphson algorithm [8, 9] to solve Eq. (8.26) with m^0 as the initial guess gives

$$m^{\ell+1} = m^\ell + \delta m^{\ell+1}, \tag{8.27}$$

where $\delta m^{\ell+1}$ is obtained by solving the linear system problem

$$J(m^\ell)\delta m^{\ell+1} = -\nabla O(m^\ell). \tag{8.28}$$

Here, ℓ is the iteration index and $J(m^\ell)$ is the Jacobian matrix evaluated at the ℓth iterate. In this setting, $J(m) = H(m)$ where the $H(m)$ denotes the Hessian matrix defined by

$$H(m) = \nabla\big((\nabla O(m))^{\mathrm{T}}\big). \tag{8.29}$$

The element in the ith row and jth column of $H(m)$ is given by

$$h_{i,j}(m) = \frac{\partial^2 O(m)}{\partial m_i \partial m_j}. \tag{8.30}$$

Eqs. (8.27) and (8.28) are applied for $\ell = 1, 2 \ldots$ until convergence. Eqs. (8.27) and (8.28) represent the Newton method for minimizing an objective function, $O(m)$, using a full Newton step. In many applications, however, the step size must be restricted at least at early iterations. Thus, a practical implementation of Newton's method calculates a search direction, $\delta m^{\ell+1}$, by solving Eq. (8.28) and then calculates the proposed update to the approximation of the minimum by replacing Eq. (8.27) by

$$m^{\ell+1} = m^\ell + \mu_\ell \delta m^{\ell+1}, \tag{8.31}$$

where μ_ℓ is the step size. Because we are seeking a minimum of $O(m)$, μ_ℓ should be chosen to obtain a sufficient decrease in the objective function. Later in this chapter, we discuss both line search methods and restricted-step methods for choosing μ_ℓ.

For minimizing the objective function of Eq. (8.23), the gradient and Hessian of the objective function are given respectively by

$$\nabla O(m) = C_M^{-1}(m - m_{\mathrm{uc}}) + G^{\mathrm{T}} C_D^{-1}(g(m) - d_{\mathrm{uc}}) \tag{8.32}$$

and

$$H = C_M^{-1} + G^{\mathrm{T}} C_D^{-1} G + (\nabla G^{\mathrm{T}}) C_D^{-1}(g(m) - d_{\mathrm{uc}}). \tag{8.33}$$

Here, G is the sensitivity matrix defined by

$$G = G(m) = \begin{bmatrix} (\nabla_m g_1)^{\mathrm{T}} \\ (\nabla_m g_2)^{\mathrm{T}} \\ \vdots \\ (\nabla_m g_{N_d})^{\mathrm{T}} \end{bmatrix} = (\nabla_m g^{\mathrm{T}})^{\mathrm{T}} \tag{8.34}$$

and

$$G^{\mathrm{T}} = G(m)^{\mathrm{T}} = \begin{bmatrix} \nabla_m g_1 & \nabla_m g_2 & \cdots & \nabla_m g_{N_d} \end{bmatrix} = \nabla_m(g^{\mathrm{T}}). \tag{8.35}$$

The individual elements of the sensitivity matrix are

$$g_{ij}(m) = \frac{\partial g_i(m)}{\partial m_j}, \tag{8.36}$$

for $i = 1, 2, \ldots, N_d$ and $j = 1, 2, \ldots, N_m$. The $g_{ij}(m)$s are referred to as the sensitivities or sensitivity coefficients. Note from a first-order Taylor expansion, $g_{ij}(m)$ gives the sensitivity of the ith (predicted or calculated) data, $d_i = g_i(m)$, to the jth model parameter m_j at a particular m and gives a measure of the change in the ith predicted data that results from a unit change in m_j keeping all other entries of m fixed. Procedures for calculating the sensitivity coefficients will be in the next chapter.

8.3.2   Gauss–Newton method

Here, we consider the Gauss–Newton method for minimizing the objective function of Eq. (8.23). In Gauss–Newton, we simply ignore the term involving the second derivatives of g in Eq. (8.33) and replace the Newton Hessian matrix by the Gauss–Newton Hessian matrix given by

$$H = C_M^{-1} + G^{\mathrm{T}} C_D^{-1} G. \tag{8.37}$$

With this replacement, the Gauss–Newton algorithm is given by Eqs. (8.28) and (8.31) or

$$\left(C_M^{-1} + G_\ell^{\mathrm{T}} C_D^{-1} G_\ell\right) \delta m^{\ell+1} = -\left(C_M^{-1}(m^\ell - m_{\mathrm{uc}}) + G_\ell^{\mathrm{T}} C_D^{-1}(g(m^\ell) - d_{\mathrm{uc}})\right) \tag{8.38}$$

and

$$m^{\ell+1} = m^{\ell} + \mu_{\ell} \delta m^{\ell+1}, \tag{8.39}$$

for $\ell = 0, 1, 2, \ldots$ where the subscript ℓ denotes evaluation at m^{ℓ}, m^0 is the initial guess for the minimizing model, and μ_{ℓ} is the step size.

Replacing the Newton method by the Gauss–Newton method often arises by necessity. For many problems of interest, an explicit formula for g is not available so that computation of the gradient of g must be done numerically and computation of the second derivatives is not feasible. If at a minimum, we get a perfect data match, i.e. $g(m) - d_{\text{obs}} = 0$, then as we approach a minimum the last term in Eq. (8.33) becomes negligible. Thus, near a minimum the Gauss–Newton method and Newton method should behave similarly. For history matching with noisy data, however, $g(m) - d_{\text{obs}} = 0$ does not normally apply. If the objective function is equal to a positive-definite quadratic function in a neighborhood of a minimum, i.e. $g(m)$ is a linear function of m in this neighborhood, then the second derivative of $g(m)$ is zero in this neighborhood and once the iterations produce a model in this neighborhood, the Gauss–Newton Hessian and Newton Hessian are identical. Because the above conditions can sometimes approximately hold, the Gauss–Newton method may sometimes behave like Newton's method near a minimum and display quadratic convergence; see Section 8.5.2.

The covariance matrices C_D and C_M are real symmetric and positive definite, so their inverses are real symmetric and positive definite. It follows that $C_M^{-1} + G_{\ell}^{\mathrm{T}} C_D^{-1} G_{\ell}$ is real symmetric. Moreover, for any nonzero real N_m-dimensional column vector x

$$x^{\mathrm{T}} \left(C_M^{-1} + G_{\ell}^{\mathrm{T}} C_D^{-1} G_{\ell} \right) x = x^{\mathrm{T}} C_M^{-1} x + \left(G_{\ell} x \right)^{\mathrm{T}} C_D^{-1} \left(G_l x \right) > 0. \tag{8.40}$$

It follows that $C_M^{-1} + G_{\ell}^{\mathrm{T}} C_D^{-1} G_{\ell}$ is real symmetric positive definite and hence non-singular. This is another advantage that the Gauss–Newton method has; the Hessian matrix for the Newton method with all second derivatives included may not be positive definite, but the Gauss–Newton Hessian matrix for constructing the MAP estimate is always real symmetric positive definite. Because of this, it can be shown that the search direction, $\delta m^{\ell+1}$, obtained by solving Eq. (8.38) is guaranteed to be a down-hill direction and the step size μ_{ℓ} can be found by one a directional line search algorithm or a restricted-step method as discussed later. On the other hand, the Hessian for Newton's method is not guaranteed to be positive definite and it may be necessary to modify it at early iterations to produce a down-hill search direction.

In the case of a pure least-squares problem, the Gauss–Newton Hessian is given by

$$H = G^{\mathrm{T}} C_D^{-1} G \tag{8.41}$$

and the basic iterative equation is

$$G_{\ell}^{\mathrm{T}} C_D^{-1} G_{\ell} \delta m^{\ell+1} = -G_{\ell}^{\mathrm{T}} C_D^{-1} (g(m^{\ell}) - d_{uc}), \tag{8.42}$$

where the subscript ℓ denotes evaluation at m^{ℓ}.

Eq. (8.42) has a unique solution if and only if the $N_m \times N_m$ Gauss–Newton Hessian, $G_\ell^T C_D^{-1} G_\ell$ is nonsingular. The data covariance matrix, C_D, and its inverse are real-symmetric positive-definite matrices, so the Gauss–Newton Hessian matrix is non-singular if and only if the rank of G_ℓ is equal to N_m. Because the rank of a matrix must be less than or equal to the minimum of the number of rows and the number of columns [12], $G_\ell^T C_D^{-1} G_\ell$ can not be nonsingular unless the number of observed data to be matched is greater than or equal to the number of model parameters to be estimated, which is typically not the case for automatic history-matching problems. Even if $N_m \ll N_d$, as is the case for the analysis of well test pressure data by nonlinear regression, the matrix $G_\ell^T C_D^{-1} G_\ell$ may still be singular or so nearly singular that obtaining an accurate numerical solution of Eq. (8.42) is difficult. Whenever this matrix is singular or approximately singular, a particular solution can be chosen by applying singular value decomposition [8]. Alternately, as discussed in Section 8.4, we can replace the Gauss–Newton method by a Levenberg–Marquardt algorithm.

8.3.3   Gauss–Newton for generating the MAP estimate

If we wish to generate the MAP estimate, the objective function to be minimized is given by Eq. (8.2) with its gradient given by

$$\nabla O(m) = C_M^{-1}(m - m_{\text{prior}}) + G^T C_D^{-1}(g(m) - d_{\text{obs}}), \tag{8.43}$$

and the Gauss–Newton Hessian is given by

$$H(m) = C_M^{-1} + G^T C_D^{-1} G. \tag{8.44}$$

The Gauss–Newton algorithm is as follows: solve the $N_m \times N_m$ matrix problem

$$\left(C_M^{-1} + G_\ell^T C_D^{-1} G_\ell\right) \delta m^{\ell+1} = - \left(C_M^{-1}(m^\ell - m_{\text{prior}}) + G_\ell^T C_D^{-1}(g(m^\ell) - d_{\text{obs}})\right) \tag{8.45}$$

for $\delta m^{\ell+1}$, calculate the step size μ_ℓ and update the approximation to the MAP estimate by

$$m^{\ell+1} = m^\ell + \mu_\ell \delta m^{\ell+1}. \tag{8.46}$$

For now we assume it is feasible to calculate the sensitivity coefficient matrix, G_ℓ, at each iteration. Then, if the dimension of m is small, it is feasible to calculate and store C_M^{-1} and the solution of Eq. (8.45) can be obtained by using a direct method for very small N_m or an iterative method such as the preconditioned conjugate gradient method using C_M^{-1}, or an incomplete LU decomposition of $H(m^\ell)$, as the preconditioning matrix.

If the number of model parameters N_m is much larger than the number of observed conditioning data, N_d, then it is usually more efficient to apply the matrix inversion lemmas of Section 7.4 to rewrite the Gauss–Newton method. Using Eqs. (7.44) and

(7.49), we can rewrite Eq. (8.45) as

$$
\begin{aligned}
\delta m^{\ell+1} &= -(C_M^{-1} + G_\ell^T C_D^{-1} G_\ell)^{-1} C_M^{-1}(m^\ell - m_{\text{prior}}) \\
&\quad - (C_M^{-1} + G_\ell^T C_D^{-1} G_\ell)^{-1} G_\ell^T C_D^{-1}(g(m^\ell) - d_{\text{obs}}) \\
&= -\left(I - C_M G_\ell^T [C_D + G_\ell C_M G_\ell^T]^{-1} G_\ell\right)(m^\ell - m_{\text{prior}}) \\
&\quad - C_M G_\ell^T (C_D + G_\ell C_M G_\ell^T)^{-1}(g(m^\ell) - d_{\text{obs}}),
\end{aligned}
\tag{8.47}
$$

or, equivalently,

$$
\begin{aligned}
\delta m^{\ell+1} &= -(m^\ell - m_{\text{prior}}) \\
&\quad - C_M G_\ell^T (C_D + G_\ell C_M G_\ell^T)^{-1}(g(m^\ell) - d_{\text{obs}} - G_\ell(m^\ell - m_{\text{prior}})).
\end{aligned}
\tag{8.48}
$$

As in the original Gauss–Newton formulation, after computing the search direction $\delta m^{\ell+1}$, the step size μ_ℓ is calculated and the approximation to the MAP estimate is updated with Eq. (8.46).

To calculate $\delta m^{\ell+1}$ from Eq. (8.45) requires solution of a linear system with coefficient matrix given by the $N_m \times N_m$ matrix $(C_M^{-1} + G_\ell^T C_D^{-1} G_\ell)$, whereas, the only matrix problem that must be solved to apply the reformulated Gauss–Newton method of Eq. (8.48) is the $N_d \times N_d$ matrix problem given by

$$
(C_D + G_\ell C_M G_\ell^T)x = g(m^\ell) - d_{\text{obs}} - G_\ell(m^\ell - m_{\text{prior}}).
\tag{8.49}
$$

Clearly the reformulated Gauss–Newton method will be more computationally efficient if $N_d \ll N_m$, whereas the original formulation is preferable if $N_m \ll N_d$. If both the number of model parameters and the number of data are large, it may not be feasible to apply either formulation. In this case, alternate optimization algorithms such as quasi-Newton or nonlinear conjugate gradient must be employed, or some form of reparameterization must be applied to reduce the size of the problem.

8.3.4   Restricted-step method for calculation of step size

Here, the restricted-step method is discussed for calculation of the step size. Although methods based on approximate line searches (Section 8.7) are more commonly used, we have used the restricted-step method successfully for history-matching problems. Moreover, the development of the restricted-step method provides an alternate derivation and enhanced understanding of the Gauss–Newton method. In this discussion $O(m)$ denotes an arbitrary objective function that has at least two continuous derivatives with respect to m.

Let m^ℓ denote the latest approximation to the minimum. Defining $\delta m = m - m^\ell$, we approximate $O(m)$ by the quadratic $Q(m)$ which represents a second-order Taylor series expansion of $O(m)$ about m^ℓ, i.e.

$$
O(m) \approx Q(m) \equiv O(m^\ell) + \nabla_m O(m^\ell)^T \delta m + \frac{1}{2} \delta m^T H_\ell \delta m,
\tag{8.50}
$$

where H_ℓ is the Hessian evaluated at m^ℓ and is given by

$$H_\ell = \nabla_m \left[\left(\nabla_m O(m^\ell) \right)^T \right]. \tag{8.51}$$

The quadratic $Q(m)$ has a unique minimum if and only if H_ℓ is a positive-definite matrix. In this case, the minimum of $Q(m)$ can be obtained by solving

$$\nabla_m Q(m) = \nabla_m O(m^\ell) + H_\ell \delta m = 0, \tag{8.52}$$

or equivalently,

$$H_\ell \delta m = -\nabla O(m^\ell). \tag{8.53}$$

Denoting the solution of Eq. (8.53) by $\delta m^{\ell+1} = m^{\ell+1} - m^\ell$, the updated approximation of the minimum of $O(m)$ is given by

$$m^{\ell+1} = m^\ell + \delta m^{\ell+1}. \tag{8.54}$$

The preceding derivation of Newton's method is quite instructive. First, it shows that if H_l is not positive definite, $Q(m)$ may not have a minimum so the Newton procedure may not give an improved estimate of a minimum of $O(m)$. Second, it suggests that if $O(m)$ is equal to the quadratic $Q(m)$ in a neighborhood of m^ℓ that includes a local minimum of $O(m)$ and H_l is positive definite, then this minimum of $O(m)$ is equal to the minimum of $Q(m)$ so Newton's method will converge in one iteration to a model that minimizes $O(m)$ locally. More generally, we can be reasonably confident that the model that minimizes Q will give an improved approximation to a minimizer of O only if Q gives a good approximation to O in a neighborhood of m^ℓ ($\{m| \parallel m - m^\ell \parallel < \alpha\}$) which contains the minimum of Q. This suggests that we should restrict the step size μ_l in the search direction $\delta m^{\ell+1}$ so that the new updated model is contained within the region where O is well approximated by the quadratic Q. This is the basis of the restricted-step method. Finally, we should note that if we replace the Newton Hessian of Eq. (8.51) by the Gauss–Newton Hessian, then the resulting iteration matrix will always be real symmetric positive definite for the objective functions of Eqs. (8.2) and (8.23) so that the quadratic function $Q(m)$ will always have a unique minimum.

Restricted-step algorithm

1. Set $\ell = 0$ and determine the initial guess for a model that corresponds to a minimum of $O(m)$. Determine h^1 which represents the maximum allowable step at the first iteration. One possibility is to choose $h^1 = \epsilon \|m^0\|$ where ϵ is on the order of 0.05 or smaller. (Alternately, one can choose h^1 based on prior variances if a Bayesian formulation is available.) Although the choice of ϵ is problem dependent, we do know from experience in automatic history-matching problems that large changes in the model at early iterations may lead to convergence to a model that does not give an acceptable match of production data [63].

2. Solve Eq. (8.53) to obtain the search direction $\delta m^{\ell+1}$.

3. If

$$\|\delta m^{\ell+1}\| > h^{\ell+1},\tag{8.55}$$

then modify the search direction as follows:

$$\delta m^{\ell+1} = \frac{h^{\ell+1}}{\|\delta m^{\ell+1}\|}\delta m^{\ell+1}.\tag{8.56}$$

4. Define a temporary value of the new iterate as

$$m_c^{\ell+1} = m^\ell + \delta m^{\ell+1},\tag{8.57}$$

and calculate the ratio

$$r_{\ell+1} = \frac{O(m^\ell) - O(m_c^{\ell+1})}{O(m^\ell) - Q(m_c^{\ell+1})}.\tag{8.58}$$

Because $Q(m^\ell) = O(m^\ell)$ (see Eq. 8.50), we see that $r_{\ell+1}$ represents the ratio of the decrease in the objective function to the predicted decrease based on the quadratic approximation.

5. If $r_{\ell+1} < 0$, then set $h^{\ell+1} = \|\delta m^{\ell+1}\|/4$, replace $\delta m^{\ell+1}$ by $\delta m^{\ell+1}/4$ and return to step 4.

6. If

$$r_{\ell+1} \leq 0.25,\tag{8.59}$$

then set $h^{\ell+2} = \|\delta m^{\ell+1}\|/4$. If

$$r_{\ell+1} \geq 0.75 \quad \text{and} \quad \|\delta m^{\ell+1}\| = h^{\ell+1}\tag{8.60}$$

then set $h^{\ell+2} = 2h^{\ell+1}$. If both Eqs. (8.59) and (8.60) fail to hold, set $h^{\ell+2} = h^{\ell+1}$.
7. Set $m^{\ell+1} = m_c^{\ell+1}$.
8. Check for convergence. If the algorithm has not converged and the maximum number of iterations allowed has not been reached, add 1 to the iteration index and return to step 2. Convergence criteria for optimization algorithms are discussed in Section 8.5.

The denominator of Eq. (8.58) represents the change in Q and will always be positive if H_l is positive definite and m^ℓ does not correspond to a minimum of Q. Thus, $r_{\ell+1} < 0$ is equivalent to $O(m_c^{\ell+1}) > O(m^\ell)$, i.e. the objective function has increased. In this case, step 5 indicates we reduce the step size in the direction of δm^{l+1} by a factor of 4 and also reduce the maximum allowable step for the subsequent iteration. Also note that if $O(m)$ is equal to the quadratic approximation, then Eq. (8.58) will give $r_{\ell+1} = 1$. In this case, it would be desirable to take a full step in the direction $\delta m^{\ell+1}$ with no restriction. According to the algorithm, when $r_{l+1} \geq 0.75$, O is well approximated by a quadratic in a region containing both $m_c^{\ell+1}$ and m^ℓ. In this case, if a restriction on the step size has been imposed at the current step, we double the allowable step size at the next iteration. As we approach a minimum, we expect $h^{\ell+1}$ to become sufficiently large so that $\delta m^{\ell+1}$ computed by solving Eq. (8.53) will be unmodified. The constants 0.25 and 0.75 that appear in Eqs. (8.59) and (8.60) are ad hoc but as noted by Fletcher [64], the algorithm is not very sensitive to small changes in these parameters.

8.4   Levenberg–Marquardt algorithm

The basic Levenberg–Marquardt algorithm for minimizing an arbitrary objective function $O(m)$ simply replaces the $N_m \times N_m$ Newton Hessian matrix H_ℓ at the ℓth iteration by $H_\ell + \lambda_\ell I$ where I is the $N_m \times N_m$ identity matrix and $\lambda_\ell \geq 0$ is chosen to obtain an improved down-hill search direction. The parameter λ_ℓ may vary from iteration to iteration. In the basic algorithm, $\delta m^{\ell+1}$ is obtained by solving

$$(\lambda_\ell I + H_\ell)\delta m^{\ell+1} = -\nabla O(m^\ell) \tag{8.61}$$

and we set

$$m_c^{\ell+1} = m^\ell + \delta m^{\ell+1}. \tag{8.62}$$

If $O(m_c^{\ell+1}) < O(m^\ell)$, then set $m^{\ell+1} = m_c^{\ell+1}$, set $\lambda_{\ell+1} = \lambda_\ell/a_1$, where $a_1 > 1$ and go to the next iteration. Otherwise, increase λ_ℓ by some factor a_2 where $a_2 > 1$ and repeat the iteration. Choosing both $a_1 = a_2 \approx 10$ is common but the optimal choice is difficult to determine for large-scale history-matching problems [65]. Although this simple procedure has been successfully applied to automatic history-matching problems, more general trust region methods are available in the optimization literature [66, 67].

In the Levenberg–Marquardt algorithm, the value λ_ℓ controls both the search direction and step size; no line search is done. As λ_ℓ becomes large, the Levenberg–Marquardt iteration is similar to taking a small step in the steepest descent direction whereas if λ_ℓ is small, the ℓth iteration in the Levenberg–Marquardt iteration becomes similar to a Newton iteration. In general, the steepest descent method converges relatively slowly (linear convergence) whereas the Newton method often tends to converge quite rapidly (quadratic convergence) once m^ℓ is close to a minimum provided that the Newton iteration matrix (Hessian) is positive definite and the Hessian satisfies appropriate continuity conditions (Lipschitz continuous) in a neighborhood of the minimum [67]. For history-matching problems, H_ℓ in the Levenberg–Marquardt algorithm will be chosen as the Gauss–Newton Hessian. Thus, we do not expect to obtain quadratic convergence; for example, the rate of convergence will depend primarily on the magnitude of the residuals as we approach the minimum as these terms were discarded to obtain the Gauss–Newton Hessian from the Newton Hessian.

The choice of λ^0 depends on the problem. For large-scale automatic history matching, the initial guess for the reservoir model m is typically far from the minimum and a large starting value of λ is often required; λ^0 on the order of 10^5 or greater is often appropriate, but the initial value depends on the problem. If the objective function decreases by a factor of 10 or greater at the first iteration, the initial value of λ may be too small, If the initial value of λ is too small, one sometimes obtains large changes in model parameters that result in unreasonable reservoir properties, e.g. permeability fields that contain very high values close to very low values. When this occurs, our experience is that it happens at the end of the first iteration so one can often detect that λ^0 is too small by examining the model obtained after one iteration rather than waiting until convergence is obtained. If we wish to construct the MAP estimate or the maximum likelihood estimate, then experience suggests that an initial value of λ between $\sqrt{O(m^0)/N_d}$ and $O(m^0)/N_d$ may be appropriate, but we have no theoretical basis for this choice or any guarantee that it will always provide a good starting value. A better choice might be to choose λ to ensure that the condition number of the Levenberg–Marquardt iteration matrix is not too large, but this requires that we compute the maximum and minimum eigenvalues of this iteration matrix and selection of an upper bound on the condition number is not an exact science.

The Levenberg–Marquardt algorithm is closely linked to the trust region method, which is popular in the optimization literature. Here, we consider only the case where $H(m)$ represents the Gauss–Newton Hessian so the Levenberg–Marquardt iteration matrix is positive definite for every $\lambda_\ell > 0$. From the trust region viewpoint, we consider the problem of minimizing

$$\hat{Q}(\delta m) = O(m^\ell) + \nabla_m O(m^\ell)^{\mathrm{T}} \delta m + \frac{1}{2} \delta m^{\mathrm{T}} H_\ell \delta m, \tag{8.63}$$

subject to the constraint that

$$\| \delta m \| \le h^{\ell+1}. \tag{8.64}$$

If the Gauss–Newton search direction satisfies Eq. (8.64), then it provides the solution of the constrained minimization problem, otherwise there exists $\lambda_\ell > 0$ such that the solution of

$$(\lambda_\ell I + H_\ell)\delta m^{\ell+1} = -\nabla O(m^\ell), \tag{8.65}$$

satisfies

$$\| \delta m^{\ell+1} \| = h^{\ell+1}. \tag{8.66}$$

Unlike the original Levenberg–Marquardt algorithm, if the trust region method is used, it is necessary to find λ_ℓ so that Eq. (8.66) is satisfied, before we accept $\delta m^{\ell+1}$ and update the model by

$$m^{\ell+1} = m^\ell + \delta m^{\ell+1}. \tag{8.67}$$

Procedures for calculating λ_ℓ and adjusting $h^{\ell+1}$ can be found in Dennis and Schnabel [66] and Nocedal and Wright [67], and modern implementations of Levenberg–Marquardt use the approaches given there. To date, the trust region method has not been applied to large history-matching problems.

8.4.1   Spectral analysis of Levenberg–Marquardt algorithm

Throughout this subsection, we assume that the $N_m \times N_m$ Hessian matrix H_ℓ is real symmetric positive semidefinite so its eigenvalues are nonnegative and may be ordered as $\beta_1 \geq \beta_2 \geq \cdots \geq \beta_{N_m} \geq 0$. Moreover, H_ℓ has a corresponding set of orthonormal eigenvectors, x^j, $j = 1, 2, \ldots, N_m$, [12]. The eigenvalue/eigenvector pairs of the Levenberg–Marquardt matrix

$$A_\ell = \lambda_\ell I + H_\ell \tag{8.68}$$

are given by $\mu_j = \lambda_\ell + \beta_j$ and x^j, $j = 1, 2, \ldots, N_m$. Thus, we have

$$A_\ell x^j = \mu_j x^j, \tag{8.69}$$

for $j = 1, 2, \ldots, N_m$ where

$$(x^i)^T x^j = \delta_{i,j}, \tag{8.70}$$

for $i = 1, 2, \ldots, N_m$ and $j = 1, 2, \ldots, N_m$. Throughout, $\delta_{i,j}$ is the Kronecker delta function. Defining X to be the matrix with jth column equal to x^j, it follows from Eqs. (8.69) and (8.70) that

$$X^T A_\ell X = \Lambda_\ell, \tag{8.71}$$

where Λ_ℓ is a diagonal matrix with jth diagonal entry given by μ_j. Because $\{x^j\}_{j=1}^{N_m}$ is linearly independent, there exist real constants a_j^ℓ, $j = 1, 2, \ldots, N_m$ such that

$$\delta m^{\ell+1} = \sum_{j=1}^{N_m} a_j^\ell x^j = \begin{bmatrix} x^1 & x^2 & \cdots & x^{N_m} \end{bmatrix} \begin{bmatrix} a_1^\ell \\ a_2^\ell \\ \vdots \\ a_{N_m}^\ell \end{bmatrix} = X a^\ell, \tag{8.72}$$

where the N_m-dimensional column vector a^ℓ is defined by

$$a^\ell = \begin{bmatrix} a_1^\ell & a_2^\ell & \cdots & a_{N_m}^\ell \end{bmatrix}^T. \tag{8.73}$$

Using Eq. (8.72) in Eq. (8.61), multiplying the resulting equation by X^T and using Eq. (8.71) gives

$$\Lambda_\ell a^\ell = -X^T \nabla O(m^\ell), \tag{8.74}$$

so the jth component of a is given by

$$a_j^\ell = -\frac{1}{\mu_j}(x^j)^T \nabla O(m^\ell) = -\frac{1}{\beta_j + \lambda_\ell} \nabla O(m^\ell)^T x^j. \tag{8.75}$$

Recalling that $\beta_1 \geq \beta_2 \geq \cdots \geq \beta_{N_m} \geq 0$, we see that if λ_ℓ and β_j are both close to zero, and $\nabla O(m_\ell)$ has even a small component in the direction of the x^j, then a_j^ℓ may be very large which will result in a large change in the component of m that lies in the direction of x^j, see Eqs. (8.72). It is important to note that even if the exact value of the projection of $\nabla O_d(m^\ell)$ onto x^j is zero, because gradients are usually computed numerically, the calculated value of $\nabla O(m^\ell)$ will often have at least a small component in the direction of x^j. In this situation, too small a value of λ_ℓ can yield a large value of a_j^ℓ even though this value is supposed to be zero. In history-matching problems, the result can be rock property fields that have unrealistically large or small values, or unreasonable spatial variability.

More generally, it is well known that if the condition number of the Levenberg–Marquardt iteration matrix is large, then small inaccuracies in $\nabla O(m^\ell)$ can result in large changes in $\delta m^{\ell+1}$ even when a stable algorithm is used for the numerical solution of Eq. (8.61). For a real-symmetric positive-definite matrix, it is convenient to define the condition number as the ratio of the maximum eigenvalue to the minimum eigenvalue. A brief description of the significance of the condition number follows. Let A be a real-symmetric positive-definite matrix and let x be the solution of

$$Ax = b. \tag{8.76}$$

Let δA represent a change or error in A so that $A + \delta A$ is still positive definite and let δb represent a change in b. Let $\kappa(A)$ represent the condition number of A. If $\tilde{x}$ is the solution of

$$(A + \delta A)\tilde{x} = b + \delta b, \tag{8.77}$$

then using the vector and matrix two norms, it can be shown that [68]

$$\frac{\|\tilde{x} - x\|}{\|x\|} \approx \kappa(A)\left(\frac{\|\delta A\|}{\|A\|} + \frac{\|\delta b\|}{\|b\|}\right). \tag{8.78}$$

Thus, if the condition number is large, then small changes in A and b can lead to large changes in the solution of Eq. (8.76). Matrices which have a "large" condition number are said to be ill-conditioned and matrices that have a condition number "close to" unity are said to be well-conditioned.

For the Levenberg–Marquardt algorithm, Eq. (8.65) provides the theoretical equation for calculation of $\delta m^{\ell+1}$, but in practical history-matching problems, the gradient and Hessian in this equation must be computed numerically, for example, by using the adjoint method discussed in Chapter 9. Although his study was not completely definitive, the results of Li [69] suggest that the adjoint method yields sensitivities that are in error by 0.1 percent. Assuming the same relative error in calculating the gradient and the Hessian, then Eq. (8.78) suggests that the relative error in $\delta m^{\ell+1}$ could be on the order of $\kappa(\lambda_\ell I + H_\ell) \times 0.002$. Thus, if $\kappa(\lambda_\ell I + H_\ell) = 100$, the solution of the approximation of Eq. (8.61) could differ from the solution of the correct equation by around 20 percent.

For the Levenberg–Marquardt iteration matrix of Eq. (8.62), the condition number κ_ℓ is given by

$$\kappa_\ell = \frac{\lambda_\ell + \beta_1}{\lambda_\ell + \beta_{N_m}}. \tag{8.79}$$

Note that if $\beta_{N_m} = 0$ then $\kappa_\ell \to \infty$ as $\lambda_\ell \to 0$. More generally, as $\lambda_\ell \to 0$, the condition number of the Levenberg–Marquardt iteration matrix approaches the condition number of the Gauss–Newton iteration matrix, H_ℓ. On the other hand, if the Hessian matrix is small enough so its minimum and maximum eigenvalues can be estimated, then we can choose λ_ℓ to obtain any κ_ℓ that we desire. In one history-matching example, Aanonsen et al. [70] found it desirable to choose λ_ℓ to keep the condition number of the Levenberg–Marquardt algorithm equal to 10 in a history-matching problem based on the PUNQ model [71] where they did not use a prior covariance matrix and used the grad zone method [72] to construct the history match. The grad zone method is a parameterization method whereby the reservoir is divided into zones and all values of a variable in a zone (e.g. transmissibilities) are multiplied by a constant to do the history match. In this approach, the set of multipliers are the parameters that are adjusted during the history match.

8.4.2   Effects of ill-conditioning

We consider a two-dimensional flow problem where water is injected into an oil reservoir with initial pressure well above bubble-point pressure. Water is injected via a well near the center of the reservoir and four wells near the corners of the reservoir are produced at a constant oil rate. All fluid and rock properties are assumed known except for the gridblock values of absolute permeability. Observed data consist of eight values of wellbore pressure at each well uniformly spaced in time with the first pressure measurements occurring at 30 days. We wish to history match these pressure data to generate the MAP estimate of the log-permeability field. The model consists of the gridblock values of $\ln(k)$ on a $21 \times 21 \times 1$ reservoir simulation grid. The grid is areally uniform, i.e. $\Delta x = \Delta y$. In the history-matching procedure, the covariance matrix C_M was generated using an isotropic spherical variogram with a sill of 0.5 and a range equal to about $4\Delta x$.

History matching was done by minimizing the objective function of Eq. (8.2) using a Levenberg–Marquardt (LM) algorithm and with the Gauss–Newton method. Figure 8.2 shows the behavior of the normalized objective function during the iterations of the two optimization algorithms. Here the normalized objective function is defined by

$$O_N(m) = 2\frac{O(m)}{N_d}, \tag{8.80}$$

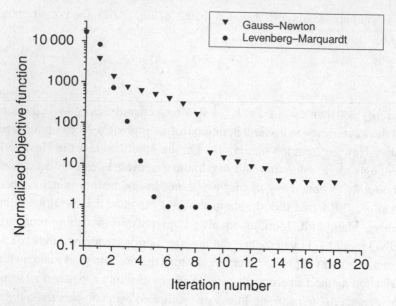

Figure 8.2. Behavior of the normalized objective function for Gauss–Newton and Levenberg–Marquardt algorithms. (Adapted from Li *et al.* [63], copyright SPE.)

where the objective function $O(m)$ is given by Eq. (8.2). Because $N_d = 32$, Eq. (8.6) and the discussion surrounding the equation indicates that expected value of the normalized objective function evaluated at the MAP estimate should be around unity and should satisfy

$$O_N(m_{\text{MAP}}) \leq 1 + 5\sqrt{\frac{2}{N_d}} = 2.25. \tag{8.81}$$

The results of Fig. 8.2 indicate that Eq. (8.81) is satisfied by the MAP estimate obtained with the LM algorithm but is not satisfied by the MAP estimate obtained with the Gauss–Newton method even though more iterations were done with the Gauss–Newton algorithm. Much more importantly, the MAP estimate of $\ln(k)$ obtained from the Gauss–Newton method is rough in the sense that small values of $\ln(k)$ on the order of 3.0 or less occur in gridblocks close to gridblocks where $\ln(k)$ is 4.5 or greater. This roughness is a consequence of obtaining an overly rough model at the first iteration where the data mismatch was large. By damping the change in the model at the first iteration by using the LM algorithm, we avoid this roughness and obtain a MAP estimate which is much smoother or more continuous (Fig. 8.3b). (In the LM algorithm, the value of λ_ℓ was decreased by a factor of 10 at each iteration.)

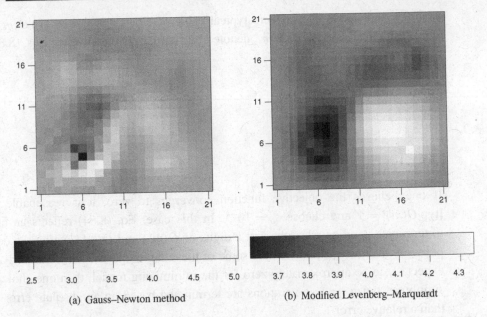

(a) Gauss–Newton method   (b) Modified Levenberg–Marquardt

Figure 8.3. MAP estimates of log-permeability field. (Adapted from Li *et al.* [63], copyright SPE.)

8.5   Convergence criteria

An optimization algorithm for minimization of an objective function $O(m)$ requires a convergence criteria test for termination of the algorithm. Natural convergence tests are based on stopping when the objective function stops changing or the model stops changing, i.e.

$$\frac{|O(m^\ell) - O(m^{\ell+1})|}{|O(m^{\ell+1})|} < \epsilon_o, \tag{8.82}$$

or

$$\max_{1 \leq i \leq N_m} \left(\frac{|m_i^{\ell+1} - m_i^\ell|}{|m_i^{\ell+1}|} \right) < \epsilon_m. \tag{8.83}$$

The convergence tests of Eqs 8.82 and 8.83 are based on relative error. If the goal is to terminate iteration when the first k digits of $O(m^{\ell+1})$ and $O(m^\ell)$ are in agreement, then we would set $\epsilon_o = 10^{-k}$. If our goal is to require that the first k digits of $m^{\ell+1}$ and m^ℓ are in agreement, we would iterate until Eq. (8.83) holds with $\epsilon_m = 10^{-k}$. When the iteration index ℓ is such that the selected convergence criterion (either Eq. (8.82), (8.83), or both equations) is satisfied, the approximate model that minimizes $O(m)$ is given by $m^{\ell+1}$.

Eq. (8.82) can not be applied if $O(m^{\ell+1}) = 0$ and Eq. (8.83) can not be applied if some component of the model becomes zero. One way to avoid this problem [66] assumes knowledge of a nonzero typical value of $O(m)$ denoted by typ $O(m)$ and a nonzero typical value of each m_i, denoted by typ m_i. We then replace Eqs. (8.82) and (8.83), respectively, by

$$\frac{|O(m^\ell) - O(m^{\ell+1})|}{\max\{|O(m^{\ell+1})|, |\text{typ } O(m)|\}} < \epsilon_o \tag{8.84}$$

and

$$\max_{1 \leq i \leq N_m} \left(\frac{|m_i^{\ell+1} - m_i^\ell|}{\max\{|m_i^{\ell+1}|, |\text{typ } m_i|\}} \right) < \epsilon_m. \tag{8.85}$$

If one believes the objective function converges to zero, it is reasonable to set $|\text{typ } O(m)| = 1$ and choose $\epsilon_o = 10^{-k}$. In this case, Eq. (8.84) reflects an attempt to terminate iteration when the first k digits of $O(m^\ell)$ and $O(m^{\ell+1})$ after the decimal point are in agreement where here some of these digits may be zero. Similarly, if we expect m_i to be approximately zero for the minimizing model, then one choice is to simply set typ $m_i = 1$ so iterations are terminated based on an absolute error rather than a relative error.

If the model m^* minimizes a smooth objective function $O(m)$, then

$$\nabla_m O(m^*) = 0. \tag{8.86}$$

This suggests applying the following stopping condition:

$$\|\nabla_m O(m^*)\|_\infty < \epsilon_g, \tag{8.87}$$

where $\|\cdot\|_\infty$ denotes the infinity vector norm. Unfortunately, the proper choice of ϵ_g can depend on the scale of O and is difficult to determine. Because of this, Dennis and Schnabel [66] suggest using a relative gradient in place of the gradient in Eq. (8.87). Gao [73] found the relative gradient to be inefficient for determining convergence in large-scale history-matching problems although his computational experiments were not very extensive. In general, he found that we obtain reasonable estimates of the minimizing model at far fewer iterations than are needed to obtain a small value of the relative gradient and that determining an appropriate convergence tolerance based on the relative gradient is difficult. Nevertheless, if the gradient becomes zero, the minimization algorithm should terminate. One appropriate condition is to set a convergence criterion based on

$$\frac{\|\nabla O\|_2}{\max\{\|m\|_2, 1\}} < \epsilon_g. \tag{8.88}$$

Applying Eq. (8.84) is natural because we are trying to minimize the objective function. But, this criterion is not in and of itself sufficient. In fact, it is easy to argue that Eq. (8.85) should also be satisfied before terminating the algorithm. Here, we follow the discussion of this aspect given in Gill *et al.* [74]. One situation that can occur is the generation of a search direction along which the objective function decreases only slightly. This can occur if the search direction is almost orthogonal to the steepest descent direction, or if we are in a region where the objective function is relatively flat. In either situation, the objective function might decrease so little that Eq. (8.84) is satisfied, but if Eq. (8.85) is not satisfied, then large changes in the model occurred over the iteration and the algorithm should not be terminated.

8.5.1 Convergence tolerance for history-matching problems

In the history-matching examples presented in this book and related papers, we have typically followed the following termination criteria.

If

$$\frac{\|\nabla O\|_2}{\max\{\|m\|_2, 1\}} < 10^{-8}. \tag{8.89}$$

the algorithm is terminated.

If both of the following two conditions are satisfied, we also terminate the algorithm:

$$\frac{|O(m^\ell) - O(m^{\ell+1})|}{\max\{|O(m^{\ell+1})|, |1|\}} < \epsilon_o \tag{8.90}$$

and

$$\max_{1 \le i \le N_m} \left(\frac{|m_i^{\ell+1} - m_i^\ell|}{\max\{|m_i^{\ell+1}|, |\mathrm{typ}\, m_i|\}} \right) < \epsilon_m, \tag{8.91}$$

where $\mathrm{typ}\, m_i$ is set equal to 1 if m_i corresponds to permeability or log-permeability and is set equal to 0.1 if m_i corresponds to porosity. Although, one can find some theoretical justification for choosing $\epsilon_o \approx \epsilon_m^2$, the typical tolerance values we have used are $\epsilon_o = 10^{-4}$ and $\epsilon_m = 10^{-3}$, but in many cases it is possible to get good results with larger tolerances.

8.5.2 Convergence of algorithms

Ignoring memory requirements, the total computational cost of an algorithm is the average time required per iteration times the number of iterations required to obtain convergence. Thus, we wish convergence to be as rapid as possible.

Definition 8.1. *Suppose an iterative algorithm converges to m^*. The algorithm is said to be linearly convergent if there exists a constant c with $0 < c < 1$ and a positive integer k_c such that for all $k \ge k_c$*

$$\|m^{k+1} - m^*\| \le c\|m^k - m^*\|. \tag{8.92}$$

The algorithm is said to be quadratically convergent if there exists a constant $b > 0$ such that

$$\|m^{k+1} - m^*\| \le b\|m^k - m^*\|^2 \tag{8.93}$$

for all $k \ge k_c$.

Note in the definition of quadratic convergence, it is not necessary that b be less than unity. Also, we should remark that currently it is common to use the terminology Q-linearly convergent and Q-quadratically convergent and write Eqs. (8.92) and (8.93), respectively, as

$$\frac{\|m^{k+1} - m^*\|}{\|m^k - m^*\|} \le c \quad \text{and} \quad \frac{\|m^{k+1} - m^*\|}{\|m^k - m^*\|^2} \le b. \tag{8.94}$$

The "Q" added to the terminology reflects the fact that the conditions involve quotients.

The norm used in the preceding equations can be any real vector norm defined on the set of all real N_m-dimensional column vectors including the one defined by

$$\|x\|_R = \sqrt{x^T R x}, \tag{8.95}$$

where R is any real-symmetric positive-definite matrix. If R is the $N_m \times N_m$ identity matrix, Eq. (8.96) becomes the standard two or Euclidean norm.

There is a slightly stronger form of linear convergence known as Q-superlinear convergence.

Definition 8.2. *Suppose an iterative algorithm converges to m^*. The algorithm is said to converge superlinearly if*

$$\lim_{k \to \infty} \frac{\|m^{k+1} - m^*\|}{\|m^k - m^*\|} = 0. \tag{8.96}$$

Clearly Q-quadratic convergence implies superlinear convergence and superlinear convergence implies Q-linear convergence.

The following theoretical result is given in Luenberger [75].

Proposition 8.1. *If m^* corresponds to a local minimum of an objective function $O(m)$ which is three times continuously differentiable, the Hessian is real symmetric positive definite in some neighborhood of m^* and the initial guess m^0 is sufficiently close to m^*, then Newton's method with a full step ($\alpha_k = 1$) for all k converges Q-quadratically.*

It is important to note that the theoretical result indicates that we will obtain quadratic convergence in Newton's method if we use a full step from some point on in the iterations, i.e. once we are close to the minimum. Far from the minimum, we would always use a line search or restricted-step method. When using a line search, it is recommended that one should always try a full step at the first iteration of the line search procedure because the conditions of the theorem are based on a full step.

Also note that far from the minimum, there is no guarantee that the Hessian will be positive definite or even nonsingular so some modification of Newton's method may be necessary. The Levenberg–Marquardt algorithm provides one way to avoid such problems provided one ensures that λ_k is chosen such that $\lambda_k I + H_k$ is positive definite where H_k denotes the Hessian at the kth iteration.

For a least-squares problem where the number of data is greater than the number of parameters estimated, it is possible that the Gauss–Newton Hessian will be nonsingular. If, however, the residual term (the second derivative term neglected in deriving the Gauss–Newton method from Newton's method) is large in a neighborhood of the minimum, the Gauss–Newton method may not converge. When the Gauss–Newton method converges, convergence will only be Q-linear unless the residual term is zero in which case, Gauss–Newton and Newton become identical so the convergence rate near a minimum may be Q-quadratic. When the Gauss–Newton Hessian is used in the Levenberg–Marquardt algorithm, the modified Hessian will always be positive definite so we will always obtain a down-hill direction. Although the Levenberg–Marquardt method tends to be a robust algorithm, like the Gauss–Newton method only Q-linear convergence can generally be established. A careful discussion of the preceding comments on convergence properties is given in Chapter 6 of Fletcher [64] and Chapter 10 of Dennis and Schnabel [66].

The simplest descent algorithm is steepest descent, which is discussed in the next section on scaling. Although this method typically converges Q-linearly, for poorly scaled problems, the convergence rate is well known to be excessively slow.

8.6 Scaling

As before, m^* denotes the model that minimizes a particular objective function. The scaling of the model parameters can have a significant effect on the performance of an optimization algorithm applied to estimate m^* but the effect of poor scaling is more significant for some algorithms than others. The steepest descent algorithm is particularly sensitive to scaling and converges very slowly when the model is poorly scaled. To illustrate this consider the simple objective function,

$$O(m) = \frac{1}{2}\left(\frac{m_1 - a_1}{\sigma_1}\right)^2 + \frac{1}{2}\left(\frac{m_2 - a_2}{\sigma_2}\right)^2, \tag{8.97}$$

where a_1, a_2, σ_1, and σ_2 are constants. The steepest descent algorithm is given by

$$m^{\ell+1} = m^\ell + \alpha_\ell d^\ell, \tag{8.98}$$

where the search direction d^ℓ is given by

$$d^\ell = -\nabla O(m^\ell) \tag{8.99}$$

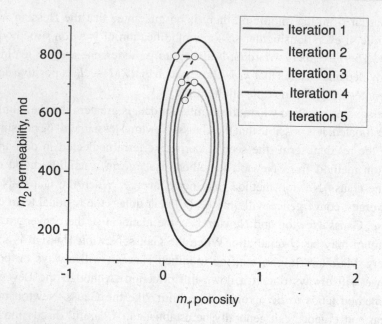

Figure 8.4. Behavior of steepest descent for a poorly scaled problem.

and the step size α_ℓ is obtained by a line search. If an exact line search is used, then α_ℓ is the value of α that minimizes $O(m^\ell + \alpha d^\ell)$, i.e.

$$\alpha_\ell = \underset{\alpha}{\operatorname{argmin}} \, O(m^\ell + \alpha d^\ell). \tag{8.100}$$

Consider the simple case where $a_1 = 0.3$, $a_2 = 500$, $\sigma_1 = 0.05$, and $\sigma_2 = 500$. In this case, we may think of m_1 and m_2, respectively as representing porosity and permeability. With a starting value of $m^0 = [0.1, 800]^T$, the steepest descent algorithm with an exact line search required 39 iterations to obtain the correct minimum with three digit accuracy. Figure 8.4 shows the initial guess and the results of the first four iterations. The elliptical contours represent level curves, i.e. on each elliptical curve, the value of the objective function is constant. The minimizing model is at the center of the concentric ellipses. Note the iterates proceed in a zig-zag path down the long narrow valley. This is the standard behavior of the steepest descent algorithm when the lengths of principle axes of hyperellipses are radically different. (For the same problem, the Newton method with a unit step converges in one iteration for any initial guess.)

For the objective function of Eq. (8.97), we can rescale the model by defining a new model by

$$\tilde{m} = \begin{bmatrix} 1/\sigma_1 & 0 \\ 0 & 1/\sigma_2 \end{bmatrix} \begin{bmatrix} m_1 - a_1 \\ m_2 - a_2 \end{bmatrix}. \tag{8.101}$$

In terms of the rescaled model, the objective function of Eq. (8.97) becomes

$$\tilde{O}(\tilde{m}) = \frac{1}{2}\tilde{m}^{\mathrm{T}}\tilde{m} \tag{8.102}$$

and

$$\nabla_{\tilde{m}}\tilde{O}(\tilde{m}) = \tilde{m}. \tag{8.103}$$

Thus, for any initial guess $\tilde{m}^0$, Eqs. (8.100) and (8.98), respectively give at the first iteration

$$\alpha_\ell = \underset{\alpha}{\mathrm{argmin}}\ \tilde{O}(\tilde{m}^0 - \alpha\tilde{m}^0) = 1 \tag{8.104}$$

and

$$\tilde{m}^1 = \tilde{m}^0 - \tilde{m}^0 = 0, \tag{8.105}$$

i.e. for the rescaled problem, steepest descent converges to the model that minimizes $O(\tilde{m})$ in one iteration. The minimizing model, $\tilde{m} = 0$, is equivalent to $m = [a_1, a_2]^{\mathrm{T}}$ which is the model corresponding to the minimum of the original objective function of Eq. (8.97). Moreover, since the Hessian for the rescaled problem is equal to the identity matrix, steepest descent for the rescaled problem is equivalent to Newton's method with a unit step.

The scaling procedure used in the preceding example is general and can be applied to any quadratic objective function. If

$$O(m) = \frac{1}{2}(m - m_{\mathrm{prior}})^{\mathrm{T}} A (m - m_{\mathrm{prior}}), \tag{8.106}$$

where A is a real-symmetric positive-definite matrix, and m_{prior} is a constant N_m-dimensional column vector, then a rescaled model is given by

$$\tilde{m} = A^{1/2}(m - m_{\mathrm{prior}}). \tag{8.107}$$

Here $A^{1/2}$ is a symmetric square root of A and can be obtained by spectral decomposition. Alternatively, we may set $A^{1/2} = L^{\mathrm{T}}$ where $A = LL^{\mathrm{T}}$ is the Cholesky decomposition of A. In terms of the rescaled model, $O(\tilde{m})$ is still given by Eq. (8.102). Thus, for any initial guess, the steepest descent algorithm with an exact line search will converge to the model that minimizes $O(\tilde{m})$ in one iteration.

For the quadratic objective function of Eq. (8.106), one could simply apply the Gauss–Newton method with a unit step and obtain convergence in one iteration for any initial guess using the original formulation in terms of m or the transformed problem. This result is a reflection of the fact that Newton's method is invariant under scaling even in the case of a general scaling. This means a unit Newton step in terms of the rescaled variable $\tilde{m}$ is identical to a unit step in terms of the variable m where

$$\tilde{m} = Dm, \tag{8.108}$$

where D is a nonsingular $N_m \times N_m$ matrix with entries independent of m. Given an arbitrary objective function $O(m)$, we define the objective function in terms of the new variable $\tilde{m}$ by

$$\tilde{O}(\tilde{m}) = O(D^{-1}\tilde{m}). \tag{8.109}$$

It is easy to show that the following chain rule holds:

$$\nabla_{\tilde{m}} \tilde{O}(\tilde{m}) = (\nabla_{\tilde{m}} m^{\mathrm{T}})(\nabla_m O(m)). \tag{8.110}$$

Using

$$\nabla_{\tilde{m}} m^{\mathrm{T}} = \nabla_{\tilde{m}} (D^{-1}\tilde{m})^{\mathrm{T}} = (\nabla_{\tilde{m}} \tilde{m}^{\mathrm{T}}) D^{-\mathrm{T}} = D^{-\mathrm{T}}, \tag{8.111}$$

in Eq. (8.110) gives

$$\nabla_{\tilde{m}} \tilde{O}(\tilde{m}) = D^{-\mathrm{T}} \nabla_m O(m), \tag{8.112}$$

where $D^{-\mathrm{T}}$ denotes the inverse of the transpose of D. Using the same type of chain rule, the Hessian for the transformed problem is given by

$$H(\tilde{m}) = \nabla_{\tilde{m}} [(\nabla_{\tilde{m}} \tilde{O}(\tilde{m}))^{\mathrm{T}}] = (\nabla_{\tilde{m}} m^{\mathrm{T}}) \nabla_m [(\nabla_{\tilde{m}} \tilde{O}(\tilde{m}))^{\mathrm{T}}]. \tag{8.113}$$

Using Eqs. (8.112) and (8.111) in Eq. (8.113) gives

$$\tilde{H}(\tilde{m}) = D^{-\mathrm{T}} \nabla_m [(D^{-\mathrm{T}} \nabla_m O(m))^{\mathrm{T}}] = D^{-\mathrm{T}} H(m) D^{-1}, \tag{8.114}$$

where $H(m)$ is the Hessian in terms of the original variable. From Eqs. (8.112) and (8.113), a unit Newton step is given by

$$\delta\tilde{m} = -(\tilde{H}(\tilde{m}))^{-1} \nabla_{\tilde{m}} \tilde{O}(\tilde{m}) = -DH(m)^{-1} \nabla_m O(m). \tag{8.115}$$

Eq. (8.108) implies that $\delta\tilde{m} = D\delta m$, so Eq. (8.115) is equivalent to

$$\delta m = -H(m)^{-1} \nabla_m O(m). \tag{8.116}$$

Thus, Newton's method is scale invariant. For steepest descent, however, it is easy to see from Eq. (8.112) that

$$\delta\tilde{m} = -D^{-1} D^{-\mathrm{T}} \delta m; \tag{8.117}$$

i.e. steepest descent is not scale invariant.

Although Newton's method and the Gauss–Newton method are both scale invariant, when applying the methods computationally, the methods are scale invariant only if infinite precision arithmetic is used, which is never the case. Thus, it is possible that rescaling the model may have some beneficial effect on Newton's method if $H(\tilde{m})$ has a significantly lower condition number than $H(m)$; see the discussion of subsection 8.3.2.

For the quadratic objective function of Eq. (8.106), one can rescale the model parameters by using the transformation of Eq. (8.107). However, for a nonquadratic objective function, the Hessian is not known a priori and changes with each iteration so determining the proper transformation for rescaling is considerably more difficult. In fact, when the number of model parameters is large, it is not feasible to even calculate the Hessian. As Eq. (8.107) indicates, scaling transforms a model m to a model $\tilde{m}$ where near the minimizing model, all components of $\tilde{m}$ are expected to be of roughly the same order of magnitude. As our approach to data integration and automatic history matching is rooted in Bayesian probability, a prior PDF for m is available. If this PDF is approximated by a multi-variate Gaussian, then the objective function minimized to match dynamic data always has the form

$$O(m) = \frac{1}{2}(m - m_0)^{\mathrm{T}} C_M^{-1}(m - m_0) + O_d(m), \tag{8.118}$$

where $O_d(m)$ includes all data mismatch terms and is given by Eq. (8.15). If one is computing the MAP estimate, m_0 is the vector of prior means, whereas, if one is using the randomized maximum likelihood estimate to generate a realization of m conditional to observed data, then m_0 is an unconditional realization generated from the prior PDF. The jth diagonal entry of the covariance matrix C_M is the prior variance of m_j and is denoted by $\sigma_{m,j}^2$. If we let D be the diagonal matrix with jth diagonal entry equal to $\sigma_{m,j}$ then

$$C_M = D\tilde{C}_M D, \tag{8.119}$$

where $\tilde{C}_M$ represents the prior correlation matrix. The element in the ith row and jth column of $\tilde{C}_M$ is given by

$$\tilde{c}_{i,j} = \frac{c_{i,j}}{\sigma_{m,i}\sigma_{m,j}}, \tag{8.120}$$

where $c_{i,j}$ is the corresponding entry of C_M. Defining a new set of variables by the transformation

$$\tilde{m} = D^{-1}(m - m_0), \tag{8.121}$$

the objective function is

$$O(m) = \frac{1}{2}\tilde{m}^{\mathrm{T}}\tilde{C}_M^{-1}\tilde{m} + O_d(m). \tag{8.122}$$

If the prior model is multi-variate Gaussian and $m_0 = m_{\mathrm{prior}}$, then $\tilde{m}$ is multi-variate Gaussian with mean equal to the zero vector and all components of $\tilde{m}$ have unit variance. Gao [73] and Gao and Reynolds [76] used the scaling of Eq. (8.121) and found that it could improve the rate of convergence of a quasi-Newton method (Section 8.7) when history matching production data for some example problems. However, the experiments were fairly cursory and there is no indication that this result is general. In fact, the condition number of the correlation matrix is not always less than or equal to the condition number of the covariance matrix except for the 2×2 case. One can construct examples where the condition number of the correlation matrix is greater than the condition number of the covariance matrix. Despite this, using the correlation matrix has the advantage that its entries are dimensionless, whereas different entries of the covariance matrix may have different units. In this case, generating eigenvalues and eigenvectors of the covariance matrix does not make sense.

8.7   Line search methods

As discussed briefly in Section 8.5.2, to theoretically establish the optimal convergence rates for Newton algorithms, we need to ensure that a full step is proposed at some point in a Newton or quasi-Newton algorithm and also ensure that the Wolfe or strong Wolfe conditions be satisfied at termination of the line search algorithm [64, 67]. Here we assume that we simply always try a full step at the first iteration of the line search algorithm.

Assume we wish to minimize $O(m)$, that m_k is our current estimate of the minimum, that d_k is the down-hill search direction and that we wish to find $\alpha_k > 0$ so that

$$m_{k+1} = m_k + \alpha_k d_k, \tag{8.123}$$

where k denotes the iteration index, gives an improved estimate of the model which minimizes O. If α_k were obtained by an exact line search, we would have

$$\alpha_k = \underset{\alpha}{\mathrm{argmin}}\ O(m_k + \alpha d_k). \tag{8.124}$$

For history-matching problems, each iteration of a line search algorithm requires a run of the simulator so we wish to minimize the number of iterations of the line search algorithm.

8.7.1   Strong Wolfe conditions

An exact line search is neither necessary nor desirable. To show theoretical convergence to a local minimum, assuming one exists, we need only require that the α_k used to update the model by Eq. (8.123) satisfies the strong Wolfe conditions which are given by the following two equations:

$$O(m_k + \alpha_k d_k) \leq O(m_k) + c_1 \alpha_k (\nabla O(m_k))^T d_k \tag{8.125}$$

and

$$\left| \left(\nabla O(m_k + \alpha_k d_k) \right)^T d_k \right| \leq c_2 \left| \left(\nabla O(m_k) \right)^T d_k \right|, \tag{8.126}$$

where c_1 and c_2 are positive constants that should satisfy

$$0 < c_1 < 0.5 \tag{8.127}$$

and

$$c_1 < c_2 < 1. \tag{8.128}$$

In our history-matching examples, we use $c_1 = 10^{-4}$ and $c_2 = 0.9$. Once we obtain an α_k which satisfies the strong Wolfe conditions, we update the model with Eq. (8.123) and proceed to the next iteration of the optimization algorithm. The statement that d_k is a down-hill direction from m_k is equivalent to

$$\left(\nabla O(m_k)\right)^T d_k < 0. \tag{8.129}$$

We see that Eq. (8.125) guarantees that $O(m_k + \alpha_k d_k)$ is strictly smaller than $O(m_k)$, but to guarantee convergence of the minimization algorithm to a local minimum, it is necessary to ensure a sufficient decrease in the objective function at each iteration. The first strong Wolfe condition, Eq. (8.125), does not guarantee this because it does not ensure that the step sizes are bounded away from zero. Thus, we need a second Wolfe condition.

As opposed to the strong Wolfe conditions, the Wolfe conditions still require that Eq. (8.125) be satisfied, but the second Wolfe condition simply requires

$$\left(\nabla O(m_k + \alpha_k)\right)^T d_k \geq c_2 \left(\nabla O(m_k)\right)^T d_k. \tag{8.130}$$

Although theoretical convergence can be obtained by only enforcing the first and second Wolfe conditions, our preference is to require that the strong Wolfe conditions be satisfied as this restricts acceptable step sizes to a smaller interval. It is possible that the second strong Wolfe condition can on occasion require more iterations of the line search algorithm for history-matching problems, although Nocedal and Wright [67] report that the difference in computational work is minimal based on their experience.

For further discussion and a geometric interpretation of the Wolfe conditions, it is convenient to define a function of the step size α by

$$\phi(\alpha) = O(m_k + \alpha d_k), \tag{8.131}$$

with the derivative of $\phi(\alpha)$ given by

$$\phi'(\alpha) = \left(\nabla O(m_k + \alpha d_k)\right)^T d_k. \tag{8.132}$$

With this definition, a line search algorithm seeks to find an α_k sufficiently close to a minimum of $\phi(\alpha)$ such that the strong Wolfe conditions are satisfied. Moreover, the first and second strong Wolfe conditions, respectively, can be rewritten as

$$\phi(\alpha_k) \leq \phi(0) + c_1 \alpha_k \phi'(0) \tag{8.133}$$

and

$$|\phi'(\alpha_k)| \leq c_2 |\phi'(0)|. \tag{8.134}$$

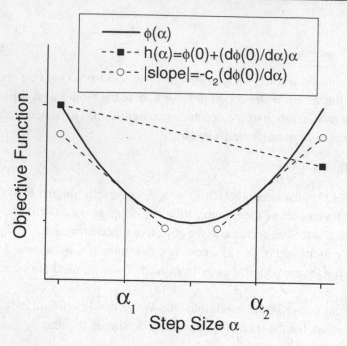

Figure 8.5. A graphical interpretation of strong Wolfe conditions.

But Eq. (8.129) implies that $\phi'(0) < 0$ so Eq. (8.134) is equivalent to

$$c_2\phi'(0) \leq \phi'(\alpha_k) \leq -c_2\phi'(0). \tag{8.135}$$

Figure 8.5 shows the behavior of an objective function in the direction of d_k, i.e. represents a plot of $\phi(\alpha)$. The top dotted line represents the upper bound from the first strong Wolfe conditions:

$$h(\alpha) = \phi(0) + c_1\phi'(0)\alpha. \tag{8.136}$$

Note the first Wolfe condition, Eq. (8.133), requires that the point $(\alpha_k, \phi(\alpha_k))$ on the curve defined by $\phi(\alpha)$ must be on or below the line, $h(\alpha)$, i.e. $\phi(\alpha_k) \leq h(\alpha_k)$. The left dashed line through two circles has a slope $c_2\phi'(0)$ and is tangent to $\phi(\alpha)$ at $\alpha = \alpha_1$. The right dashed line through two circles has slope $-c_2\phi'(0)$ and is tangent to $\phi(\alpha)$ at $\alpha = \alpha_2$. Thus Eq. (8.135) is equivalent to

$$\phi'(\alpha_1) \leq \phi'(\alpha_k) \leq \phi'(\alpha_2) \tag{8.137}$$

and any α such that $\alpha_1 \leq \alpha \leq \alpha_2$ will satisfy the second strong Wolfe condition. Note Fig. 8.5 displays a very well behaved convex objective function. It is possible that $\phi(\alpha)$ could have multiple minima in the direction of α and there could be several subintervals on which the second strong Wolfe condition will be satisfied. Also note that if we require only the Wolfe conditions of Eqs. (8.125) and (8.130) be satisfied, then α_k must be between α_1 and the value of α corresponding to the intersection of $\phi(\alpha)$ and $h(\alpha)$.

If we reduce c_2, we reduce the absolute values of the slopes of the two lines representing the second strong Wolfe condition in Fig. 8.5. In this case, the second strong Wolfe condition would require that we find an α_k closer to the true minimum before terminating the line search. While this may seem desirable, doing a more accurate line search requires more iterations of the line search algorithm, and as each iteration requires a simulation run to evaluate the objective functions, small values of c_2 are not desirable. Although the optimal value of c_2 has not to our knowledge been carefully investigated for automatic history-matching problems, our experience suggests the commonly recommended value of $c_2 = 0.9$ works well.

8.7.2   Quadratic and cubic line search algorithms

The most commonly used line search algorithms are either based on a backtracking strategy or a bracketing strategy (pages 34–39 of Fletcher [64]). Both methods use quadratic and/or cubic interpolation along the lines given in Dennis and Schnabel [66] where detailed line search algorithms that can easily be converted to code can be found. Nocedal and Wright [67] give a clear discussion of the bracketing strategy. At iteration k of the optimization algorithm, we wish to find an α_k such that the strong Wolfe conditions are satisfied.

In history matching, we are minimizing objective functions that are nonnegative and include data mismatch terms squared so if we take a large enough step, $\bar{\alpha}$, we will always obtain a model such that $O(m_k + \bar{\alpha}d_k)$ does not satisfy the first Wolfe condition. In this case, the $\phi(\alpha)$ must intersect the line $h(\alpha)$ at some $\alpha < \bar{\alpha}$ (see Fig. 8.5) so there exists a point in the interval $[0, \bar{\alpha}]$ where the strong Wolfe conditions are satisfied. The procedure is to generate a sequence of iterates, $\alpha_\ell^*, \ell = 1, 2, \ldots$, until we find α_ℓ such that the strong Wolfe (or Wolfe) conditions are satisfied by

$$m_{k+1}^c = m_k + \alpha_\ell d_k. \tag{8.138}$$

Then m_{k+1} is set equal to m_{k+1}^c and the iteration of the optimization is complete. At each iteration of the line search algorithm, the procedure uses points in the α_ℓ^* sequence to determine a new subinterval of smaller width that contains points where the strong Wolfe conditions are satisfied.

Here, we assume we always try a full step at the first iteration so $\alpha_1^* = 1$. If at the first iteration of the line search algorithm, the first Wolfe condition is not satisfied, there is no need to compute the derivative of ϕ necessary to check the second Wolfe condition. Instead, we simply do a quadratic fit and minimize the quadratic to estimate a new trial step size. If the first Wolfe condition (Eq. 8.133) is not satisfied, then

$$\phi(\alpha_\ell^*) > \phi(0) + c_1 \alpha_\ell^* \phi'(0) > \phi(0) + \alpha_\ell^* \phi'(0). \tag{8.139}$$

In this situation, $\phi(\alpha) = O(m_k + \alpha d_k)$ is approximated by a quadratic $q(\alpha)$ such that

$$q(0) = \phi(0),$$
$$q'(0) = \phi'(0), \qquad (8.140)$$
$$q(\alpha_\ell^*) = \phi(\alpha_\ell^*),$$

where at the first iteration of the line search algorithm, $\ell = 1$ and $\alpha_\ell^* = 1$. It is easy to show that $q(\alpha)$ is given by

$$q(\alpha) = \frac{\phi(\alpha_\ell^*) - \phi'(0)\alpha_\ell^* - \phi(0)}{(\alpha_\ell^*)^2} \alpha^2 + \phi'(0)\alpha + \phi(0). \qquad (8.141)$$

Letting $\alpha_{\ell+1}^*$ denote the value of α at which the derivative of q is zero, we find that

$$\alpha_{\ell+1}^* = -\frac{\phi'(0)(\alpha_\ell^*)^2}{2\big(\phi(\alpha_\ell^*) - \phi'(0)\alpha_\ell^* - \phi(0)\big)}. \qquad (8.142)$$

Taking the second derivative of q and using Eq. (8.139) gives

$$q''(\alpha_{\ell+1}^*) = 2\frac{\phi(\alpha_\ell^*) - \phi'(0)\alpha_\ell^* - \phi(0)}{(\alpha_\ell^*)^2} > 0; \qquad (8.143)$$

thus, Eq. (8.142) gives the unique value of α which minimizes q. Because $\phi'(0) < 0$, Eq. (8.139) also implies that Eq. (8.142) gives a positive value of $\alpha_{\ell+1}^*$. From Eqs. (8.142) and (8.139), it follows that

$$\alpha_{\ell+1}^* < \frac{-\phi'(0)(\alpha_\ell^*)^2}{2\big(\phi(0) + c_1\alpha_\ell^*\phi'(0) - \phi'(0)\alpha_\ell^* - \phi(0)\big)} = \frac{\alpha_\ell^*}{2(1 - c_1\alpha_\ell^*)} \approx \frac{\alpha_\ell^*}{2}, \qquad (8.144)$$

where the last approximate equality follows from the fact that c_1 is typically chosen very small and $\alpha_\ell^* \leq 1$. The preceding equation indicates Eq. (8.142) gives a proposed step size reduction on the order of 1/2 or more. On the other hand, if $\phi(\alpha_\ell^*) \gg \phi(0)$, then the value of $\alpha_{\ell+1}^*$ calculated from Eq. (8.142) may be close to zero. This can sometimes indicate that $\phi(\alpha)$ is not well approximated by a quadratic in the interval $0 \leq \alpha \leq 1$ and that $\alpha_{\ell+1}^*$ is a poor approximation to the minimum of $\phi(\alpha)$. Because of this, it is common [66] to specify a maximum reduction, $\alpha_{\min}$, in the step size over an iteration, i.e. require that

$$\alpha_{\ell+1}^* \geq \alpha_{\min}\alpha_\ell^*. \qquad (8.145)$$

If the value of $\alpha_{\ell+1}^*$ calculated from 8.142 is below the lower bound in Eq. (8.145), then we set $\alpha_{\ell+1}^* = \alpha_{\min}\alpha_\ell^*$. A typical choice is to set $\alpha_{\min} = 0.1$ so that the step size is not reduced by more than a factor of 10.

If we have generated points $\alpha_{\ell-1}^*$ and α_ℓ^* and neither result in a model which satisfies the strong Wolfe conditions, $\alpha_{\ell+1}^*$ is estimated by cubic interpolation because a cubic can model changes in concavity whereas a quadratic can not. Letting $q(\alpha)$ denote the cubic polynomial such that

$$
\begin{aligned}
q(0) &= \phi(0), \\
q'(0) &= \phi'(0), \\
q(\alpha_{\ell-1}^*) &= \phi(\alpha_{\ell-1}^*), \\
q(\alpha_\ell^*) &= \phi(\alpha_\ell^*).
\end{aligned}
\tag{8.146}
$$

The minimizing point is

$$
\alpha_{\ell+1}^* = \frac{-b + \sqrt{b^2 - 3a\phi'(0)}}{3a}.
\tag{8.147}
$$

Defining δ_c by

$$
\delta_c = \frac{1}{\alpha_\ell^* - \alpha_{\ell-1}^*},
\tag{8.148}
$$

a and b are given respectively by

$$
a = \delta_c \left(\frac{1}{(\alpha_\ell^*)^2} [\phi(\alpha_\ell^*) - \phi(0) - \phi'(0)\alpha_\ell^*] - \frac{1}{(\alpha_{\ell-1}^*)^2} [\phi(\alpha_{\ell-1}^*) - \phi(0) - \phi'(0)\alpha_{\ell-1}^*] \right)
\tag{8.149}
$$

and

$$
b = \delta_c \left(-\frac{\alpha_{\ell-1}^*}{(\alpha_\ell^*)^2} [\phi(\alpha_\ell^*) - \phi(0) - \phi'(0)\alpha_\ell^*] + \frac{\alpha_\ell^*}{(\alpha_{\ell-1}^*)^2} [\phi(\alpha_{\ell-1}^*) - \phi(0) - \phi'(0)\alpha_{\ell-1}^*] \right).
\tag{8.150}
$$

As in the quadratic case, it is customary to bound the step size away from zero. Unlike the quadratic case, however, one is no longer guaranteed that Eq. (8.147) will reduce the step size by a factor of about 2 or more. Thus, it is customary to impose both a lower and upper limit on the reduction on step size, i.e. require that

$$
\alpha_{\min}\alpha_\ell^* \leq \alpha_{\ell+1}^* \leq 0.5\alpha_\ell^*,
\tag{8.151}
$$

where as in the quadratic case, we normally set $\alpha_{\min} \approx 0.1$. If the calculated step size is not within the bounds, we simply set $\alpha_{\ell+1}^* = 0.5\alpha_\ell^*$. Although the preceding procedure usually works well, using a bracketing strategy is more robust.

To present a bracketing strategy, we suppose we have located an interval $[\lambda_1, \lambda_2]$ and it is known that this interval contains a minimum of $\phi(\alpha)$. We also suppose we have derivative information available which we use to compute the cubic interpolate. (If not, the cubic interpolation formula of Eq. (8.146) can be used.) If derivative information has been computed, we determine a cubic interpolate $q(\alpha)$ from the conditions

$$q(\lambda_1) = \phi(\lambda_1),$$
$$q'(\lambda_1) = \phi'(\lambda_1),$$
$$q(\lambda_2) = \phi(\lambda_2), \tag{8.152}$$
$$q'(\lambda_2) = \phi'(\lambda_2).$$

The cubic interpolate is given by

$$q(\alpha) = a(\alpha - \lambda_1)^3 + b(\alpha - \lambda_1)^2 + c(\alpha - \lambda_1) + d, \tag{8.153}$$

where

$$a = \frac{-2(\phi(\lambda_2) - \phi(\lambda_1)) + (\phi'(\lambda_2) + \phi'(\lambda_1))(\lambda_2 - \lambda_1)}{(\lambda_2 - \lambda_1)^2}, \tag{8.154}$$

$$b = \frac{3(\phi(\lambda_2) - \phi(\lambda_1)) - (\phi'(\lambda_2) + 2\phi'(\lambda_1))(\lambda_2 - \lambda_1)}{(\lambda_2 - \lambda_1)^2}, \tag{8.155}$$

$$c = \phi'(\lambda_1), \tag{8.156}$$

and

$$d = \phi(\lambda_1). \tag{8.157}$$

If $a \neq 0$ and $b^2 - 3ac > 0$, this cubic has a minimum in the interior of $[\lambda_1, \lambda_2]$ given by

$$\alpha_{\ell+1}^* = \lambda_1 + \frac{-b + \sqrt{b^2 - 3ac}}{3a}, \tag{8.158}$$

or the minimum occurs at one of the endpoints of the interval $[\lambda_1, \lambda_2]$. If $a = 0$, then the cubic reduces to a quadratic which has a zero derivative at

$$\alpha_{\ell+1}^* = \lambda_1 - \frac{c}{2b}. \tag{8.159}$$

A line search strategy which uses interpolation to refine the $[\lambda_1, \lambda_2]$ interval can now be easily described. At the first iteration of the line search, set $\ell = 1$, $\alpha_{\ell-1}^* = 0$ and $\alpha_\ell^* = 1$. If the strong Wolfe conditions hold with the unit step size, we would terminate the line search. However, in all history-matching cases we have tried, one or more of the following typically hold when $\ell = 1$:

$$\phi(\alpha_\ell^*) > \phi(\alpha_{\ell-1}^*), \tag{8.160}$$

$$\phi(\alpha_\ell^*) > \phi(0) + c_1\phi'(0)\alpha_\ell^*, \tag{8.161}$$

or

$$\phi'(\alpha_\ell^*) > 0. \tag{8.162}$$

If any of the preceding equations hold, then the strong Wolfe conditions must hold on some subinterval of $[0, 1]$ and we let $\lambda_1 = 0$ and $\lambda_2 = 1$. If none of these conditions hold, we simply increase α_1^* by a factor ρ until we find a value of α_1^* such that one of the preceding three conditions holds. Then λ_2 is set equal to this value of α_1^*.

If Eq. (8.160) holds when $\ell = 1$, then Eq. (8.161) also holds, i.e. the first Wolfe condition is violated. When Eq. (8.161) holds, $\alpha_{\ell+1}^*$ is computed from Eq. (8.142). If the first Wolfe condition is satisfied, but Eq. (8.162) holds, $\alpha_{\ell+1}^*$ is computed by the cubic interpolation formula of Eq. (8.158) (or Eq. 8.147), or if $a = 0$, from Eq. (8.159). If the computed value of $\alpha_{\ell+1}^*$ is not in the interior of the interval $[\lambda_1, \lambda_2]$, we set

$$\alpha_{\ell+1}^* = \frac{\lambda_1 + \lambda_2}{2}. \tag{8.163}$$

Next we evaluate $\phi(\alpha_{\ell+1}^*) = O(m_k + \alpha_{\ell+1}^* d_k)$ and check the first Wolfe condition. If the first Wolfe condition is not satisfied, i.e.

$$\phi(\alpha_{\ell+1}^*) > \phi(0) + c_1 \phi'(0) \alpha_{\ell+1}^* \tag{8.164}$$

or if

$$\phi(\alpha_{\ell+1}^*) > \phi(\lambda_1) \tag{8.165}$$

then set $\lambda_2 = \alpha_{\ell+1}^*$, compute $\phi'(\lambda_2)$, add one to the iteration index ℓ, and compute $\alpha_{\ell+1}^*$ from Eq. (8.158), or if $a = 0$, from Eq. (8.159), but enforcing the requirement that $\alpha_{\ell+1}^*$ be contained in the interior of the interval $[\lambda_1, \lambda_2]$; which can be enforced by using Eq. (8.163) if necessary. Then we check again the validity of Eqs. (8.164) and (8.165) and repeat the process. Eventually, we must find a $\alpha_{\ell+1}^*$ which does not satisfy either Eq. (8.164) or (8.165).

Then, we check the second strong Wolfe condition. If it is satisfied, then the line search is terminated and we set $\alpha_k = \alpha_{\ell+1}^*$ and update the model. Otherwise check the sign of ϕ'. If

$$\phi'(\alpha_{\ell+1}^*) \geq 0, \tag{8.166}$$

then, the strong Wolfe conditions must hold on some subinterval of $[\lambda_1, \alpha_{\ell+1}^*]$ so we set $\lambda_2 = \alpha_{\ell+1}^*$. If Eq. (8.166) does not hold, then we must have

$$\phi'(\alpha_{\ell+1}) < 0, \tag{8.167}$$

then the strong Wolfe conditions must hold on some subinterval of $[\alpha_{\ell+1}^*, \lambda_2]$ so we set $\lambda_1 = \alpha_{\ell+1}^*$. Note this procedure guarantees that the derivative of ϕ is negative at λ_1. At $\alpha = \lambda_2$, ϕ' is positive or $\phi(\lambda_2) > \phi(\lambda_1)$ or the first Wolfe condition is not satisfied. In any of these situations, the interval contains a subinterval on which the strong Wolfe conditions are satisfied, in fact, contains a point where $\phi' = 0$.

Now we add 1 to the line search iteration index ℓ and repeat the process.

For slightly more sophisticated procedures in algorithmic form, see Dennis and Schnabel [66].

8.7.3   Alternate line search method

The line search procedures discussed above are well-established methods that can be used with Newton, Gauss–Newton, quasi-Newton and nonlinear conjugate gradient optimization algorithms. For large-scale history-matching problems, however, each iteration of the line search algorithm requires checking the strong Wolfe conditions. This requires evaluation of $\phi(\alpha_{\ell+1}^*) = O(m_k + \alpha_{\ell+1}^* d_k)$ which requires a forward run of the reservoir as well as calculation of $\phi'(\alpha_{\ell+1}^*) = \left(\nabla O(m_k + \alpha_{\ell+1}^*)\right)^T d_k$, which requires calculation of the gradient of the objective function. Calculation of this gradient would normally be done with the adjoint method discussed in Chapter 9.

Because of the high computational expense, some effort has been invested in finding alternate line search algorithms [73, 77] including experiments with algorithms which do not always require that the strong Wolfe conditions are satisfied at every iteration. Our experience with quasi-Newton methods suggests that if the strong Wolfe conditions are not satisfied, one invariably generates a direction which is not down hill and the computational effort for correcting this problem offsets the computational gain achieved by not requiring the strong Wolfe conditions be satisfied.

One promising line search algorithm for reducing the computational cost was presented by Gao *et al.* [78]. This algorithm uses a normalized search direction and estimates an initial step size based on the previous step size and updates the step sizes based on a fairly complex algorithm which considers the local behavior of the objective function. Although this algorithm proved to be very efficient for some automatic history-matching problems based on using a limited-memory BFGS algorithm to minimize an appropriate objective function, as presented in Gao *et al.* [78], the algorithm does not try a full step at the first iteration. Thus, the method is very efficient compared to the quadratic/cubic fitting algorithm when the step size must be very small to satisfy the strong Wolfe conditions. For simple optimization problems, however the Gao–Reynolds line search algorithm is less efficient than the one based on quadratic and/or cubic interpolation.

8.8   BFGS and LBFGS

For large-scale problems, construction of the Hessian matrix is not computationally feasible. For such problems, quasi-Newton methods appear to be the most promising alternative. We may think of a quasi-Newton method as a method in which we generate an initial approximation of the inverse Hessian $\tilde{H}_0^{-1}$ and then, using a simple updating formula, generate an improved approximation of the inverse Hessian at each iteration.

The actual implementation, however, avoids direct generation and storage of all approximate inverse Hessians except for $\tilde{H}_0^{-1}$. Thus, when the Hessian is large, the computer memory requirements are quite small particularly if one uses a sparse matrix, such as a diagonal matrix, for $\tilde{H}_0^{-1}$. Although there exist a large number of quasi-Newton methods, a large body of computational evidence suggests that the Broyden–Fletcher–Goldfarb–Shanno (BFGS) is the most reliable quasi-Newton method [67, 79]. Thus, our focus is on the BFGS algorithm.

The search direction in the Newton method is obtained by solving

$$H_k \delta m^{k+1} = -g_k, \tag{8.168}$$

where k is the iteration index, g_k denotes the gradient of the objective function evaluated at m^k, and H_k denotes the Hessian matrix. Letting $d_k = \delta m^{k+1}$, it follows that the search direction at the kth iteration is given by the solution of Eq. (8.168), i.e. by

$$d_k = -H_k^{-1} g_k. \tag{8.169}$$

In quasi-Newton methods, H_k^{-1} is approximated by a symmetric positive-definite matrix $\tilde{H}_k^{-1}$, which is corrected or updated from iteration to iteration so the search direction is approximated by

$$d_k = -\tilde{H}_k^{-1} g_k. \tag{8.170}$$

As discussed in more detail later, it is desirable that a quasi-Newton method satisfies the following three properties.

1. Because we are seeking a minimum, we wish d_k to represent a down-hill direction, i.e.

$$d_k^T g_k = g_k^T d_k < 0. \tag{8.171}$$

Note that if $\tilde{H}_k^{-1}$ is a real-symmetric positive-definite matrix, then by multiplying Eq. (8.170) by g_k^T, it follows that d_k will be a down-hill direction unless g_k is zero, i.e. m_k minimizes the objective function.

2. The second desirable property is that it should be relatively simple to calculate $\tilde{H}_{k+1}^{-1}$ from $\tilde{H}_k^{-1}$, i.e. the formula for updating the inverse Hessian from iteration to iteration does not require excessive computational time.

3. The matrix $\tilde{H}_{k+1}^{-1}$ should satisfy the quasi-Newton condition given by

$$\tilde{H}_{k+1}^{-1}(g_{k+1} - g_k) = m^{k+1} - m^k. \tag{8.172}$$

Defining

$$y_k = g_{k+1} - g_k \tag{8.173}$$

and

$$s_k = m^{k+1} - m^k, \tag{8.174}$$

the quasi-Newton condition (Eq. 8.172) can be written as

$$\tilde{H}_{k+1}^{-1} y_k = s_k. \tag{8.175}$$

The quasi-Newton condition is often referred to as the secant equation.

For a quadratic objective function with a constant invertible Hessian matrix, H, taking the gradient of the Taylor series expansion of $O(m)$ about m^k gives

$$\nabla O(m) = g_k + H(m - m^k). \tag{8.176}$$

Evaluating at $m = m^{k+1}$, multiplying by H^{-1} and using the definitions of s_k and y_k gives

$$H^{-1} y_k = s_k. \tag{8.177}$$

Since the quasi-Newton condition is satisfied by a quadratic objective function with a constant invertible Hessian, it is reasonable to enforce this condition on any quasi-Newton method.

By writing the second property of quasi-Newton methods as

$$\tilde{H}_{k+1}^{-1} = \tilde{H}_k^{-1} + E_k, \tag{8.178}$$

it is possible to derive a formula for a general family (the Huang family) of quasi-Newton methods. A very detailed derivation of the Huang family is given in Appendix B of Zhang [80]. The Broyden family of quasi-Newton methods is a subfamily of the Huang family and the BFGS algorithm emphasized here is a special case of the Broyden family. The update formula for calculating $\tilde{H}_{k+1}^{-1}$ from the real-symmetric positive-definite matrix $\tilde{H}_k^{-1}$ in the BFGS algorithm is the solution of the problem:

$$\text{minimize } \| H - \tilde{H}_k^{-1} \|_F \tag{8.179}$$

subject to the constraints that

$$H = H^{\mathrm{T}} \quad \text{and} \quad \tilde{H}^{-1} y_k = s_k, \tag{8.180}$$

where $\| \cdot \|_F$ denoted a weighted Frobenius norm, see, for example Nocedal and Wright [67]. Using different weighting matrices to define the Frobenius norm, yields different quasi-Newton methods. By using a weighting defined by an average Hessian [66, 67], the unique solution of the preceding minimization problem is

$$\tilde{H}_{k+1}^{-1} = V_k^{\mathrm{T}} \tilde{H}_k^{-1} V_k + \rho_k s_k s_k^{\mathrm{T}}, \tag{8.181}$$

where

$$\rho_k = \frac{1}{y_k^{\mathrm{T}} s_k}, \tag{8.182}$$

$$V_k = I - \rho_k y_k s_k^{\mathrm{T}}, \tag{8.183}$$

and I denotes the identity matrix having the same dimensions as $\tilde{H}_{k+1}^{-1}$. Eq. (8.181) represents the BFGS updating equation.

An outline of the BFGS algorithm for minimizing a continuously differentiable objective function $O(m)$ is given immediately below.

BFGS algorithm outline

1. Select an initial guess, m^0, define $\tilde{H}_0^{-1}$ as a real-symmetric positive-definite approximation of the inverse Hessian evaluated at m_0. If such an approximation is unavailable, one can set $\tilde{H}_0^{-1}$ to the $N_m \times N_m$ identity matrix, I_{N_m} or a scaled version of this matrix as discussed later. Set the iteration index k equal to zero.

2. Compute the search direction

$$d_k = -\tilde{H}_k^{-1} g_k, \tag{8.184}$$

where g_k is the gradient of the objective function $O(m)$ evaluated at model iterate m^k.

3. Use a line search algorithm to find the step size $\alpha_k \geq 0$ such that

$$\alpha_k \approx \underset{\alpha}{\operatorname{argmin}} \, O(m^k + \alpha d_k). \tag{8.185}$$

In practice, an exact line search is not done. Thus, the α_k generated is only approximately equal to the right-hand side of Eq. (8.185). As discussed in Section 8.7, when using an approximate line search, it is important to require that the strong Wolfe conditions (or Wolfe conditions) are satisfied.

4. Update the minimizing model by

$$m^{k+1} = m^k + \alpha_k d_k. \tag{8.186}$$

5. Check for convergence. If the convergence criteria are satisfied, the approximation of the minimizing model is set equal to m^{k+1}. If not, calculate $g_{k+1} = \nabla O(m^{k+1})$, set

$$y_k = g_{k+1} - g_k \quad \text{and} \quad s_k = m^{k+1} - m^k = \alpha_k d_k \tag{8.187}$$

and calculate $\tilde{H}_{k+1}^{-1}$ from Eq. (8.181).

6. $k \leftarrow k + 1$ and go to step 2.

The rest of this section is devoted to comments on the theoretical underpinnings for the BFGS algorithm and its practical implementation.

8.8.1   Theoretical overview of BFGS

$\tilde{H}_{k+1}$ is clearly real symmetric if $\tilde{H}_k^{-1}$ is real symmetric, but we also wish $\tilde{H}_{k+1}$ to be positive definite so that d_k calculated from Eq. (8.184) gives a down-hill search direction.

Proposition 8.2. *If*

$$y_k^T s_k > 0 \tag{8.188}$$

and $\tilde{H}_k^{-1}$ is real symmetric positive definite, then $\tilde{H}_{k+1}^{-1}$ is real symmetric positive definite.

A straightforward proof of this result can be found in Luenberger [75]. The preceding proposition indicates that if we use a real-symmetric positive-definite matrix for $\tilde{H}_0^{-1}$, then, for all k, the updating formula of Eq. (8.181) will give a real-symmetric positive-definite matrix provided that Eq. (8.188) holds.

It is easy to show that Eq. (8.188) holds if the line search of step 3 is done accurately enough so that the Wolfe conditions or strong Wolfe conditions hold and $\tilde{H}_k^{-1}$ is real symmetric positive definite. As noted before, if $\tilde{H}_k^{-1}$ is real symmetric positive definite, then d_k computed from Eq. (8.184) is a down-hill direction, i.e. $g_k^T d_k < 0$. If either the second Wolfe condition (Eq. 8.130) or the second strong Wolfe condition (Eq. 8.126) hold, it follows that

$$g_{k+1}^T d_k \geq c_2 g_k^T d_k. \tag{8.189}$$

Using the definitions of y_k and s_k (Eq. 8.187) and Eq. (8.189), it follows that for $0 < c_2 < 1$,

$$y_k^T s_k = (g_{k+1} - g_k)^T \alpha_k d_k = \alpha_k(g_{k+1}^T d_k - g_k^T d_k) \geq \alpha_k(c_2 - 1)g_k^T d_k > 0. \tag{8.190}$$

Thus, we have established the following result:

Proposition 8.3. *If $\tilde{H}_0^{-1}$ for the BFGS algorithm is chosen to be real symmetric positive definite and, at each iteration, the line search algorithm is done sufficiently accurately so that the Wolfe (or strong Wolfe) conditions are satisfied, then for all nonnegative integers k, $y_k^T s_k > 0$ and $\tilde{H}_{k+1}^{-1}$ computed from the BFGS updating formula of Eq. (8.181) is real symmetric positive definite.*

8.8.2 Convergence rate of BFGS

The BFGS algorithm builds up an approximation to the Hessian using information on the descent directions and gradients at previous iterations. Significant insight on the BFGS algorithm and other quasi-Newton methods can be obtained by considering the analysis for the case of a strictly convex quadratic objective function. In particular the following results are established in [75].

Proposition 8.4. *Let $O(m)$ be a quadratic objective function which has a real-symmetric positive-definite matrix H as its Hessian. Then, starting with any real-symmetric positive-definite matrix $\tilde{H}_0^{-1}$ as the initial guess for the inverse Hessian, any*

Broyden algorithm (including BFGS) with an exact line search will converge to the model which corresponds to the unique minimum of the objective function in at most N_m iterations and $\tilde{H}_{N_m}^{-1} = H^{-1}$.

Of course for such an objective function, Newton's method with a full step would converge to the minimizing model in one iteration for any initial guess. Nonetheless, as for Newton's method, much theoretical motivation for quasi-Newton algorithms in general, and the BFGS method in particular, is provided by the quadratic case. After all, the underlying motivation for Newton's method is that in a sufficiently small neighborhood of a local minimum, the objective function is well approximated by a quadratic.

Fundamental convergence results on the BFGS algorithm were established long ago [81, 82]. Although the theorems are normally stated based on the assumptions that the objective function is twice continuously differentiable and that the Hessian is Lipschitz continuous in some neighborhood of a model m* which represents a minimum of the objective function, we state the theoretical result under the stronger condition that the objective function is three times continuously differentiable.

Proposition 8.5. *If m* corresponds to a local minimum of an objective function $O(m)$ which is three times continuously differentiable, the Hessian is real symmetric positive definite in some neighborhood of m*, the initial guess m^0 is sufficiently close to m* and $\tilde{H}_0^{-1}$ is sufficiently close to the initial approximation to the inverse Hessian evaluated at m*, then the BFGS algorithm with an exact line search converges to m* superlinearly.*

While the assumptions of the theorem are not practically verifiable, the result is worthwhile in that it indicates that in some cases, it is possible to obtain faster convergence with the BFGS algorithm than with the Gauss–Newton method, or the Levenberg–Marquardt modification of the Gauss–Newton method, which may only give Q-linear convergence; see Section 8.5.2. However, this is an asymptotic result. As one often chooses $\tilde{H}_0^{-1}$ as an identity matrix, BFGS behaves like a steepest descent method at early iterations. For comparisons we have done with history-matching examples, BFGS and LBFGS always required far more iterations to obtain convergence than the Levenberg–Marquardt (LM) algorithm although LBFGS typically requires far less computer time to reach convergence than does LM.

It may be unsettling to note that one assumption of Proposition 8.5 is that an exact line search is done which is not feasible for the large-scale problems of interest here. However, under appropriate conditions, it is possible to show superlinear convergence holds with an approximate line search algorithm that satisfies the Wolfe or strong Wolfe conditions provided one always tries a full step in the line search algorithm as we approach convergence [67]. That one should try a full step at some point in the line search is not totally surprising in light of the convergence theorem for Newton's method.

Superlinear convergence proofs for any gradient based optimization algorithm rest on the assumption that the Dennis–Moré condition

$$\lim_{k \to \infty} \frac{\|H(m^*)d_k + g_k\|}{\|d_k\|} = 0 \tag{8.191}$$

holds where d_k is the down-hill search direction at the kth iteration, $H(m^*)$ is the true Hessian at the minimizing model m^* and g_k is the gradient of the objective function at the kth iteration. The exact line search condition of Proposition 8.5 can be replaced by the following two assumptions: (i) Eq. (8.191) holds and (ii) the strong Wolfe conditions (or Wolfe conditions) with $c_1 \le 1/4$ hold in the line search procedure. These conditions ensure that the full step side ($\alpha_k = 1$) will be accepted and satisfy the Wolfe conditions [67] once we are sufficiently close to the minimum. Note the condition on c_1 is no restriction in practice because, as noted in Section 8.7, one normally chooses c_1 very near zero in the Wolfe conditions.

8.8.3   Scaling of BFGS

The performance of BFGS can be highly susceptible to the scaling of the objective function. In the discussion of scaling, it is convenient to write the BFGS updating formula of Eq. (8.181) as

$$\tilde{H}_{k+1}^{-1} = \left(\tilde{H}_k^{-1} - \frac{\tilde{H}_k^{-1} y_k y_k^T \tilde{H}_k^{-1}}{y_k^T \tilde{H}_k^{-1} y_k} + v_k v_k^T \right) + \frac{s_k s_k^T}{s_k^T y_k}, \tag{8.192}$$

where

$$v_k = (y_k^T \tilde{H}_k^{-1} y_k)^{1/2} \left(\frac{s_k}{s_k^T y_k} - \frac{\tilde{H}_k^{-1} y_k}{y_k^T \tilde{H}_k^{-1} y_k} \right). \tag{8.193}$$

It is well known that changing the starting inverse Hessian from H_0^{-1} to $\gamma_o \tilde{H}_0^{-1}$ for some real constant $\gamma_0 \ne 1$ can significantly change the performance of the algorithm. For example, if H_0^{-1} is such that $\|d_0\|$ is very large, it can sometimes take several iterations of the line search algorithm to obtain a step size that satisfies the strong Wolfe conditions.

As BFGS was derived as an approximation to Newton's method and Newton's method may exhibit quadratic convergence (Proposition 8.1), it is desirable to try to introduce scaling into the BFGS algorithm or any quasi-Newton method in a way so that the $N_m \times N_m$ matrix $\tilde{H}_k^{-1} H(m^*)$ approaches the $N_m \times N_m$ identity matrix as k increases. As the condition number of the identity matrix is unity, one way to attempt to achieve this goal is by ensuring that the condition number decreases from iteration to iteration, i.e. to require that

$$\kappa\left(\tilde{H}_{k+1}^{-1} H(m^*)\right) < \kappa\left(\tilde{H}_k^{-1} H(m^*)\right), \tag{8.194}$$

for all k. Complete theoretical results for such a procedure are given in Luenberger [75] but only for the case where the objective function is a quadratic with a real-symmetric positive-definite Hessian H. In this simple case, if the eigenvalues of the real-symmetric positive-definite matrix $\tilde{H}_k^{-1} H$ has eigenvalues $\lambda_1 \leq \lambda_2 \leq \cdots \leq \lambda_{N_m}$, and

$$\lambda_1 \leq 1 \leq \lambda_{N_m}, \tag{8.195}$$

then it can be shown that

$$\kappa\left(\tilde{H}_{k+1}^{-1} H(m^*)\right) \leq \kappa\left(\tilde{H}_k^{-1} H(m^*)\right). \tag{8.196}$$

The strict inequality of Eq. (8.194) can not be proved. Moreover, this result has only been established for a strictly convex quadratic objective function. Since $\tilde{H}_k^{-1}$ is real symmetric positive definite, its eigenvalues are all positive. Moreover, for any positive real number γ_k, the eigenvalues of $\gamma_k \tilde{H}_k^{-1}$ are given by $\mu_j = \gamma_k \lambda_j$, $j = 1, 2, \ldots, N_m$. Thus if γ_k is chosen such that

$$\lambda_1 \leq \frac{1}{\gamma_k} \leq \lambda_{N_m}, \tag{8.197}$$

then

$$\mu_1 \leq 1 \leq \mu_{N_m}. \tag{8.198}$$

Thus, at the kth iteration, to ensure that Eq. (8.196) holds, we simply need to replace $\tilde{H}_k^{-1}$ by $\gamma_k \tilde{H}_k^{-1}$ when applying Eqs. (8.192) and (8.193). This is equivalent to replacing Eq. (8.192) by

$$\tilde{H}_{k+1}^{-1} = \gamma_k \left(\tilde{H}_k^{-1} - \frac{\tilde{H}_k^{-1} y_k y_k^T \tilde{H}_k^{-1}}{y_k^T \tilde{H}_k^{-1} y_k} + v_k v_k^T \right) + \frac{s_k s_k^T}{s_k^T y_k}. \tag{8.199}$$

For large-scale optimization problems, it is not feasible to compute the largest and smallest eigenvalues of the approximate inverse Hessian at each iteration. In fact, as shown later, for large-scale problems, we avoid direct calculation of the approximate inverse Hessian. Results presented in Luenberger [75] indicate that

$$\gamma_k = \frac{s_k^T y_k}{y_k^T \tilde{H}_k^{-1} y_k} \tag{8.200}$$

is an appropriate choice. This scaling factor is also the one derived by Oren and Spedicato [83] for the BFGS algorithm. With the modified updating formula of Eq. (8.199), BFGS becomes what is known as a "self-scaling" quasi-Newton method or variable metric method. The principal objection to self-scaling variable metric algorithms, which were developed by Oren and Luenberger [84] and Oren [85], that has been raised is that the sequence $\tilde{H}_{k+1}$ does not converge to the true inverse Hessian in n iterations in the case where the objective function is an n-dimensional quadratic function, i.e. Proposition 8.4 is not valid. On the other hand, quasi-Newton methods from the Broyden family satisfy the "quadratic termination property" (Proposition 8.4) if $\gamma_k = 1$ for all $k > 0$. Motivated by this reasoning, Shanno and Phua [86] suggested that one should only scale at the first iteration and then do no further scaling.

As discussed in the next section, Nocedal [87] has proposed an implementation of BFGS which is far more efficient than an implementation based on the updating formula of Eq. (8.192). His method avoids the explicit computation of $\tilde{H}_k^{-1}$ for $k \geq 1$. Only the initial approximation, $\tilde{H}_0^{-1}$, needs to be known explicitly. Since $\tilde{H}_k^{-1}$ for $k \geq 1$ is not known explicitly, Eq. (8.200), or other scaling algorithms based on knowledge of the approximate Hessian and/or its inverse [83], can not be applied for $k \geq 1$. Thus, if one wishes to scale at all iterations, the methods must be modified. In place of Eq. (8.200), Liu and Nocedal [88] suggest using

$$\gamma_k = \frac{s_k^T y_k}{y_k^T \tilde{H}_0^{-1} y_k},$$
(8.201)

for $k \geq 1$ if one wishes to scale at all iterations.

Many other scaling methods have been proposed for the BFGS method as well as other quasi-Newton methods. Oren and Spedicato [83] proposed an "optimal" conditioning of variable metric methods and proposed four different scaling methods. According to their results, one should actually consider switching between different updating formulas from the general Broyden family from iteration to iteration. Again, however, their results assume that the quasi-Newton method is applied to a strictly convex quadratic objective function and that an exact line search is performed at each iteration. From computational experiments on history-matching problems, Zhang [80] and Zhang and Reynolds [77] found that the general switching rule proposed by Oren and Spedicato exhibits poorer convergence properties than are obtained by applying the BFGS update at every iteration and then computing the optimal γ_k from Eq. (8.200). Like Eq. (8.200), the Oren–Spedicato scaling procedures can not be applied with the Nocedal BFGS implementation because $\tilde{H}_k^{-1}$ is not computed for $k \geq 1$. Because of this, Zhang [80] tried several modifications of recommended scaling factors that could be applied at each iteration with Nocedal's BFGS and limited-memory BFGS implementations.

The basic scaling procedures investigated by Zhang [80] and Zhang and Reynolds [77] can be written in the general form,

$$\gamma_k = \begin{cases} \tau_k & \text{if } \tau_k < 1, \\ \sigma_k = \frac{s_k^T y_k}{y_k^T \tilde{H}_0^{-1} y_k} & \text{if } \tau_k \geq 1. \end{cases}$$
(8.202)

Note that the formula for σ_k is identical to the Liu and Nocedal [88] procedure for scaling at every iteration, Eq. (8.201). Zhang [80] also tried setting $\gamma_k = \sigma_k$ at all iterations as well as setting $\gamma_k = \tau_k$ at all iterations. Three different definitions τ_k were tried:

$$\tau_{1k} = \frac{s_k^T \tilde{H}_0 s_k}{s_k^T y_k}, \tag{8.203}$$

$$\tau_{2k} = -\alpha_k \frac{g_k^T s_k}{s_k^T y_k}, \tag{8.204}$$

and

$$\tau_{3k} = \frac{s_k^T g_k}{g_k \tilde{H}_0^{-1} y_k}. \tag{8.205}$$

It is easy to show that at the first iteration all formulas are equivalent, but they are not equivalent at later iterations. The equation for τ_{2k} is actually the equation for τ_k used in one of the Oren–Spedicato switching formulas. Some specific computational results comparing scaling methods will be presented after considering Nocedal's implementation of the BFGS and LBFGS algorithms.

Remark 8.1. For history-matching production data, experiments summarized in Zhang and Reynolds [77] indicate that the preferred method is to use Eq. (8.202) with $\tau_k = \tau_{1k}$, although in some cases alternate switching formulas gave comparable results. The two most important results are (i) scaling at only the first iteration often results in a significantly faster rate of convergence than is achieved with no scaling; (ii) scaling at all iterations using Eq. (8.202) with $\tau_k = \tau_{1k}$ typically gives faster convergence than scaling at only the first iteration. They found only one case where no scaling and scaling gave almost identical rates of convergence and history matches of production data.

8.8.4 Efficient implementation of BFGS

Note that Eq. (8.181) suggests that the inverse Hessian at each iteration is computed and stored. As noted by Nocedal [87], however, applying Eq. (8.181) recursively gives

$$\begin{aligned}
\tilde{H}_{k+1}^{-1} =\ & V_k^T V_{k-1}^T \cdots V_0^T \tilde{H}_0^{-1} V_0 \cdots V_{k-1} V_k \\
& + V_k^T \cdots V_1^T \rho_0 s_0 s_0^T V_1 \cdots V_k \\
& \ \vdots \\
& + V_k^T \rho_{k-1} s_{k-1} s_{k-1}^T V_k \\
& + \rho_k s_k s_k^T,
\end{aligned} \tag{8.206}$$

where

$$\rho_k = 1/y_k^T s_k, \tag{8.207}$$

and the matrix V_k for $k = 0, 1, \ldots$ is defined by

$$V_k = I - \rho_k y_k s_k^{\mathrm{T}}.$$

(8.208)

In this form, it is clear that the only inverse Hessian that needs to be explicitly known to apply the BFGS updating formula is H_0^{-1}. In fact, we need only compute and store products of $\tilde{H}_0^{-1}$ and vectors. For example, at the second iteration of the BFGS algorithm, we need to compute the search direction

$$d_1 = \tilde{H}_1^{-1} g_1 = \left(V_0^{\mathrm{T}} \tilde{H}_0^{-1} V_0 + \rho_0 s_0 s_0^{\mathrm{T}}\right) g_1,$$

(8.209)

where g_1 is the gradient of the objective function evaluated at m^1. The second term on the right-hand side of Eq. (8.209) can be computed by taking the inner product of s_0 and g_1 to find the scalar $s_0^{\mathrm{T}} g_1$ and then multiplying the column vector s_0 by the number $\rho_0 s_0^{\mathrm{T}} g_1$. To complete the calculation for d_1, we need to compute

$$\begin{aligned}
V_0^{\mathrm{T}} \tilde{H}_0^{-1} V_0 g_1 &= \left(I - \rho_0 s_0 y_0^{\mathrm{T}}\right) \tilde{H}_0^{-1} \left(I - \rho_0 y_0 s_0^{\mathrm{T}}\right) g_1 \\
&= \tilde{H}_0^{-1} g_1 - \rho_0 s_0 y_0^{\mathrm{T}} \tilde{H}_0^{-1} g_1 - \rho_0 \tilde{H}_0^{-1} y_0 s_0^{\mathrm{T}} g_1 + \rho_0 s_0 y_0^{\mathrm{T}} \tilde{H}_0^{-1} \rho_0 y_0 s_0^{\mathrm{T}} g_1.
\end{aligned}$$

(8.210)

Now it is clear that evaluation of the right-hand side of Eq. (8.210) requires only the multiplication of column vectors by $\tilde{H}_0^{-1}$ and the calculation of vector inner products. Moreover, it is clear that this process can be extended recursively so that the calculation of $d_k = -\tilde{H}_k^{-1} g_k$ at iteration k requires only vector inner products and the multiplication of column vectors by $\tilde{H}_0^{-1}$. Thus, $\tilde{H}_k^{-1}$ does not need to ever be explicitly calculated or stored. At iteration k, the vector pairs y_j and s_j, $j = 0, 1, 2, \ldots, k$ and g_k need to be in storage. In addition, a way to compute $\tilde{H}_0^{-1}$ times a vector must be available, but one often chooses $\tilde{H}_0^{-1}$ as a sparse matrix such as a diagonal matrix so we have direct knowledge of $\tilde{H}_0^{-1}$ and may be able to store it conveniently.

If scaling is done only at the first iteration, then we simply replace Eq. (8.206) by

$$\begin{aligned}
\tilde{H}_{k+1}^{-1} &= V_k^{\mathrm{T}} V_{k-1}^{\mathrm{T}} \cdots V_0^{\mathrm{T}} \left(\gamma_0 \tilde{H}_0^{-1}\right) V_0 \cdots V_{k-1} V_k \\
&\quad + V_k^{\mathrm{T}} \cdots V_1^{\mathrm{T}} \rho_0 s_0 s_0^{\mathrm{T}} V_1 \cdots V_k \\
&\quad \vdots \\
&\quad + V_k^{\mathrm{T}} \rho_{k-1} s_{k-1} s_{k-1}^{\mathrm{T}} V_k \\
&\quad + \rho_k s_k s_k^{\mathrm{T}}.
\end{aligned}$$

(8.211)

For scaling at all iterations, Liu and Nocedal [88] suggest that an effective strategy is to replace Eq. (8.211) by

$$
\begin{aligned}
\tilde{H}_{k+1}^{-1} = {} & V_k^T V_{k-1}^T \cdots V_0^T \left(\gamma_k \tilde{H}_0^{-1} \right) V_0 \cdots V_{k-1} V_k \\
& + V_k^T \cdots V_1^T \rho_0 s_0 s_0^T V_1 \cdots V_k \\
& \vdots \\
& + V_k^T \rho_{k-1} s_{k-1} s_{k-1}^T V_k \\
& + \rho_k s_k s_k^T .
\end{aligned}
\tag{8.212}
$$

Note that the preceding updating scheme is not equivalent to the updating formula of Eq. (8.199) which can be rewritten as

$$
\begin{aligned}
\tilde{H}_{k+1}^{-1} = {} & \gamma_k^{1/2} V_k^T \gamma_{k-1}^{1/2} V_{k-1}^T \cdots \gamma_0^{1/2} V_0^T \tilde{H}_0^{-1} \gamma_0^{1/2} V_0 \cdots \gamma_{k-1}^{1/2} V_{k-1} \gamma_k^{1/2} V_k \\
& + \gamma_k^{1/2} V_k^T \cdots \gamma_1^{1/2} V_1^T \rho_0 s_0 s_0^T \gamma_1^{1/2} V_1 \cdots \gamma_k^{1/2} V_k \\
& \vdots \\
& + \gamma_k^{1/2} V_k^T \rho_{k-1} s_{k-1} s_{k-1}^T \gamma_k^{1/2} V_k \\
& + \rho_k s_k s_k^T .
\end{aligned}
\tag{8.213}
$$

8.8.5 Limited-memory Broyden–Fletcher–Goldfarb–Shanno algorithm

The limited-memory Broyden–Fletcher–Goldfarb–Shanno (LBFGS) algorithm is simply an implementation of BFGS where we limit the amount of the preceding gradient and curvature information retained when applying the updating formula. Our discussion essentially follows the one given in Nocedal [87].

If the optimization problem is truly large scale, then when the iteration index becomes large, it may not be feasible to store all y_j and s_j for $j = 0, 1, 2, \ldots$. In this case the limited-memory version can be used.

To motivate LBFGS, Nocedal rewrote Eq. (8.212) as

$$
\begin{aligned}
\tilde{H}_{k+1}^{-1} = {} & V_k^T V_{k-1}^T \cdots V_{k-p+1}^T \left(\gamma_k \tilde{H}_0^{-1} \right) V_{k-p+1} \cdots V_{k-1} V_k \\
& + V_k^T \cdots V_{k-p+2}^T \rho_{k-p+1} s_{k-p+1} s_{k-p+1}^T V_{k-p+2} \cdots V_k \\
& \vdots \\
& + V_k^T \rho_{k-1} s_{k-1} s_{k-1}^T V_k \\
& + \rho_k s_k s_k^T ,
\end{aligned}
\tag{8.214}
$$

where $p = \min\{L, k+1\}$ and the parameter L is an integer chosen by the user. Although we have used the same notation for the scaling parameter in Eqs. (8.214) and (8.199), we emphasize again that the two equations are identical if and only if $\gamma_k = 1$ for all $k > 0$. If L is greater than the total number of iterations allowed, then

LBFGS becomes equivalent to BFGS if the efficient implementation of Nocedal is used for both schemes. If L is too small, however, the number of iterations required for convergence is increased. For the history-matching problems he considered, Zhang [80] found $L = 30$ to be a good choice; larger values of L resulted in only a small improvement in convergence properties, but using values of L smaller than 20 resulted in a noticeable degradation in the rate of convergence and resulted in higher values of the objective function at convergence. There is no reason to expect, however, that $L = 30$ is a universally good choice and there may be problems where substantially larger values of L are needed.

The LBFGS algorithm given by Nocedal [87] for computing the search direction at the kth iteration is as follows.

1. If $k \le L$, then $inc = 0$ and $ibd = k$, else $inc = k - L$ and $ibd = L$.
2. $q_{ibd} = g_k$.
3. For $i = ibd - 1, ibd - 2, \ldots 0$ set

$$j = i + inc,$$
$$\alpha_i = \rho_j s_j^\mathsf{T} q_{i+1},$$
$$q_i = q_{i+1} - \alpha_i y_j.$$

4. Set $r_0 = \gamma_k \tilde{H}_0^{-1} q_0$.
5. For $i = 0, 1, \ldots, ibd - 1$, set

$$j = i + inc,$$
$$\beta_j = \rho_j y_j^\mathsf{T} r_i,$$
$$r_{i+1} = r_i + s_j(\alpha_i - \beta_i).$$

In this algorithm, k is the iteration index and the final r_{i+1} represents the search direction for the next iteration, i.e. d_k.

8.9   Computational examples

In many of the history-matching examples, we either minimize Eq. (8.2) to obtain a MAP estimate or minimize

$$O_r(m) = \frac{1}{2}(m - m_{\mathrm{uc}})^\mathsf{T} C_M^{-1}(m - m_{\mathrm{uc}}) + \frac{1}{2}(g(m) - d_{\mathrm{uc}})^\mathsf{T} C_D^{-1}(g(m) - d_{\mathrm{uc}}), \quad (8.215)$$

when applying RML to sample the a posteriori PDF of Eq. (8.1). In either case, it is reasonable to use the prior covariance matrix as the initial approximation to the inverse Hessian, i.e. set $\tilde{H}_0^{-1} = C_M$. A somewhat simpler procedure, which yields a less robust algorithm, is to set $\tilde{H}_0^{-1}$ equal to the diagonal matrix $\tilde{D}$ obtained from C_M by setting all off diagonal elements equal to zero. Even when $\tilde{H}_0^{-1} = C_M$, Zhang et al. [89] found

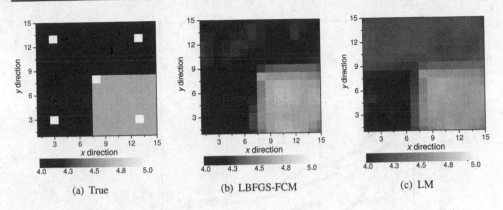

Figure 8.6. The log-permeability field.

it is usually satisfactory to compute τ_{1k} of Eq. (8.203) from the approximation

$$\tau_{1k} = \frac{s_k^T \tilde{D}^{-1} s_k}{s_k^T y_k},$$ (8.216)

instead of

$$\tau_{1k} = \frac{s_k^T \tilde{C}_M^{-1} s_k}{s_k^T y_k}.$$ (8.217)

Applying Eq. (8.216) avoids solving the matrix problem $C_M x_k = s_k$ for x_k at each iteration.

8.9.1   2D three-phase example

This synthetic example pertains to a two-dimensional, three-phase flow problem simulated on a $15 \times 15 \times 1$ grid. The gridblock porosities are fixed. The truth case, from which synthetic production data were generated, is shown in Fig. 8.6 (a). Note that there are three distinct zones with log-permeability uniform in each zone. This example has the advantage that the problem is small, so all methods require only modest computer resources. Moreover, because only three log-permeability values are involved, it is easy to visualize the quality of the MAP estimate of log-permeability. (The other plots of Fig. 8.6 are discussed later.) The prior covariance matrix is generated from an isotropic spherical variogram with range equal to six gridblocks. This covariance would suggest that a realization with the prior model should be much smoother than the true model of Fig. 8.6. Thus, our MAP estimate will not be able to capture the distinct boundaries between the three permeability zones. Four producers are located near the four corners of the reservoir and one injector is located at the center. GOR, WOR, and flowing bottomhole pressure (p_{wf}) data from the four producers and p_{wf} data from the injector are used as observed conditioning data. The total number of data are 364. As noted in the discussion of Figs. 8.9 and 8.10, the observed data were set equal to the true

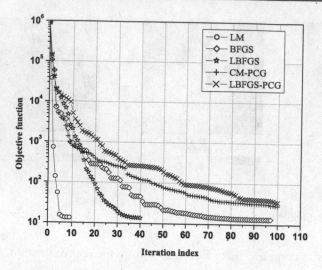

Figure 8.7. Behavior of the objective function [80].

production data generated from the reservoir simulator without adding measurement error.

Comparison of optimization algorithms [80].

Because the nonlinear conjugate gradient (NL-CG) algorithm [64] represents a potential competitor to the LBFGS algorithm, we will make some comparisons between nonlinear conjugate gradient results and those obtained with LBFGS in this example. A NL-CG algorithm requires even less memory than LBFGS and also only requires computation of the gradient. Unfortunately, results for this example show NL-CG to be fairly inefficient compared to LBFGS, which is consistent with early work on using NL-CG for history matching by Makhlouf *et al.* [90]. In one example they tried, Makhlouf *et al.* [90] found that NL-CG required over 200 iterations to obtain convergence. As the efficiency of any conjugate gradient algorithm depends largely on the preconditioner used, Zhang [80] attempted to find a suitable preconditioner for the Polak–Ribière form of the nonlinear conjugate gradient algorithm [64], but was unable to develop an algorithm as efficient and robust as LBFGS.

Figure 8.7 shows the behavior of the objective function as a function of the iteration number obtained for five different optimization algorithms, Levenberg–Marquardt (LM), two preconditioned conjugate gradient algorithms (CM-PCG and LBFGS-PCG), BFGS, and LBFGS. The two preconditioned NL-CG algorithms, which are described in more detail in Zhang and Reynolds [77] performed relatively poorly, failing to converge in 100 iterations. Clearly, one needs to find a better preconditioner to render NL-CG viable. Of the four remaining algorithms, BFGS required the most iterations (96) to obtain convergence and at convergence the value of the objective function is 16.2. The BFGS algorithm used the implementation of Eq. (8.199) with scaling only

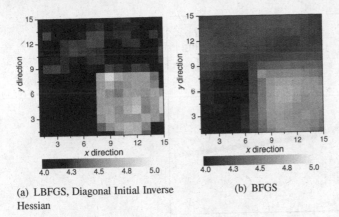

(a) LBFGS, Diagonal Initial Inverse
Hessian

(b) BFGS

Figure 8.8. The log-permeability field from LBFGS with diagonal initial Hessian (a) and BFGS with initial scaling (b) [80].

at the first iteration with $\gamma_0 = \sigma_0$; see Eq. (8.202). The LBFGS algorithm uses scaling at all iterations based on the scaling factor of Eq. (8.202) with $\tau_k = \tau_{1k}$ where τ_{1k} is calculated from the approximation of Eq. (8.216) and the initial inverse Hessian is approximated by the inverse of the diagonal of C_M. Nocedal's efficient implementation with $L = 30$ is used.

For LM, the initial value of λ was set equal to 10^5 and we simply increased or decreased λ by a factor of 10 based on requiring that a new model is not accepted unless it results in a decrease in the objective function. As shown in Fig. 8.7, LM converged in nine iterations to a MAP estimate, m_∞, such that $O(m_\infty) = 13.3$. The BFGS algorithm converged in 66 iterations to an estimate m_∞ such that $O(m_\infty) = 16.2$; LBFGS converged in 40 iterations to an estimate m_∞ such that $O(m_\infty) = 13.7$.

Although not shown, when the prior covariance matrix, C_M, was used as the initial inverse Hessian, LBFGS converged in 33 iterations to an estimate m_∞ such that $O(m_\infty) = 14.0$. Although LM required far fewer iterations than LBFGS, Levenberg–Marquardt actually required more than ten times the CPU time required by LBFGS to achieve convergence. This was true whether the full C_M was used as the initial inverse Hessian and in computing τ_{1k} or whether only the diagonal of C_M was used. For significantly larger problems, the cost of Levenberg–Marquardt would be prohibitive due to the high cost of computing sensitivity coefficients. In subsequent figures, we refer to the case where $\tilde{H}_0$ is set equal to the diagonal of the prior covariance as LBFGS-DCM and refer to the case where $\tilde{H}_0^{-1} = C_M$ as LBFGS-FCM.

Figure 8.6(b) shows the log-permeability field obtained by LBFGS-FCM. We can see this model is very close to the log-permeability field obtained by the LM method (Fig. 8.6c) and is similar to the true model. The log-permeability field obtained where we approximate the initial inverse Hessian by $\tilde{H}_0^{-1} = \tilde{D}^{-1}$ (see Fig. 8.8a) is much rougher than those shown in Figs. 8.6(b) and 8.6(c) because the smoothing effect

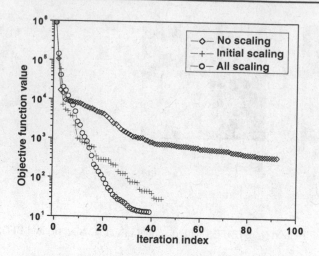

Figure 8.9. Scaling versus no scaling for LBFGS [80].

of multiplication by C_M is lost when we use only the diagonal of C_M as the initial approximate inverse Hessian. The estimated model obtained with BFGS is shown in Fig. 8.8(b) and is quite similar to the models obtained with LM and LBFGS with $\tilde{H}_0^{-1} = C_M$.

Comparison of scaling procedures

Here, we consider the results obtained with Nocedal's LBFGS algorithm with different scaling schemes. The results pertain to history matching WOR, GOR, and pressure data for a 2D three-phase flow example discussed in more detail later. A total of 392 production data were history matched. In this example, $L = 30$ where L is the number of previous sets of vectors stored in the Nocedal updating scheme. First, consider the case where the scaling factor γ_k in LBFGS is given by Eq. (8.202) with $\tau_k = \tau_{1k}$ computed from Eq. (8.203) and $\tilde{H}_0^{-1}$ set equal to the inverse of the diagonal of C_M, where C_M is the prior covariance matrix. The model m represents gridblock log-permeabilities on a uniform $15 \times 15 \times 1$ grid and the covariance was generated from a spherical isotropic variogram with range equal to the width of six gridblocks. The model was purposely chosen very small so that the sensitivity coefficients needed for the Gauss–Newton (GN) and Levenberg–Marquardt algorithm could be quickly computed. In this case, we minimized the objective function of Eq. (8.2) to obtain a MAP estimate of the log-permeabililty field. In all optimization algorithms, a maximum of 100 iterations was allowed.

In Fig. 8.9, the diamonds represent the case where LBFGS was applied without scaling. The plus signs represent the case where LBFGS was scaled only at the first iteration. The circles represent the case where LBFGS were scaled at all iterations. The results clearly demonstrate the advantages of scaling; i.e. convergence is obtained in

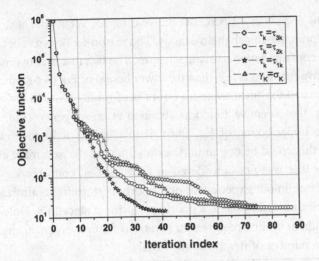

Figure 8.10. Comparison of scaling procedures for LBFGS [80].

significantly fewer iterations, and at convergence, a much lower value of the objective function is obtained. Also note that for this example, scaling at all iterations is superior to scaling only at the first iteration.

For Nocedal's LBFGS algorithm, the scaling factor recommended by Liu and Nocedal [88] is the one derived by Oren and Spedicato [83] and is given by $\gamma_k = \sigma_k$ for all k where σ_k is defined by the second equality of Eq. (8.202). In Fig. 8.10, we compare the performance of LBFGS with this choice with the performance obtained with scaling factors based on simply setting $\gamma_k = \tau_k$ for all k, where we try three different values of τ_k, namely, τ_{1k}, τ_{2k}, and τ_{3k} given respectively by Eqs. (8.203)–(8.205). Note that LBFGS with $\gamma_k = \tau_{1k}$ requires the fewest iterations to obtain convergence and obtains the lowest value of the objective function at convergence. This result can not be generally established, but based on computational experiments, Zhang and Reynolds [77], Zhang [80] concluded that this choice gave the best overall performance for history-matching problems although the choice of $\gamma_k = \sigma_k$ is often recommended.

Remark 8.2. With $N_d = 392$, the criteria of Eq. (8.6) indicates that at convergence we should have

$$126 = 0.5\left[N_d - 5\sqrt{2N_d}\right] \le O(m_{\mathrm{MAP}}) \le 0.5\left[N_d + 5\sqrt{2N_d}\right] = 266. \qquad (8.218)$$

Note that for all LBFGS with scaling results shown in Figs. 8.9 and 8.10, we obtained a value of the objective function below 126. This occurred because the data we matched were generated from the reservoir simulation with the true model and no noise (measurement error) was added to the data, but the data covariance matrix, C_D, was based on having measurement errors with a variance of 1 psi^2 for pressure,

25 $(scf/STB)^2$ for GOR and about a 1 percent relative error for WOR. Because data are exact, we could in principle match them exactly. The reason we do not is because the objective function contains a model mismatch term. The fact that at convergence the value of the objective function is less than the lower bound of Eq. (8.6) indicates that we have overestimated the "measurement errors" in forming C_D. On the other hand, without scaling, the objective function obtained at convergence (Fig. 8.9) is far greater than the upper bound of Eq. (8.6). If this were a field case rather than a synthetic case, this could be due to underestimating the measurement error, due to modeling error that was not considered in forming C_D, convergence to a local minimum (quite rare in our experience) or a poorly performing optimization algorithm. In this specific case, it is due to a poorly performing algorithm. Note that in Fig. 8.9, LBFGS without scaling required 96 iterations for convergence, slightly less than the maximum number of iterations allowed.

8.9.2   3D three-phase example

In this example, a 3D three-phase flow problem on a $40 \times 40 \times 6$ grid is considered. The true log-permeability field is an unconditional realization generated by Gaussian co-simulation. The top layer of the true log-permeability field is shown in Fig. 8.11(c). The porosity field is fixed. Six producers and four water injection wells are completed in the reservoir. Wellbore pressure (p_{wf}), GOR and WOR data from the producers, and p_{wf} data from the injectors are used as conditioning data. Figure 8.11(a) shows the top layer of the unconditional realization of the model which was used as the initial guess in the history-matching process. Fig. 8.11(b) shows the corresponding layer of the model obtained by history matching the production data by minimizing the objective function of Eq. (8.215). Optimization was done with the LBFGS-FCM version of

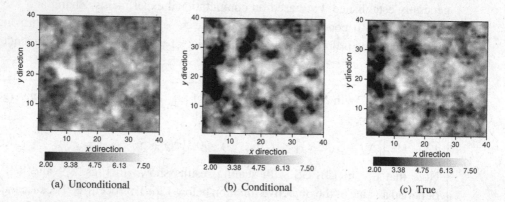

(a) Unconditional   (b) Conditional   (c) True

Figure 8.11. Unconditional, conditional, and true realizations of the log-permeability field in the top layer [80].

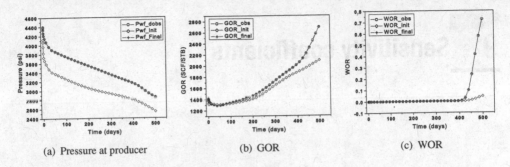

Figure 8.12. Data matches [80].

LBFGS presented in the previous example. Note that to match data, the permeability had to be increased in much of the reservoir. Also note that the conditional realization obtained by history matching captures the main features of the true model much better than the unconditional realization (Fig. 8.11a).

Figures 8.12(a)–(c) show the data match for the pressure, GOR, and WOR from one producer. In all of these figures, diamonds represent the data obtained from the unconditional realization, i.e. the initial model, circles represent the observed data and the plus signs represent the data obtained from the model which was obtained by history matching all observed data. Note that good matches were obtained. Matches of similar quality were obtained at all wells. It is important to note that the size of the problem precluded the application of Levenberg–Marquardt, Gauss–Newton, and the standard BFGS algorithm with the personal computer used for the study.

9 Sensitivity coefficients

Efficient inverse theory methods based on the optimization methods presented earlier require knowledge of the sensitivity coefficients, i.e. the relationship between small changes in model variables (such as permeability at a point or within a gridblock) and changes in the state variables such as pressure or saturation. For linear problems, this relationship is often clear from first principles, but for nonlinear flow and transport problems, the relationship must be derived from the partial differential equations that govern flow and transport.

With the Bayesian approach we prefer, generating a single history-matched realization using randomized maximum likelihood requires the minimization of the objective function of Eq. (8.23). The optimization procedures that we have discussed involve sensitivity coefficients, i.e. the derivatives of predicted data with respect to model parameters. However, one does not necessarily have to calculate all individual elements of the sensitivity matrix G. For example, we have seen that the BFGS and LBFGS algorithms simply require the computation of the gradient of the objective function, which can be done without explicitly computing the sensitivity matrix.

In the next few sections we will develop the basic ideas through simple examples involving steady flow, then apply the methods to realistic problems involving transient multi-phase flow for history matching.

We will start the discussion of sensitivity coefficients for the case where the property to be varied is a function instead of a set of discrete properties such as the gridblock permeabilities in a reservoir simulation model. In some respects, the continuous problem is easier and it is sometimes fairly easy to treat inverse problems with an infinite number of unknowns. To deal with functional inverse problems, we must discuss functional differentiation. We note, however, that the subsections that involve Fréchet derivative in their title are not needed elsewhere in this book.

9.1 The Fréchet derivative

In working with nonlinear inverse problems, it is often useful to linearize the problem so that standard minimization methods can be used. Normally, when we think of linearization, we think of throwing away the quadratic and higher-order terms in a Taylor series expansion. When the dependent variable m is a vector and f is a real-valued function of m, the standard Taylor series expansion is

$$f(m + \delta m) = f(m) + \delta m^T \nabla f(m) + \frac{1}{2} \delta m^T \nabla [\nabla f(m)]^T \delta m + \cdots . \tag{9.1}$$

A functional g is simply a mapping from a linear space (vector space) into the scalar field associated with the vector space. We will assume that the scalar field is real, that the linear space has a real inner product defined on it which defines a norm and that the linear space is complete. This means the linear space, denoted in this subsection by M, is a Hilbert space and the functional is a mapping from M into the set of real numbers. In this case, we say that the functional g is Fréchet differentiable at the element m if

$$g[m + \delta m] = g[m] + (D, \delta m) + R[\delta m] \tag{9.2}$$

and

$$\lim_{\|\delta m\| \to 0} \frac{R[\delta m]}{\|\delta m\|} = 0. \tag{9.3}$$

Here $(D, \delta m)$ denotes the inner product of D and δm and the element D is called the Fréchet derivative of g at m [see 3, p. 304]. Note that the Taylor expansion of $f(m + \delta m)$ in Eq. (9.1) is of exactly this form. Thus, the ordinary gradient at m is a Fréchet derivative for the vector space consisting of column vectors with the inner product defined by the dot product. From Eq. (9.2), it is clear that the Fréchet derivative of "data" can be obtained through a perturbation analysis of the forward problem. We would simply determine the resulting change in the data (which is a functional of the model) when the model is changed by a small amount. This also explains why we often refer to D as the *sensitivity* of the data to the model.

An example that arises often is the calculation of the Fréchet derivative of a linear functional. Suppose that $d = g[m] = (G, m)$, then

$$g[m + \delta m] = (G, m + \delta m) = (G, m) + (G, \delta m) \tag{9.4}$$
$$= g[m] + (G, \delta m)$$

and the remainder term is zero in this case. For a linear problem, the Fréchet derivative is the data kernel G.

The chain rule can be used with Fréchet derivatives in the same way that it is used with normal derivatives. Suppose that we want to calculate the Fréchet derivative of a function S of a functional $g[m]$, i.e.

$$S[m] = f(g[m]). \tag{9.5}$$

If D_S is the Fréchet derivative of $S[m]$ and D_g is the Fréchet derivative of $g[m]$, then

$$D_S = f'(g[m])D_g. \tag{9.6}$$

This formula would be useful for determining the sensitivity of the "objective function" with respect to the model parameters.

9.1.1 Sensitivity of steady pressure to permeability

One of the simplest inverse problems with relevance to reservoir engineering involves estimation of permeability from measurements of steady-state pressure in a linear, 1D system. This is a problem that we have used several times previously in this book. In this section, we will consider the calculation of sensitivity coefficients. The equation for one-dimensional steady flow in porous media is

$$\frac{d}{dx}\left(k\frac{dp}{dx}\right) = 0,$$

(9.7)

with boundary conditions

$$\frac{kA}{\mu}\frac{dp}{dx}\bigg|_{x=0} = -q_0$$

(9.8)

$$p(L) = p_e.$$

(9.9)

If the permeability is known everywhere, then Eq. (9.7) can be solved explicitly for $p(x)$:

$$\begin{aligned}
p(x) &= p_e + \frac{q_0\mu}{A}\int_x^L \frac{dx'}{k(x')} \\
&= p_e + \frac{q_0\mu}{A}\int_0^L \frac{H(x'-x)}{k(x')}\,dx',
\end{aligned}$$

(9.10)

where L is the length of the domain. The Heaviside unit step function $H(x)$ is equal to 1 when $x > 0$ and equal to 0 otherwise.

9.1.2 Fréchet derivative approach

Suppose that we wanted to know how the pressure at some location would change if the permeability field were changed from $k_0(x)$ to $k_0(x) + \delta(x)$. After perturbing the permeability field, the pressure field would change from $p_0(x)$ to $p_0(x) + \delta p(x)$. The Fréchet derivative approach requires we calculate the perturbation in pressure that results from a perturbation in permeability. To compute the Fréchet derivative, we recognize that p is a function of k and x and write Eq. (9.10) as

$$p(x, k) = p_e + \frac{q_0\mu}{A}\int_0^L \frac{H(x'-x)}{k(x')}\,dx',$$

(9.11)

so if the permeability $k(x)$ is perturbed to $k(x) + \delta k(x)$, then it follows easily that

$$
p(x, k + \delta k) - p(x, k) = \frac{q_0 \mu}{A} \int_0^L H(x' - x) \left(\frac{1}{k(x') + \delta k(x')} - \frac{1}{k(x')} \right) dx'
$$

$$
= -\frac{q_0 \mu}{A} \int_0^L \frac{H(x' - x)}{k(x')} \left(1 - \frac{1}{1 + (\delta k(x')/k(x'))} \right) dx'.
$$

$$(9.12)$$

Since we assume the perturbation in the permeability field is very small compared to the permeability, the following geometric expansion holds:

$$
\frac{1}{1 + (\delta k(x')/k(x'))} = 1 - \frac{\delta k(x')}{k(x')} + \frac{(\delta k(x'))^2}{(k(x'))^2} + \cdots .
$$

$$(9.13)$$

Using Eq. (9.13) in (9.12), it follows that

$$
p(x, k + \delta k) - p(x, k) = -\frac{q_0 \mu}{A} \int_0^L \frac{H(x' - x)}{k^2(x')} \delta k(x') \, dx' + R[\delta k],
$$

$$(9.14)$$

where the remainder term, $R[\delta k]$, involves quadratic and higher powers of δk so that

$$
\lim_{\|\delta k\| \to 0} \frac{R[\delta k]}{\|\delta k\|} = 0.
$$

$$(9.15)$$

Thus, comparing Eq. (9.14) with Eq. (9.2), it follows that the Fréchet derivative or the sensitivity of the pressure at x to a perturbation in the permeability at x' is given by

$$
G_k = \frac{-q_0 \mu H(x' - x)}{A k^2(x')}.
$$

$$(9.16)$$

To see if this makes sense, consider the calculation of the sensitivity of pressure at the inlet ($x = 0$) to a change in the permeability at some location x', where $0 < x' < L$. If the injection rate q_0 is positive, then the pressure is highest at the inlet and lowest at the outlet. An increase in permeability would cause the inlet pressure to decrease, in accordance with Eq. (9.16).

9.1.3   Adjoint approach to calculate Fréchet derivative

We consider the same problem of calculating the sensitivity of the steady-state pressure at some particular location in a one-dimensional domain $[0, L]$ to permeabilities at any other location in the domain, but in this subsection we introduce the adjoint method to calculate the Fréchet derivative [91, 92]. Assume that the pressure satisfies the equation

$$
k(x) \frac{dp}{dx} = -\frac{q_0 \mu}{A},
$$

$$(9.17)$$

with boundary condition

$$
p(L) = p_e.
$$

$$(9.18)$$

Again q_0 is constant and independent of x. Let J be a functional for which we want to find the derivative. For the specific case in which we want to calculate the sensitivity of pressure at a particular location to changes in permeability, we can write

$$J = p(x') = \int_0^L p(x)\delta(x - x')\,dx. \tag{9.19}$$

Here we use the notation $dJ/dk(x)$ to denote the Fréchet derivative of J with respect to the function k evaluated at x. In this case, the sensitivity of J to permeability is

$$
\begin{aligned}
\frac{dJ}{dk(x'')} &= \int_0^L \left[\frac{\partial[p(x)\delta(x - x')]}{\partial k(x'')} + \frac{\partial[p(x)\delta(x - x')]}{\partial p(x)} \frac{d[p(x)]}{dk(x'')} \right] dx \\
&= \int_0^L \delta(x - x') \frac{d[p(x)]}{dk(x'')}\,dx,
\end{aligned}
\tag{9.20}
$$

where the first term in the first integral represents the "direct effect" of k on the "objective function" J (there is no effect in this case so the first term is zero) and the second term represents the "indirect effect" due to the implicit dependence of J on permeability through the dependence of pressure on permeability. Let $\psi(x, x'') = dp(x)/dk(x'')$ represent this dependence and write Eq. (9.20) as

$$\frac{dJ}{dk(x'')} = \int_0^L \delta(x - x')\psi(x, x'')\,dx. \tag{9.21}$$

At this point, it appears that nothing has really been gained since we have almost identical terms on both sides of the equation. In fact, Eq. (9.21) just says that if we knew the sensitivity of pressures at all locations to permeability at all locations, we could obtain the sensitivity of pressure at a particular location by selecting that location with the delta function.

In order to use Eq. (9.21), therefore, we need to evaluate $\psi(x, x'')$. An equation for $\psi(x, x'')$ can be obtained by differentiating Eq. (9.17) with respect to $k(x'')$ to obtain

$$\frac{dk(x)}{dk(x'')}\frac{dp}{dx} + k(x)\frac{d}{dx}\left(\frac{dp(x)}{dk(x'')} \right) = 0, \tag{9.22}$$

or, using the notation for $dp(x)/dk(x'')$ that was introduced earlier,

$$\delta(x - x'')\frac{dp}{dx} + k(x)\frac{d\psi(x, x'')}{dx} = 0. \tag{9.23}$$

The boundary condition for this equation is obtained by differentiating the boundary condition Eq. (9.18) with respect to $k(x'')$ to obtain

$$\psi(x, x'') = 0 \qquad \text{at } x = L. \tag{9.24}$$

Note that the direct use of Eq. (9.23) to obtain $\psi(x, x'')$ would require us to solve a differential equation with a source function at every value of x''. For this simple problem, we could write the solution directly, but in general, the sensitivity equation must be solved numerically, and this is not a practical approach. Instead, we will use Eq. (9.23) to eliminate $\psi(x, x'')$ from Eq. (9.21). We begin by multiplying Eq. (9.23) by an arbitrary function ψ^*, and then integrating over the entire domain. This gives

$$\int_0^L \psi^*(x)\left[\delta(x - x'')\frac{dp}{dx} + k(x)\frac{d\psi(x, x'')}{dx}\right] dx = 0. \tag{9.25}$$

Integrate the second term in Eq. (9.25) by parts to obtain

$$\int_0^L \psi^*(x)k(x)\frac{d\psi(x, x'')}{dx} dx = \left.\psi^*(x)k(x)\psi(x, x'')\right|_0^L$$
$$- \int_0^L \psi(x, x'')\frac{d}{dx}[k(x)\psi^*(x)] dx. \tag{9.26}$$

Because the terms in Eq. (9.25) sum to zero, they can be added to Eq. (9.21) without affecting the equality. We then have

$$\frac{dJ}{dk(x'')} = \int_0^L \left[\delta(x - x')\psi(x, x'') - \psi(x, x'')\frac{d}{dx}[k(x)\psi^*(x)]\right.$$
$$\left. + \psi^*(x)\delta(x - x'')\frac{dp}{dx}\right] dx + \left.\psi^*(x)k(x)\psi(x, x'')\right|_0^L. \tag{9.27}$$

Gathering terms that multiply $\psi(x, x'')$ yields

$$\frac{dJ}{dk(x'')} = \int_0^L \left[\psi(x, x'')\left[\delta(x - x') - \frac{d}{dx}[k(x)\psi^*(x)]\right] + \psi^*(x)\delta(x - x'')\frac{dp}{dx}\right] dx$$
$$+ \left.\psi^*(x)k(x)\psi(x, x'')\right|_0^L. \tag{9.28}$$

From Eq. (9.24), we see that if we require ψ^* to be equal to zero at $x = 0$ the boundary term vanishes.

We do not want to have to evaluate $\psi(x, x'')$. We see now that it can be eliminated from Eq. (9.28) if we require that ψ^* satisfy the equation,

$$\frac{d}{dx}[k(x)\psi^*(x)] - \delta(x - x') = 0 \tag{9.29}$$

in which case Eq. (9.28) simplifies to

$$\frac{dJ}{dk(x'')} = \int_0^L \psi^*(x)\delta(x - x'')\frac{dp}{dx} dx = \psi^*(x'')\frac{dp(x'')}{dx''}. \tag{9.30}$$

Although Eqs. (9.21) and (9.30) look superficially similar, and it might appear that Eq. (9.30) is more complex, in most practical cases it requires far less work to obtain the sensitivity from Eq. (9.30) than from Eq. (9.21). This is because obtaining ψ^* in Eq. (9.30) only requires that Eq. (9.29) be solved with one source at $x = x'$, the location at which the pressure is observed.

Solution of the Adjoint System

Integrate Eq. (9.29) from 0 to x to obtain

$$k(x)\psi^*(x) - H(x - x') = 0, \tag{9.31}$$

where we used the fact that $\psi^* = 0$ at $x = 0$. Rearranging Eq. (9.31) gives

$$\psi^*(x) = \frac{H(x - x')}{k(x)}. \tag{9.32}$$

To obtain the sensitivity of pressure at x' with respect to permeability at x'', we need only substitute Eq. (9.32) into Eq. (9.30) to obtain

$$\frac{dJ}{dk(x'')} = \frac{H(x'' - x')}{k(x'')} \frac{dp(x'')}{dx''}. \tag{9.33}$$

In this relatively simple example, we can simplify further by recalling that

$$\frac{dp}{dx} = -\frac{q_0\mu}{k(x)A} \tag{9.34}$$

so we finally write

$$\frac{dJ}{dk(x'')} = -\frac{q_0\mu H(x'' - x')}{k(x'')^2 A}, \tag{9.35}$$

which is the same result that we derived in Eq. (9.16) using a perturbation expansion to calculate the Fréchet derivative. In practice, this would all be done numerically on a discrete grid. We would first solve the forward system Eq. (9.17) for the pressure field, and Eq. (9.29) for the adjoint solution, then apply Eq. (9.30) to calculate the sensitivity.

9.2   Discrete parameters

At some point, almost all inverse problems have to be discretized in order to be solved. Often the discretization is necessary for the solution of the forward problem. In reservoir engineering, the PDEs for flow and transport are converted to discrete equations which are then solved numerically. Instead of estimating permeability and porosity as functions of position, we estimate their values in every gridblock of the reservoir model. The number of parameters is now finite, but still too large for the problem to be well posed. In this section, methods are developed for calculation of sensitivity in discrete problems.

9.2.1   Generic finite-dimensional problem

Consider the N-dimensional column vector of dependent variables y (e.g. pressure or saturation), and the M-dimensional vector of model variables m (e.g. permeability and porosity), which are implicitly related by the following set of model equations:

$$f(y, m) = 0. \tag{9.36}$$

Here f is an N-dimensional column vector, i.e. Eq. (9.36) represents a system of N equations that we can solve for y assuming m is known. We are often only interested in a small subset of dependent variables instead of the value of the dependent variable in every gridblock. Let J be a real-valued function of y such as an objective function or a result (e.g. a water–oil ratio) that is related to y as follows

$$J = h(y). \tag{9.37}$$

Also define generic parameters, $\{\alpha_i, i = 1, \ldots, N_i\}$. These could be components of m, coefficients defining a reparameterization of m, or other parameters we want to estimate that are directly related to m. The goal of a sensitivity calculation is to calculate the rate of change of J with respect to the parameters α_i, i.e. the total derivative of J with respect to each α_i. From Eq. (9.37), the total differential of J is seen to be

$$dJ = (\nabla_y h)^T \, dy. \tag{9.38}$$

So, the sensitivity of J with respect to the parameter α_i is

$$\frac{dJ}{d\alpha_i} = (\nabla_y h)^T \frac{\partial y}{\partial \alpha_i}. \tag{9.39}$$

A comment on notation is in order here. Recall that the gradient of a scalar function with respect to a vector is:

$$\nabla_y h = \begin{bmatrix} \frac{\partial h}{\partial y_1} \\ \vdots \\ \frac{\partial h}{\partial y_N} \end{bmatrix}. \tag{9.40}$$

On the other hand, the derivative of a vector y with respect to a scalar is

$$\frac{\partial y}{\partial \alpha} = \begin{bmatrix} \frac{\partial y_1}{\partial \alpha} \\ \vdots \\ \frac{\partial y_N}{\partial \alpha} \end{bmatrix}. \tag{9.41}$$

We will often need to calculate the gradient of a vector f of length N with respect to another vector m of length M. In this case, we mean

$$(\nabla_m f^T)^T = \begin{bmatrix} \frac{\partial f_1}{\partial m_1} & \cdots & \frac{\partial f_1}{\partial m_M} \\ \vdots & & \vdots \\ \frac{\partial f_N}{\partial m_1} & \cdots & \frac{\partial f_N}{\partial m_M} \end{bmatrix} = \begin{bmatrix} (\nabla f_1)^T \\ \vdots \\ (\nabla f_N)^T \end{bmatrix} = \begin{bmatrix} \frac{\partial f}{\partial m_1} & \frac{\partial f}{\partial m_2} & \cdots & \frac{\partial f}{\partial m_M} \end{bmatrix}. \tag{9.42}$$

The quantity $(\nabla_y h)^T$ is usually easy to evaluate because $h(y)$ is often of the form $h(y) = y_i$ or $h(y) = (y(x_0) - y_{obs})^T (y(x_0) - y_{obs})$. The rate of change of y with respect to the model parameters is more difficult to calculate, however. We begin by differentiating Eq. (9.36) with respect to a model parameter, α_i. This gives

$$(\nabla_y f^{\mathrm{T}})^{\mathrm{T}} \frac{\partial y}{\partial \alpha_i} + (\nabla_m f^{\mathrm{T}})^{\mathrm{T}} \frac{\partial m}{\partial \alpha_i} = 0. \tag{9.43}$$

Rearranging, we obtain the following system of equations for $\partial y/\partial \alpha_i$:

$$(\nabla_y f^{\mathrm{T}})^{\mathrm{T}} \frac{\partial y}{\partial \alpha_i} = -(\nabla_m f^{\mathrm{T}})^{\mathrm{T}} \frac{\partial m}{\partial \alpha_i}. \tag{9.44}$$

9.2.2   Sensitivity by the direct method

One way to obtain the sensitivity of the dependent variables to the model parameters is to directly solve Eq. (9.44) for $\partial y/\partial \alpha_i$ and use this result in Eq. (9.39). This is usually called the *direct* method. In the petroleum industry, however, it is commonly called the gradient simulator method [93]. If f represents the simulator equations, then the dimension of the system to be solved in Eq. (9.44) is the same as the dimension of the simulator system. If there are very few model parameters this can be an efficient method for calculating derivatives. This might be the case when the goal is to calculate permeability multipliers for layers in a layered reservoir simulation model.

Quite often, however, large numbers of parameters are required to adequately describe the distribution of petrophysical properties in a reservoir. A typical reservoir simulator requires at least a porosity and a permeability for every gridblock. If there are many gridblocks, the number of model parameters can be quite large. Note that Eq. (9.44) has as many right-hand sides as parameters so if N_i is large, the computation of sensitivities using the direct method can be computationally demanding. Of course the system is linear, and the coefficient matrix on the left-hand side only needs to be factored once, so some increase in efficiency can be gained through a clever algorithm [94].

9.2.3   Sensitivity by the adjoint method

If the Gauss–Newton or Levenberg–Marquardt algorithm is used for minimization of an objective function, we need to be able to calculate the derivative of each predicted data with respect to all of the model parameters. While this is typically a large number of derivatives, it is often far less than the number of derivatives calculated in the direct method. This is because the direct method calculates the derivative of every dependent variable (y is usually a vector of large dimension) with respect to all of the parameters, whereas we only need the derivative of the dependent variables at locations where they are observed (typically at well locations). This is usually a much smaller number. In the adjoint method, we evaluate $dJ/d\alpha_i$ without first computing $\partial y/\partial \alpha_i$.

We begin by adjoining the simulator equations to Eq. (9.37), i.e.

$$J = h(y) + \lambda^{\mathrm{T}} f(m, y). \tag{9.45}$$

This relationship is true for arbitrary multipliers λ because $f(m, y) = 0$. As before, we compute the total differential of J:

$$dJ = (\nabla_y h)^{\mathrm{T}} dy + \lambda^{\mathrm{T}} (\nabla_m f^{\mathrm{T}})^{\mathrm{T}} dm + \lambda^{\mathrm{T}} (\nabla_y f^{\mathrm{T}})^{\mathrm{T}} dy. \tag{9.46}$$

In Section 9.1.3 we were able to choose the adjoint variable in such a way that the computation of sensitivity was greatly simplified. The choice is even easier here. We see from Eq. (9.46), that if we require λ to satisfy

$$\lambda^{\mathrm{T}} (\nabla_y f^{\mathrm{T}})^{\mathrm{T}} = -(\nabla_y h)^{\mathrm{T}} \tag{9.47}$$

then the expression for dJ of Eq. (9.46) no longer involves dy. The transpose of Eq. (9.47) is

$$\nabla_y f^{\mathrm{T}} \lambda = -\nabla_y h. \tag{9.48}$$

This system is referred to as the adjoint system or simply the *adjoint* equation.

This is the equation that must be solved for the adjoint vector λ. From Eq. (9.46), we see that we now have an expression for the derivative of J with respect to the model parameter, α_i, that is,

$$\frac{dJ}{d\alpha_i} = \lambda^{\mathrm{T}} (\nabla_m f^{\mathrm{T}})^{\mathrm{T}} \frac{\partial m}{\partial \alpha_i}. \tag{9.49}$$

The advantage of this method is that Eq. (9.48) only has to be solved once for the adjoint variables, no matter how many model parameters are needed to describe the problem. And, like the direct method, the problem to be solved (Eq. 9.48) is linear. Of course, if the intention is to use a Gauss–Newton approach with a large number of "observations" for which derivatives with respect to parameters are required, then we must solve as many adjoint systems as we have data. This number is, however, often much smaller than the number of model parameters.

9.2.4   Sensitivity by finite-difference method

Suppose a predicted data vector y depends on parameters α_i, $i = 1, 2, \ldots, N_i$. A conceptually simple way to compute the derivative of a vector y with respect to a parameter α_i is to use a finite-difference approximation to the partial derivative, i.e.

$$\frac{\partial y}{\partial \alpha_i} = \frac{1}{\delta \alpha_i} \big(y(\alpha_1, \ldots, \alpha_{i-1}, \alpha_i + \delta \alpha_i, \alpha_{i+1}, \ldots, \alpha_{N_i})$$
$$- y(\alpha_1, \ldots, \alpha_{i-1}, \alpha_i, \alpha_{i+1}, \ldots, \alpha_{N_i}) \big). \tag{9.50}$$

Here $\delta \alpha_i$ denotes a perturbation in the parameter α_i, and theoretically, as $\alpha_i \to 0$, the right-hand side of Eq. (9.50) converges to the derivative. However, in practice,

computation of predicted data y is inaccurate and if the change in y due to the perturbation α_i is too small, then the calculated value of the numerator on the right-hand side
of the preceding equation may contain no correct digits. In this case, the calculated
right-hand side of Eq. (9.50) gives a completely erroneous estimate of $\partial y / \partial \alpha_i$; in fact,
the components of the sign may not even be correct. Choosing $\delta \alpha_i$ to be equal to about
0.01 times α_i often yields good results, but this is not always the case. The only way to
be sure the results are accurate is to try a range of perturbations and find a stable range
that gives the same value.

In addition to the difficulty in choosing the magnitude of the perturbation, the finite-
difference method suffers from the fact that the number of calculations of y required
is equal to the number of parameters plus one. (In this sense, the method is similar
to the direct method (gradient simulator method).) Because of this, neither method is
feasible when calculation of y requires a reservoir simulation run unless the number
of parameters is reasonably small, say on the order of few dozen. For this reason,
in our discussion of history matching of reservoir models, we focus on the adjoint
method. The adjoint method, however, requires considerable care to implement and the
finite-difference method provides one way to check the adjoint calculations [69, 95].

9.3   One-dimensional steady-state flow

Let us apply the methods of the previous sections to the problem of calculating sensitivities of inlet pressure to gridblock permeabilities in a small one-dimensional porous
medium. The flow is governed by Eq. (9.7), which is repeated here

$$\frac{d}{dx}\left[k(x)\frac{dp}{dx}\right] = 0, \tag{9.51}$$

with boundary conditions

$$\frac{kA}{\mu}\frac{dp}{dx}\bigg|_{x=\Delta x} = -q_0, \tag{9.52}$$

$$p(L - \Delta x/2) = p_e. \tag{9.53}$$

where the slightly odd boundary conditions were chosen to simplify the calculations
of sensitivity in the discrete problem. We will replace the differential Eq. (9.51) with
a finite-difference approximation. The grid system is shown in Fig. 9.1. The finite-
difference equation is

$$k_{i+1/2}[p_{i+1} - p_i] - k_{i-1/2}[p_i - p_{i-1}] = 0, \tag{9.54}$$

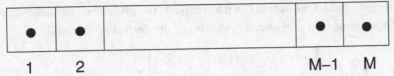

1   2   M–1   M

Figure 9.1. The grid system for the one-dimensional flow problem.

while the boundary conditions become

$$k_{3/2}[p_2 - p_1]/\Delta x = -q_0\mu/A,$$ (9.55)

$$p_M = p_e,$$ (9.56)

and

$$k_{i+1/2} = \frac{2k_i k_{i+1}}{k_i + k_{i+1}}.$$ (9.57)

For the purpose of exposition, we will treat the extremely simplified example in which the porous medium is represented by only three gridblocks. Our goal will be to calculate the sensitivity of inlet pressure to the permeability of each of the gridblocks.

The system of model equations can be written as follows

$$f \equiv \begin{bmatrix} -k_{3/2} & k_{3/2} & 0 \\ k_{3/2} & -(k_{3/2} + k_{5/2}) & k_{5/2} \\ 0 & 0 & 1 \end{bmatrix} \begin{bmatrix} p_1 \\ p_2 \\ p_3 \end{bmatrix} - \begin{bmatrix} -q_0\mu\Delta x/A \\ 0 \\ p_e \end{bmatrix} = 0,$$ (9.58)

which is in the form $f(m, y) = F(m)y - b = 0$ where $m = [k_1, k_2, k_3]^T$ and $y = [p_1, p_2, p_3]^T$. In order to calculate the sensitivities, we need to evaluate a few of the intermediate results. In particular, we need to evaluate $(\nabla_y h)^T, (\nabla_y f^T)^T, (\nabla_m f^T)^T$, and $\partial m/\partial\alpha_i$.

Derivative of objective function with respect to pressure. We assume here that we are able to observe the pressure at the center of the first gridblock, and that we wish to know how the calculated pressure would change if the permeability in the gridblocks were to change. Thus $h = p_1$ and

$$(\nabla_y h)^T = \begin{bmatrix} 1 & 0 & 0 \end{bmatrix}.$$ (9.59)

Derivative of model equations with respect to pressure. Because the model Eqs. (9.58) are linear in pressure, calculation of the derivative is trivial:

$$(\nabla_y f^T)^T = F(m) = \begin{bmatrix} -k_{3/2} & k_{3/2} & 0 \\ k_{3/2} & -(k_{3/2} + k_{5/2}) & k_{5/2} \\ 0 & 0 & 1 \end{bmatrix}$$

$$= \begin{bmatrix} -\frac{2k_1 k_2}{k_1+k_2} & \frac{2k_1 k_2}{k_1+k_2} & 0 \\ \frac{2k_1 k_2}{k_1+k_2} & -\left(\frac{2k_1 k_2}{k_1+k_2} + \frac{2k_2 k_3}{k_2+k_3}\right) & \frac{2k_2 k_3}{k_2+k_3} \\ 0 & 0 & 1 \end{bmatrix}. \tag{9.60}$$

The inverse of this matrix is

$$\left((\nabla_y f^T)^T\right)^{-1} = \begin{bmatrix} -\left(\frac{1}{2k_1} + \frac{1}{k_2} + \frac{1}{2k_3}\right) & -\frac{k_2+k_3}{2k_2 k_3} & 1 \\ -\frac{k_2+k_3}{2k_2 k_3} & -\frac{k_2+k_3}{2k_2 k_3} & 1 \\ 0 & 0 & 1 \end{bmatrix}. \tag{9.61}$$

Derivative of model equations with respect to model parameters. Assuming that the model parameters are the gridblock permeabilities, we need to evaluate $(\nabla_m f^T)^T$, which is a 3×3 matrix whose i, j entry is $\partial f_i / \partial k_j$. Clearly, the permeability only enters in the interblock transmissivity terms which can be differentiated easily.

$$\frac{\partial k_{j+1/2}}{\partial k_i} = \begin{cases} 0 & \text{if } i \neq j \text{ and } i \neq j+1, \\ \frac{2k_{i+1}^2}{(k_i+k_{i+1})^2} & \text{if } i = j, \\ \frac{2k_{i-1}^2}{(k_{i-1}+k_i)^2} & \text{if } i = j+1. \end{cases} \tag{9.62}$$

Using this result, we can write the derivatives of f with respect to each of the gridblock permeabilities.

$$\frac{\partial f}{\partial k_1} = \begin{bmatrix} -\frac{2k_2^2}{(k_1+k_2)^2} & \frac{2k_2^2}{(k_1+k_2)^2} & 0 \\ \frac{2k_2^2}{(k_1+k_2)^2} & -\frac{2k_2^2}{(k_1+k_2)^2} & 0 \\ 0 & 0 & 0 \end{bmatrix} \begin{bmatrix} p_1 \\ p_2 \\ p_3 \end{bmatrix} = \begin{bmatrix} \frac{2k_2^2(p_2-p_1)}{(k_1+k_2)^2} \\ -\frac{2k_2^2(p_2-p_1)}{(k_1+k_2)^2} \\ 0 \end{bmatrix}, \tag{9.63}$$

$$\frac{\partial f}{\partial k_2} = \begin{bmatrix} -\frac{2k_1^2}{(k_1+k_2)^2} & \frac{2k_1^2}{(k_1+k_2)^2} & 0 \\ \frac{2k_1^2}{(k_1+k_2)^2} & -\left(\frac{2k_1^2}{(k_1+k_2)^2} + \frac{2k_3^2}{(k_2+k_3)^2}\right) & \frac{2k_3^2}{(k_2+k_3)^2} \\ 0 & 0 & 0 \end{bmatrix} \begin{bmatrix} p_1 \\ p_2 \\ p_3 \end{bmatrix}$$

$$= \begin{bmatrix} \frac{2k_1^2(p_2-p_1)}{(k_1+k_2)^2} \\ -\frac{2k_1^2(p_2-p_1)}{(k_1+k_2)^2} + \frac{2k_3^2(p_3-p_2)}{(k_2+k_3)^2} \\ 0 \end{bmatrix}, \tag{9.64}$$

and

$$\frac{\partial f}{\partial k_3} = \begin{bmatrix} 0 & 0 & 0 \\ 0 & -\frac{2k_2^2}{(k_2+k_3)^2} & \frac{2k_2^2}{(k_2+k_3)^2} \\ 0 & 0 & 0 \end{bmatrix} \begin{bmatrix} p_1 \\ p_2 \\ p_3 \end{bmatrix} = \begin{bmatrix} 0 \\ \frac{2k_2^2(p_3-p_2)}{(k_2+k_3)^2} \\ 0 \end{bmatrix}. \tag{9.65}$$

From Eq. (9.42), we see that the matrix $(\nabla_m f^T)^T$ is the combination of these last three results, i.e.

$$(\nabla_m f^T)^T = \begin{bmatrix} \frac{\partial f}{\partial k_1} & \frac{\partial f}{\partial k_2} & \frac{\partial f}{\partial k_3} \end{bmatrix}$$

$$= \begin{bmatrix} \frac{2k_1^2(p_2-p_1)}{(k_1+k_2)^2} & \frac{2k_1^2(p_2-p_1)}{(k_1+k_2)^2} & 0 \\ -\frac{2k_1^2(p_2-p_1)}{(k_1+k_2)^2} & -\frac{2k_1^2(p_2-p_1)}{(k_1+k_2)^2} + \frac{2k_3^2(p_3-p_2)}{(k_2+k_3)^2} & \frac{2k_3^2(p_3-p_2)}{(k_2+k_3)^2} \\ 0 & 0 & 0 \end{bmatrix}. \tag{9.66}$$

Derivative of model vector with respect to parameters. The only remaining term that can be calculated analytically is the derivative of the vector of gridblock parameters with respect to the model parameters. In this example, the two sets of parameters are the same, so

$$\partial m/\partial \alpha_i = e_i, \tag{9.67}$$

where e_i is the unit vector for which the entry is zero at every location except the ith, at which the value is 1. This also means that we can simplify the product on the right-hand side of Eq. (9.44) to

$$(\nabla_m f^T)^T \frac{\partial m}{\partial \alpha_i} = \frac{\partial f}{\partial m_i}, \tag{9.68}$$

so that Eq. (9.44) reduces to

$$(\nabla_y f^T)^T \frac{\partial y}{\partial m_i} = -\frac{\partial f}{\partial m_i}. \tag{9.69}$$

9.3.1 Direct method

In the direct method for evaluating derivatives of the object function with respect to the parameters, we start with Eq. (9.39) which we repeat here as

$$\frac{dJ}{d\alpha_i} = (\nabla_y h)^T \frac{\partial y}{\partial \alpha_i} \tag{9.70}$$

$$= \begin{bmatrix} 1 & 0 & 0 \end{bmatrix} \frac{\partial y}{\partial \alpha_i}.$$

The simplification on the second line was a result of the fact that $J = h(y)$ is equal to pressure at the inlet; see Eq. (9.59).

The second term in Eq. (9.70) can be evaluated by solving Eq. (9.44) using the inverse from Eq. (9.61) to obtain

$$
\frac{\partial y}{\partial m_i} = -(\nabla_y f^T)^{-T} \frac{\partial f}{\partial m_i}
$$

$$
= - \begin{bmatrix} -\left(\frac{1}{2k_1} + \frac{1}{k_2} + \frac{1}{2k_3}\right) & -\frac{k_2+k_3}{2k_2k_3} & 1 \\ -\frac{k_2+k_3}{2k_2k_3} & -\frac{k_2+k_3}{2k_2k_3} & 1 \\ 0 & 0 & 1 \end{bmatrix} \frac{\partial f}{\partial m_i}. \tag{9.71}
$$

Change in permeability of first block. Using Eq. (9.63) in Eq. (9.71), it follows that

$$
\frac{\partial y}{\partial k_1} = -(\nabla_y f^T)^{-T} \frac{\partial f}{\partial k_1}
$$

$$
= - \begin{bmatrix} -\left(\frac{1}{2k_1} + \frac{1}{k_2} + \frac{1}{2k_3}\right) & -\frac{k_2+k_3}{2k_2k_3} & 1 \\ -\frac{k_2+k_3}{2k_2k_3} & -\frac{k_2+k_3}{2k_2k_3} & 1 \\ 0 & 0 & 1 \end{bmatrix} \begin{bmatrix} \frac{2k_2^2(p_2-p_1)}{(k_1+k_2)^2} \\ -\frac{2k_2^2(p_2-p_1)}{(k_1+k_2)^2} \\ 0 \end{bmatrix}
$$

$$
= - \begin{bmatrix} \frac{k_2(p_1-p_2)}{k_1(k_1+k_2)} \\ 0 \\ 0 \end{bmatrix} \tag{9.72}
$$

$$
= \begin{bmatrix} \frac{-q_0\mu\Delta x}{2Ak_1^2} \\ 0 \\ 0 \end{bmatrix}.
$$

This result indicates that the pressures in gridblocks 2 and 3 are insensitive to a change in the permeability of the first gridblock, which is correct for these particular boundary conditions since the flow rate is specified at the interface between gridblocks 1 and 2. It also says that the sensitivity is proportional to the rate and that it is inversely proportional to the square of the permeability of gridblock 1. Both these statements are physically reasonable as well as correct. Also note from Eq. (9.70) that the first entry of the vector $\partial y/\partial k_1$ represents dJ/dk_1, i.e.

$$
\frac{dJ}{dk_1} = \frac{dp_1}{dk_1} = \frac{-q_0\mu\Delta x}{2Ak_1^2}. \tag{9.73}
$$

Change in permeability of third (last) block. Similarly, the derivative of the pressure vector with respect to the permeability of the third gridblock is

$$
\frac{\partial y}{\partial k_3} = -\left(\nabla_y f^{\mathrm{T}}\right)^{-\mathrm{T}} \frac{\partial f}{\partial k_3}
$$

$$
= -\begin{bmatrix} -\left(\frac{1}{2k_1} + \frac{1}{k_2} + \frac{1}{2k_3}\right) & -\frac{k_2+k_3}{2k_2k_3} & 1 \\ -\frac{k_2+k_3}{2k_2k_3} & -\frac{k_2+k_3}{2k_2k_3} & 1 \\ 0 & 0 & 1 \end{bmatrix} \begin{bmatrix} 0 \\ \frac{2k_2^2(p_3-p_2)}{(k_2+k_3)^2} \\ 0 \end{bmatrix}
$$

$$
= -\begin{bmatrix} \frac{k_2(p_2-p_3)}{k_3(k_2+k_3)} \\ \frac{k_2(p_2-p_3)}{k_3(k_2+k_3)} \\ 0 \end{bmatrix} \tag{9.74}
$$

$$
= \begin{bmatrix} \frac{-q_0\mu\Delta x}{2Ak_3^2} \\ \frac{-q_0\mu\Delta x}{2Ak_3^2} \\ 0 \end{bmatrix}.
$$

This result shows that the pressure in gridblock 3 is insensitive to a change in the permeability of gridblock 3. This is correct for this problem because the pressure in gridblock 3 is fixed. It also says that the sensitivity of pressures in gridblocks 1 and 2 are proportional to the rate and inversely proportional to the square of the permeability of gridblock 3. These results are both physically reasonable and correct.

Remark 9.1. Note that even though we only wanted to calculate the sensitivity of pressure in the first gridblock with respect to all permeabilities, we had to calculate the sensitivity of all gridblock pressures when using the direct method.

9.3.2 The adjoint method

Computation of the sensitivities of pressure in the first gridblock with respect to the model parameters is somewhat simpler using the adjoint method. We begin by solving Eq. (9.48) for the adjoint variables. Using Eq. (9.61), this gives

$$
\lambda = -\left(\nabla_y f^{\mathrm{T}}\right)^{-1} \nabla_y h
$$

$$
= -\begin{bmatrix} -\left(\frac{1}{2k_1} + \frac{1}{k_2} + \frac{1}{2k_3}\right) & -\frac{k_2+k_3}{2k_2k_3} & 0 \\ -\frac{k_2+k_3}{2k_2k_3} & -\frac{k_2+k_3}{2k_2k_3} & 0 \\ 1 & 1 & 1 \end{bmatrix} \begin{bmatrix} 1 \\ 0 \\ 0 \end{bmatrix}
$$

$$
= \begin{bmatrix} \frac{1}{2k_1} + \frac{1}{k_2} + \frac{1}{2k_3} \\ \frac{k_2+k_3}{2k_2k_3} \\ -1 \end{bmatrix}. \tag{9.75}
$$

Using Eq. (9.68), we can rewrite Eq. (9.49) as

$$\frac{dJ}{dm_i} = \lambda^T \frac{\partial f}{\partial m_i}.$$
(9.76)

Thus, once the adjoint solution has been calculated, the sensitivities are easy to obtain. For example, the sensitivity of the "objective function" (in this case the pressure in gridblock 1) to the permeability of gridblock 1 is found by setting $m_i = m_1 = k_1$ in Eq. (9.76) and using Eqs. (9.63) and (9.75) to obtain

$$\frac{dJ}{dk_1} = \lambda^T \frac{\partial f}{\partial k_1}$$

$$= \left[\left(\frac{1}{2k_1} + \frac{1}{k_2} + \frac{1}{2k_3}\right) \quad \frac{k_2+k_3}{2k_2k_3} \quad -1\right] \begin{bmatrix} \frac{2k_2^2(p_2-p_1)}{(k_1+k_2)^2} \\ -\frac{2k_2^2(p_2-p_1)}{(k_1+k_2)^2} \\ 0 \end{bmatrix}$$
(9.77)

$$= \frac{-k_2(p_1 - p_2)}{k_1(k_1 + k_2)} = \frac{-q_0\mu\Delta x}{2Ak_1^2}.$$

This is the same answer that was obtained by the direct method; see Eq. (9.73).

Similarly, the sensitivity of pressure in gridblock 1 to the permeability of gridblock 3 is

$$\frac{dJ}{dk_3} = \lambda^T \frac{\partial f}{\partial k_3}$$

$$= \left[\left(\frac{1}{2k_1} + \frac{1}{k_2} + \frac{1}{2k_3}\right) \quad \frac{k_2+k_3}{2k_2k_3} \quad -1\right] \begin{bmatrix} 0 \\ \frac{2k_2^2(p_3-p_2)}{(k_2+k_3)^2} \\ 0 \end{bmatrix}$$
(9.78)

$$= \frac{-k_2(p_2 - p_3)}{k_3(k_2 + k_3)} = \frac{-q_0\mu\Delta x}{2Ak_3^2}.$$

Again this is the same result as was obtained by the direct method; see the first component of Eq. (9.74).

9.3.3   Finite-difference method

The inverse of the matrix in Eq. (9.58) is given by Eq. (9.61). Multiplying Eq. (9.58) by this inverse gives

$$y(k_1, k_2, k_3) \equiv \begin{bmatrix} p_1 \\ p_2 \\ p_3 \end{bmatrix} = \begin{bmatrix} -\left(\frac{1}{2k_1} + \frac{1}{k_2} + \frac{1}{2k_3}\right) & -\frac{k_2+k_3}{2k_2k_3} & 1 \\ -\frac{k_2+k_3}{2k_2k_3} & -\frac{k_2+k_3}{2k_2k_3} & 1 \\ 0 & 0 & 1 \end{bmatrix} \begin{bmatrix} -q_0\mu\Delta x/A \\ 0 \\ p_e \end{bmatrix}.$$
(9.79)

We estimate the sensitivity of y with respect to $\alpha_i = k_1$ using the finite-difference method. Upon simplification, Eq. (9.50) becomes

$$\frac{\partial y}{\partial k_1} = \frac{1}{\delta k_1} \begin{bmatrix} -\left(\frac{1}{2k_1+2\delta k_1} - \frac{1}{2k_1}\right) & 0 & 0 \\ 0 & 0 & 0 \\ 0 & 0 & 0 \end{bmatrix} \begin{bmatrix} -q_0\mu\Delta x/A \\ 0 \\ p_e \end{bmatrix}$$

$$= \frac{1}{\delta k_1} \begin{bmatrix} -(q_0\mu\Delta x/A)((2\delta k_1)/(4k_1(k_1+\delta k_1))) \\ 0 \\ 0 \end{bmatrix}. \tag{9.80}$$

As $\delta k_1 \to 0$, Eq. (9.80) gives

$$\frac{\partial y}{\partial k_1} = \begin{bmatrix} \partial p_1/\partial k_1 \\ \partial p_2/\partial k_1 \\ \partial p_3/\partial k_1 \end{bmatrix} = \begin{bmatrix} -q_0\mu\Delta x/(2Ak_1^2) \\ 0 \\ 0 \end{bmatrix}, \tag{9.81}$$

which is identical to the results obtained with the direct method (Eq. 9.72). Identical results would also be obtained with the adjoint method although previously, we only applied the adjoint to compute the sensitivity of p_1 to k_1 and k_3; see Eqs. (9.77) and (9.78).

9.4 Adjoint methods applied to transient single-phase flow

Here, we consider only single-phase flow of a slightly compressible fluid of constant viscosity and constant compressibility. We assume that the dot product of the spatial gradients of one over the formation volume (FVF) and pressure is negligible so that FVF can be deleted from the flow equation. We also assume rock compressibility is constant and neglect the variation in porosity with pressure. Gravity effects are neglected, but could be incorporated using a pseudopotential. Under these assumptions, the partial differential equation describing flow is linear as are the associated initial and boundary conditions describing flow. Under these conditions, Carter et al. [96] was able to construct the sensitivity of a gridblock pressure or wellbore pressure to gridblock permeability for two-dimensional horizontal flow. His elegant derivation was based on calculating the Fréchet derivative and provided an alternate approach to constructing sensitivities to the method based on an electric circuit analog developed by Jacquard and Jain [97]. Carter et al. [54] showed that for a two-dimensional horizontal flow problem, his procedure for constructing sensitivities is mathematically equivalent to the adjoint method as derived independently by Chen et al. [98] and Chavent et al. [99]. Unlike the adjoint method, the Carter et al. derivation does not apply to nonlinear problems and in fact, is difficult to implement efficiently even for three-dimensional linear single-phase flow problems [100]. Thus, our focus is on the adjoint method.

One can try to formulate the adjoint problem from the fully continuous problem (which requires Green's functions), the semidiscrete problem, or the fully discrete equations used in the reservoir simulator. When using the adjoint to compute derivatives needed in a Levenberg–Marquardt or quasi-Newton optimization procedure for matching dynamic data with predicted data generated from a reservoir simulator, the use of the fully discrete formulation is most straightforward and less likely to lead to errors. Nevertheless, we will first present a semidiscrete formulation as it is helpful in understanding the main concepts and difficulties.

9.4.1 Discrete and semidiscrete flow equations

The constants C_1 and C_2 represent unit conversion factors. We use oil field units throughout, in which case,

$$C_1 = 1.127 \times 10^{-3} \tag{9.82}$$

and

$$C_2 = 5.615. \tag{9.83}$$

The flow equation is written as

$$\frac{C_1}{\mu} \nabla \cdot \left([k]\nabla p(x, y, z, t)\right) = \frac{\phi c_t}{C_2}\frac{\partial p}{\partial t} + \hat{q}(x, y, z, t), \tag{9.84}$$

where $[k]$ is the permeability tensor, p is pressure in psi, ϕ is porosity and c_t is total compressibility in psi^{-1}. The source/sink term $\hat{q}(x, y, z, t)$ has units of RB/ft^3-day and is positive for a producing well, negative for an injection well and is nonzero only if the point (x, y, z) is intersected by a well. We assume that the coordinate directions are aligned with the principle permeability directions so that

$$[k] = \begin{bmatrix} k_x & 0 & 0 \\ 0 & k_y & 0 \\ 0 & 0 & k_z \end{bmatrix}. \tag{9.85}$$

We assume a rectangular parallelepiped reservoir, i.e. Eq. (9.84) applies for all $t > 0$ on

$$\Omega = \{(x, y, z)|0 < x < L_x, 0 < y < L_y, 0 < z < L_z\}, \tag{9.86}$$

where the boundary of Ω is denoted by $\partial\Omega$. We assume no flow boundary conditions at the outer edges of the reservoir. Initial conditions are given by

$$p(x, y, z, 0) = p_0(x, y, z). \tag{9.87}$$

We partition Ω into gridblocks using a standard block-centered grid and let (x_i, y_j, z_k), $i = 1, 2, \ldots, n_x$, $j = 1, 2, \ldots, n_y$, $k = 1, 2, \ldots, n_z$, denote the gridblock centers. Considering Eq. (9.84) at (x_i, y_j, z_k), we use a standard three-point finite-difference procedure to approximate spatial derivatives in Eq. (9.84) and multiply the resulting equation by $\Delta x_i \Delta y_j \Delta z_k$ to obtain the following equation:

$$T_{x,i+1/2,j,k}(p_{i+1,j,k}(t) - p_{i,j,k}(t)) - T_{x,i-1/2,j,k}(p_{i,j,k}(t) - p_{i-1,j,k}(t))$$
$$+ T_{y,i,j+1/2,k}(p_{i,j+1,k}(t) - p_{i,j,k}(t)) - T_{y,i,j-1/2,k}(p_{i,j,k}(t) - p_{i,j-1,k}(t))$$
$$+ T_{z,i,j,k+1/2}(p_{i,j,k+1}(t) - p_{i,j,k}(t)) - T_{z,i,j,k-1/2}(p_{i,j,k}(t) - p_{i,j,k-1}(t))$$
$$= \left(\frac{\Delta x_i \Delta y_j \Delta z_k \phi_{i,j,k} c_t}{C_2} \right) \frac{\partial p_{i,j,k}}{\partial t} + q_{i,j,k}(t), \tag{9.88}$$

where the Ts represent the standard transmissibilities defined on gridblock boundaries. Note that the modified source sink terms are given by

$$q_{i,j,k}(t) = \Delta x_i \Delta y_j \Delta z_k \hat{q}(x_i, y_j, z_k, t) \tag{9.89}$$

and have units of RB/D. Throughout, Δx_i, Δy_j, and Δz_k are the dimensions of the gridblock centered at (x_i, y_j, z_k). The x-direction boundaries of this gridblock are $x_{i-1/2}$ and $x_{i+1/2}$ so that $\Delta x_i = x_{i+1/2} - x_{i-1/2}$. Similar and obvious notation is used for the gridblock boundaries in the other directions. The Ts denote transmissibilites at the gridblock boundaries. Note we have left the time variable continuous in Eq. (9.88). At any time t and for any j and k,

$$T_{x,i+1/2,j,k} = \frac{C_1 \Delta y_j \Delta z_k k_{x,i+1/2,j,k}}{\mu(x_{i+1} - x_i)}, \tag{9.90}$$

for all $i = 1, 2, \ldots, n_x - 1$. To incorporate no flow boundaries, we set

$$T_{x,1/2,j,k} = T_{x,n_x+1/2,j,k} = 0. \tag{9.91}$$

Similarly,

$$T_{y,i,j+1/2,k} = \frac{C_1 \Delta x_i \Delta z_k k_{y,i,j+1/2,k}}{\mu(y_{j+1} - y_j)}, \tag{9.92}$$

for $j = 1, 2, \ldots, n_y - 1$,

$$T_{y,i,1/2,k} = T_{y,i,n_y+1/2,k} = 0, \tag{9.93}$$

$$T_{z,i,j,k+1/2} = \frac{C_1 \Delta x_i \Delta y_j k_{z,i,j,k+1/2}}{\mu(z_{k+1} - z_k)}, \tag{9.94}$$

for all $k = 1, 2, \ldots, n_z - 1$, and

$$T_{z,i,j,1/2} = T_{z,i,j,n_z+1/2} = 0. \tag{9.95}$$

> **Remark 9.2.** That transmissibilities and the volume terms are independent of time is a consequence of the simple problem we have considered. This would not be the case even for single-phase gas flow. Although pseudopressure could be used to linearize the divergence term in the gas flow equation, it is not clear that this would be advantageous. Even for single-phase flow above bubble point, we have assumed the variation in density or FVF will have a negligible effect on the results. Please note, however, that flow rate terms are in reservoir barrels per day, not STB/D. Also note we have incorporated no flow boundaries by setting transmissibilities at the boundaries to zero. We could however easily incorporate constant pressure boundaries into the finite-difference equations without encountering difficulties in the adjoint formulation for the sensitivitity coefficients. We mention this because it is unclear how to account for constant pressure boundaries in the Carter type derivation. We could even treat the specified boundary pressures as parameters and compute the sensitivity of calculated wellbore pressure data to boundary pressures.

Permeabilities at gridblock interfaces are computed as harmonic averages. Specifically, for all $1 \leq j \leq n_y$ and $1 \leq k \leq n_z$,

$$k_{x,i+1/2,j,k} = \frac{(\Delta x_i + \Delta x_{i+1})k_{x,i,j,k}k_{x,i+1,j,k}}{\Delta x_i k_{x,i+1,j,k} + \Delta x_{i+1}k_{x,i,j,k}}, \tag{9.96}$$

for $i = 1, 2, \ldots, n_x - 1$,

$$k_{x,1/2,j,k} = k_{x,1,j,k} \tag{9.97}$$

and

$$k_{x,n_x+1/2,j,k} = k_{x,n_x,j,k}. \tag{9.98}$$

Similarly,

$$k_{y,i,j+1/2,k} = \frac{(\Delta y_j + \Delta y_{j+1})k_{y,i,j,k}k_{y,i,j+1,k}}{\Delta y_j k_{y,i,j+1,k} + \Delta y_{j+1}k_{y,i,j,k}}, \tag{9.99}$$

for $j = 1, 2, \ldots, n_y - 1$,

$$k_{y,i,1/2,k} = k_{y,i,1,k}, \tag{9.100}$$

$$k_{y,i,n_y+1/2,k} = k_{y,i,n_y,k} \tag{9.101}$$

$$k_{z,i,j,k+1/2} = \frac{(\Delta z_k + \Delta z_{k+1})k_{z,i,j,k}k_{z,i,j,k+1}}{\Delta z_k k_{z,i,j,k+1} + \Delta z_{k+1}k_{z,i,j,k}}, \tag{9.102}$$

for $k = 1, 2, \ldots, n_z - 1$,

$$k_{z,i,j,1/2} = k_{z,i,j,1}, \tag{9.103}$$

and

$$k_{z,i,j,n_z+1/2} = k_{z,i,j,n_z}.$$

(9.104)

For two-dimensional problems, we simply use one gridblock in the z-direction, replace Δz_1 by the reservoir thickness h, and set all z-direction transmissibilities to zero, i.e. delete all terms that pertain to vertical flow in Eq. (9.88) and delete the z variable from the equations.

Well constraints

In the multi-phase flow simulator we use for most examples in this book, the simulator can include water and gas injection wells and multiple producing wells. At injection wells, we normally specify the water or gas injection rate or the bottomhole pressure as a function of time. The total injection rates can also be set equal to the voidage rate. Either the flowing bottomhole pressure, the total flow rate, or the oil flow rate is specified at the producing wells. The relation between a gridblock source or sink term, the gridblock pressure and flowing bottomhole pressure is specified by the equation developed by Peaceman [11]. Specific source/sink terms in a gridblock penetrated by a well must be computed from the specified well constraint. Here we focus on the single-phase flow case. Assume a well is located in a gridblock centered at (x_i, y_j, z_k). At time t^n, Peaceman's equation relates flow rate $(q_{i,j,k}^n)$ for a well to the gridblock pressure, $p_{i,j,k}^n$, and the flowing wellbore pressure, $p_{wf,i,j}^n$ by

$$q_{i,j,k}^n = WI_{i,j,k}\left(p_{i,j,k}^n - p_{wf,i,j}^n\right),$$

(9.105)

where $WI_{i,j,k}$ is the well index term. Letting

$$C_3 = (2\pi)1.127 \times 10^{-3},$$

(9.106)

$$WI_{i,j,k} = \frac{C_3 h \sqrt{k_{x,i,j,k} k_{y,i,j,k}}}{\ln(r_{o,i,j,k}/r_{w,i,j}) + s_{i,j,k}},$$

(9.107)

where

$$r_{o,i,j,k} = \frac{0.28073\Delta x_i \sqrt{1 + \frac{k_{x,i,j,k}\Delta y_j^2}{k_{y,i,j,k}\Delta x_i^2}}}{1 + \sqrt{k_{x,i,j,k}/k_{y,i,j,k}}}.$$

(9.108)

Here $r_{w,i,j}$ is the wellbore radius and $s_{i,j,k}$ is the skin factor at the well for model layer k.

Final semidiscrete difference equations, well rates specified

We now define $V_{i,j,k}$ by

$$V_{i,j,k} = \frac{\phi_{i,j,k} c_t \Delta x_i \Delta y_j \Delta z_k}{C_2}.$$

(9.109)

Using the last equation and Peaceman's equation at any time t, Eq. (9.88) can be rewritten as

$$
\begin{aligned}
T_{x,i+1/2,j,k}&(p_{i+1,j,k}(t) - p_{i,j,k}(t)) - T_{x,i-1/2,j,k}(p_{i,j,k}(t) - p_{i-1,j,k}(t)) \\
&+ T_{y,i,j+1/2,k}(p_{i,j+1,k}(t) - p_{i,j,k}(t)) - T_{y,i,j-1/2,k}(p_{i,j,k}(t) - p_{i,j-1,k}(t)) \\
&+ T_{z,i,j,k+1/2}(p_{i,j,k+1}(t) - p_{i,j,k}(t)) - T_{z,i,j,k-1/2}(p_{i,j,k}(t) - p_{i,j,k-1}(t)) \\
&- W I_{i,j,k}(p_{i,j,k}(t) - p_{wf,i,j}(t)) = V_{i,j,k}\frac{\partial p_{i,j,k}(t)}{\partial t},
\end{aligned} \tag{9.110}
$$

for $t > 0$.

For each (i, j, k) the "total" gridblock transmissibility is defined by

$$
\begin{aligned}
T_{t,i,j,k} = T_{x,i+1/2,j,k} &+ T_{x,i-1/2,j,k} + T_{y,i,j+1/2,k} + T_{y,i,j-1/2,k} \\
&+ T_{z,i,j,k+1/2} + T_{z,i,j,k-1/2}
\end{aligned} \tag{9.111}
$$

Now Eq. (9.110) can be rewritten as

$$
\begin{aligned}
T_{z,i,j,k-1/2}\,p_{i,j,k-1}(t) &+ T_{y,i,j-1/2,k}\,p_{i,j-1,k}(t) + T_{x,i-1/2,j,k}\,p_{i-1,j,k}(t) \\
- (T_{t,i,j,k} + W I_{i,j,k})\,p_{i,j,k}(t) &+ T_{x,i+1/2,j,k}\,p_{i+1,j,k}(t) + T_{y,i,j+1/2,k}\,p_{i,j+1,k}(t) \\
+ T_{z,i,j,k+1/2}\,p_{i,j,k+1}(t) &+ W I_{i,j,k}\,p_{wf,i,j}(t) = V_{i,j,k}\frac{\partial p_{i,j,k}(t)}{\partial t},
\end{aligned} \tag{9.112}
$$

for $t > 0$. We are assuming the total flow rate at each well is specified as a function of time. The flow rate at time t at a vertical well which penetrates gridblocks centered at (x_i, y_j, z_k) for certain values of k is denoted by $q_{i,j}(t)$ and is given by

$$
q_{i,j}(t) = \sum_{k=l1}^{l2} W I_{i,j,k}(p_{i,j,k}(t) - p_{wf,i,j}), \tag{9.113}
$$

where the sum is over all gridblocks penetrated by the well. The notation used is not very clear since the well could be perforated at two distinct intervals separated by a non-perforated section. At this point, we are assuming that the flow rate at each well is specified as a function of time in RB/D so that Eq. (9.113) represents another equation for the unknowns, gridblock pressures plus wellbore pressure for the particular well. Assuming flow rates are specified at N_w wells, Eq. (9.113) is applied at each well, i.e. Eq. (9.113) represents N_w equations. Thus, Eqs. (9.112) and (9.113) represent $N + N_w$ equations in $N + N_w$ unknowns. The unknowns are the N gridblock pressures and the N_w wellbore pressures. We let $p(t)$ be the N-dimensional column vector of gridblock pressures at time t ordered in any natural way and let $p_w(t)$ denote the vector of wellbore pressures ordered in any natural way. Let $A_{1,1}$ be an $N \times N$ matrix with diagonal entries given by terms $-(T_{t,i,j,k} + W I_{i,j,k})$ and the other six nonzero diagonals given by gridblock transmissibilities. Let $A_{1,2}$ be an $N \times N_w$ matrix whose entries are either zero or equal to a well index term so that Eq. (9.112) can be written as

$$A_{1,1}p(t) + A_{1,2}p_w(t) = V\frac{dp}{dt}, \tag{9.114}$$

where V is an $N \times N$ diagonal matrix with diagonal entries given by the $V_{i,j,k}$ terms. Let $Q(t)$ be an N_w-dimensional vector containing the total flow rate terms, let $A_{2,1} = A_{1,2}^T$ (an $N_w \times N$ matrix) and let $A_{2,2}$ be an $N_w \times N_w$ diagonal matrix with a particular diagonal entry corresponding to the sum of well index terms corresponding to a particular well so that Eq. (9.113) can be rewritten as

$$A_{2,1}p(t) + A_{2,2}p_w(t) = Q(t). \tag{9.115}$$

Eqs. (9.114) and (9.115) apply for all $t > 0$. The initial condition is

$$p_{i,j,k}(0) = p_0(x_i, y_j, z_k). \tag{9.116}$$

In the simulator, we of course approximate the time derivative of $p(t)$ to get the final system of finite-difference equations.

Flowing wellbore pressure specified

If the wellbore pressure is specified at a well, then it is no longer an unknown. In such a case, we would want to find the derivative of the rate with respect to the model parameters. In history matching, we require a wellbore constraint to run the simulator and we can always choose either the rate or the pressure. (We could also switch between the two types of well constraints at particular values of time.) In any case, if all wellbore pressures are specified then Eq. (9.114) has only $p(t)$ as the unknown and is solved directly for $p(t)$. Eq. (9.113) can then be applied to obtain the flow rate at the well and model layer flow rates can be calculated from Eq. (9.105). In terms of computing sensitivities by the adjoint method, it is best to work in terms of the general formulation of Eqs. (9.114) and (9.115). If a rate is specified, then the corresponding entry of $Q(t)$ is held fixed so it is independent, and hence insensitive, to model parameters. If a wellbore pressure is specified, then the corresponding entry of the vector $p_{wf}(t)$ is held fixed and is insensitive to model parameters. With such a formulation, we can switch back and forth between applying a rate constraint and a pressure constraint.

9.5   Adjoint equations

Assume we wish to compute the total derivative of some real-valued function $\hat{g}$ with respect to a vector m of model parameters. The j component of the total derivative vector is the total derivative of $\hat{g}$ with respect to the jth model parameter, i.e. the standard sensitivity coefficient. Using the adjoint approach, we need one adjoint variable λ for each equation. In the case where flow rates are specified at each well, we have two types of equations, at any time t. Thus, we let $\lambda_f = \lambda_f(t)$ be an N-dimensional column vector and $\lambda_s = \lambda_s(t)$ be an N_w-dimensional vector and adjoin the integrals of

Eqs. (9.114) and (9.115) to $\hat{g}$ to obtain the functional J given by

$$J = \hat{g} + \int_0^{t_f} \left[\lambda_f^T(t) \left(A_{1,1} p(t) + A_{1,2} p_w(t) - V \frac{dp}{dt} \right) \right.$$
$$\left. + \lambda_s^T(t)(A_{2,1} p(t) + A_{2,2} p_w(t) - Q(t)) \right] dt, \tag{9.117}$$

where $[0, t_f]$ is the time interval of interest. Since we have formed J by adjoining to $\hat{g}$ equations which are identically zero, the sensitivity coefficients of $\hat{g}$ are simply the total derivatives of J with respect to model parameters. Integration by parts gives

$$\int_0^{t_f} \lambda_f^T V \frac{dp}{dt} \, dt = \lambda_f^T(t_f) V p(t_f) - \lambda_f^T(0) V p(0) - \int_0^{t_f} \frac{d\lambda_f(t)}{dt}^T V p(t) \, dt. \tag{9.118}$$

Note that we have assumed that V is independent of time to use integration by parts in this way. Applying Eq. (9.118), Eq. (9.117) becomes

$$J = \hat{g} + \int_0^{t_f} \left[\lambda_f^T(t) \left(A_{1,1} p(t) + A_{1,2} p_w(t) \right) + \left(\frac{d\lambda_f(t)}{dt}^T V p(t) \right) \right.$$
$$\left. + \lambda_s^T(t) \left(A_{2,1} p(t) + A_{2,2} p_w(t) - Q(t) \right) \right] dt + \lambda_f^T(0) V p(0) - \lambda_f^T(t_f) V p(t_f). \tag{9.119}$$

As now expressed, the functional J can be considered as a function of $p(t)$, $p_w(t)$, k_x, k_y, k_z, ϕ. Here k_x represents the vector of N gridblock x-direction permeabilities and k_y, k_z, and ϕ should be interpreted in the same way. In taking the total differential of Eq. (9.119), we will use the fact that the A matrices are independent of ϕ and the V matrix is independent of k_x, k_y, and k_z. Moreover, $A_{1,2}$, $A_{2,1}$, and $A_{2,2}$ are independent of k_z and since Q contains specified flow rates $dQ = 0$. Finally, because $p(0)$ is specifed, $dp(0) = 0$. Thus, taking the total differential of J gives

$$dJ = d\hat{g} + \int_0^{t_f} \left[\lambda_f(t)^T \left(A_{1,1} \, dp(t) + A_{1,2} \, dp_w(t) \right) + \left(\frac{d\lambda_f(t)}{dt}^T V \, dp(t) \right) \right.$$
$$\left. + \lambda_s(t)^T \left(A_{2,1} \, dp(t) + A_{2,2} \, dp_w(t) \right) \right] dt$$
$$+ \int_0^{t_f} \left[\lambda_f(t)^T \left(\nabla_{k_x} [A_{1,1} p(t) + A_{1,2} p_w(t)]^T \right)^T dk_x \right.$$
$$\left. + \lambda_s(t)^T \left(\nabla_{k_x} [A_{2,1} p(t) + A_{2,2} p_w(t)]^T \right)^T dk_x \right] dt$$
$$+ \int_0^{t_f} \left[\lambda_f(t)^T \left(\nabla_{k_y} [A_{1,1} p(t) + A_{1,2} p_w(t)]^T \right)^T dk_y \right.$$
$$\left. + \lambda_s(t)^T \left(\nabla_{k_y} [A_{2,1} p(t) + A_{2,2} p_w(t)]^T \right)^T dk_y \right] dt$$
$$+ \int_0^{t_f} \left[\lambda_f^T(t) \left(\nabla_{k_z} [A_{1,1} p(t)]^T \right)^T dk_z \right] dt$$
$$+ \int_0^{t_f} \left[\frac{d\lambda_f(t)}{dt}^T \left(\nabla_\phi [V p(t)]^T \right)^T d\phi \right] dt - \lambda_f^T(t_f) V \, dp(t_f). \tag{9.120}$$

We now assume there is some function g such that

$$\hat{g} = \int_0^{t_f} g(t)\, dt.$$
(9.121)

For simplicity, we also assume that the functional form of g is such that g depends explicitly only on some components of $p_w(t)$ and/or $p(t)$. Then,

$$d\hat{g} = \int_0^{t_f} \left(\nabla_{p_w(t)} g(t)\right)^{\mathrm{T}} dp_w(t)\, dt + \int_0^{t_f} \left(\nabla_{p(t)} g(t)\right)^{\mathrm{T}} dp(t)\, dt.$$
(9.122)

Using Eq. (9.122) in Eq. (9.120) and rearranging the resulting equations yields

$$
\begin{aligned}
dJ = & \int_0^{t_f} \left[\lambda_f(t)^{\mathrm{T}} A_{1,1} + \lambda_s(t)^{\mathrm{T}} A_{2,1} + \frac{d\lambda_f(t)}{dt}^{\mathrm{T}} V + \left(\nabla_{p(t)} g(t)\right)^{\mathrm{T}} \right] dp(t)\, dt \\
& + \int_0^{t_f} \left[\lambda_f(t)^{\mathrm{T}} A_{1,2} + \lambda_s(t)^{\mathrm{T}} A_{2,2} + \left(\nabla_{p_w(t)} g(t)\right)^{\mathrm{T}} \right] dp_w(t)\, dt \\
& + \int_0^{t_f} \left[\lambda_f(t)^{\mathrm{T}} \left(\nabla_{k_x}[A_{1,1} p(t) + A_{1,2} p_w(t)]^{\mathrm{T}}\right)^{\mathrm{T}} dk_x \right. \\
& \qquad\quad \left. + \lambda_s(t)^{\mathrm{T}} \left(\nabla_{k_x}[A_{2,1} p(t) + A_{2,2} p_w(t)]^{\mathrm{T}}\right)^{\mathrm{T}} dk_x \right] dt \\
& + \int_0^{t_f} \left[\lambda_f(t)^{\mathrm{T}} \left(\nabla_{k_y}[A_{1,1} p(t) + A_{1,2} p_w(t)]^{\mathrm{T}}\right)^{\mathrm{T}} dk_y \right. \\
& \qquad\quad \left. + \lambda_s(t)^{\mathrm{T}} \left(\nabla_{k_y}[A_{2,1} p(t) + A_{2,2} p_w(t)]^{\mathrm{T}}\right)^{\mathrm{T}} dk_y \right] dt \\
& + \int_0^{t_f} \left[\lambda_f(t)^{\mathrm{T}} \left(\nabla_{k_z}[A_{1,1} p(t)]^{\mathrm{T}}\right)^{\mathrm{T}} dk_z \right] dt \\
& + \int_0^{t_f} \left[\frac{d\lambda_f(t)}{dt}^{\mathrm{T}} \left(\nabla_\phi[V p(t)]^{\mathrm{T}}\right)^{\mathrm{T}} \right] d\phi\, dt - \lambda_f^{\mathrm{T}}(t_f) V\, dp(t_f).
\end{aligned}
$$
(9.123)

Specifying that the coefficients of $dp(t)$ and $dp_w(t)$ must be zero for all $t > 0$ gives the adjoint system:

$$\lambda_f(t)^{\mathrm{T}} A_{1,1} + \lambda_s(t)^{\mathrm{T}} A_{2,1} + \frac{d\lambda_f(t)}{dt}^{\mathrm{T}} V + \left(\nabla_{p(t)} g(t)\right)^{\mathrm{T}} = 0,$$
(9.124)

$$\lambda_f(t)^{\mathrm{T}} A_{1,2} + \lambda_s(t)^{\mathrm{T}} A_{2,2} + \left(\nabla_{p_w(t)} g(t)\right)^{\mathrm{T}} = 0,$$
(9.125)

and

$$\lambda_f(t_f) = 0.$$
(9.126)

Note when the last three equations are satisfied, Eq. (9.123) reduces to

$$
dJ = \int_0^{t_f} \left[\lambda_f^T(t) \left(\nabla_{k_x} [A_{1,1} p(t) + A_{1,2} p_w(t)]^T \right)^T dk_x \right.
$$
$$
\left. + \lambda_s^T(t) \left(\nabla_{k_x} [A_{2,1} p(t) + A_{2,2} p_w(t)]^T \right)^T dk_x \right] dt
$$
$$
+ \int_0^{t_f} \left[\lambda_f^T(t) \left(\nabla_{k_y} [A_{1,1} p(t) + A_{1,2} p_w(t)]^T \right)^T dk_y \right.
$$
$$
\left. + \lambda_s^T(t) \left(\nabla_{k_y} [A_{2,1} p(t) + A_{2,2} p_w(t)]^T \right)^T dk_y \right] dt
$$
$$
+ \int_0^{t_f} \left[\lambda_f^T(t) \left(\nabla_{k_z} [A_{1,1} p(t)]^T \right)^T dk_z \right] dt + \int_0^{t_f} \left[\frac{d\lambda_f(t)}{dt}^T \left(\nabla_\phi [V p(t)]^T \right)^T \right] d\phi \, dt.
$$

$$(9.127)$$

The preceding represents an expression for the total differential of J considered as a function of the vectors k_x, k_y, k_z, and ϕ, which can be written as

$$
dJ = \left(\nabla_{k_x} J \right)^T dk_x + \left(\nabla_{k_y} J \right)^T dk_y + \left(\nabla_{k_z} J \right)^T dk_z + \left(\nabla_\phi J \right)^T d\phi.
$$

$$(9.128)$$

Thus, it follows that

$$
\left(\nabla_{k_x} J \right)^T = \int_0^{t_f} \left[\lambda_f^T(t) \left(\nabla_{k_x} [A_{1,1} p(t) + A_{1,2} p_w(t)]^T \right)^T \right.
$$
$$
\left. + \lambda_s^T(t) \left(\nabla_{k_x} [A_{2,1} p(t) + A_{2,2} p_w(t)]^T \right)^T \right] dt,
$$

$$(9.129)$$

$$
\left(\nabla_{k_y} J \right)^T = \int_0^{t_f} \left[\lambda_f^T(t) \left(\nabla_{k_y} [A_{1,1} p(t) + A_{1,2} p_w(t)]^T \right)^T \right.
$$
$$
\left. + \lambda_s^T(t) \left(\nabla_{k_y} [A_{2,1} p(t) + A_{2,2} p_w(t)]^T \right)^T \right] dt,
$$

$$(9.130)$$

$$
\left(\nabla_{k_z} J \right)^T = \int_0^{t_f} \left[\lambda_f^T(t) \left(\nabla_{k_z} [A_{1,1} p(t)]^T \right)^T + \lambda_s^T(t) \left(\nabla_{k_z} [A_{2,2} p_w(t)]^T \right)^T \right] dt,
$$

$$(9.131)$$

and

$$
\left(\nabla_\phi J \right)^T = \int_0^{t_f} \left[\frac{d\lambda_f(t)}{dt}^T \left(\nabla_\phi [V p(t)]^T \right)^T \right].
$$

$$(9.132)$$

Eqs. (9.129)–(9.132) give the desired sensitivity coefficients, but there is a bit of a notational problem here. Although it is common to write the equations for the sensitivities as in the preceding four equations, we would often like to reserve notation like $\nabla_{k_x} J$ to denote the vector of partial derivatives of J with respect to the components of k_x, i.e. $\partial J / \partial k_{x,i}$, $i = 1, 2, \ldots, N$. In this case, $\partial J / \partial k_{x,i} = 0$ if the expression for J does not explicitly involve $k_{x,i}$. However, in Eq. (9.129), $\nabla_{k_x} J$

really represents the total derivative of J with respect to k_x which we denote by

$$\frac{dJ}{dk_x} = \begin{bmatrix} dJ/dk_{x,1} \\ dJ/dk_{x,2} \\ \vdots \\ dJ/dk_{x,N} \end{bmatrix}, \tag{9.133}$$

so that the right-hand side of Eq. (9.129) is the transpose of dJ/dk_x. Similar comments apply to the total derivatives of J with respect to the other model parameters.

9.5.1 Solution of the adjoint equations

Taking the transpose of Eqs. (9.124) and (9.125) using the facts that $A_{1,1}$ and $A_{2,2}$ are symmetric and $A_{2,1}^T = A_{1,2}$, and transferring the "sink" terms to the right-hand sides gives

$$A_{1,1}\lambda_f(t) + A_{1,2}\lambda_s(t) + V\frac{d\lambda_f(t)}{dt} = -\big(\nabla_{p(t)}g(t)\big), \tag{9.134}$$

and

$$A_{2,1}\lambda_f(t) + A_{2,2}\lambda_s(t) = -\big(\nabla_{p_w(t)}g(t)\big). \tag{9.135}$$

The auxiliary condition for Eqs. (9.134) and (9.135) is given by Eq. (9.126) which is recorded here as

$$\lambda_f(t_f) = 0. \tag{9.136}$$

Note we could solve algebraically Eq. (9.135) for λ_s and substitute the formula into Eq. (9.134) to eliminate λ_s from Eq. (9.134). After solving for the resulting differential equation for λ_f, we could solve Eq. (9.135) for λ_s. If it is necessary to solve Eq. (9.134) numerically at discrete times using a finite-difference method, the plus sign on the time derivative and the auxiliary condition of Eq. (9.136) indicate that we should solve the system backwards in time from $t = t_f$.

It turns out that for this linear problem, we can rewrite the adjoint system in exactly the same form as the semidiscrete finite-difference equations using the change of variable $\tau = t_f - t$. In this case, for many problems, the sensitivities can be computed by simply running the reservoir simulator forward in time. This is the procedure used by Carter et al. [96] and approximately extended to three-dimensional linear problems by He et al. [100]. Unfortunately, nonlinear problems do not share this useful feature.

9.5.2   Sensitivity to subobjective functions

When using the LBFGS algorithm in history matching, we will only need the sensitivity of functions like

$$
\hat{g} = \frac{1}{2} \sum_{l=1}^{N_j} \frac{\left(p_{wf,j}(t_l) - p_{wf,obs,j}(t_l)\right)^2}{\sigma_j^2}
$$

$$
= \frac{1}{2} \int_0^{t_f} \sum_{l=1}^{N_j} \delta(t - t_l) \frac{\left(p_{wf,j}(t) - p_{wf,obs,j}(t)\right)^2}{\sigma_j^2} \, dt, \tag{9.137}
$$

with respect to the model parameters. Here, $p_{wf,j}(t_l)$ denotes the pressure at well j at time t_l obtained from the simulator for the current estimate of model parameters and $p_{wf,obs,j}(t_l)$ denotes the corresponding measured wellbore pressure that we wish to history match by adjusting the model parameters. Based on the characteristics of the pressure gauge, we assume the variation of the measurement error is σ_j^2. In discussing LBFGS, it is customary to refer to computing the gradient of an objective function with respect to model parameters. One should always bear in mind, however, that when computing this gradient, we must consider the fact that primary variables such as pressure must also be considered as functions of the model parameters, i.e. the components of this gradient represent total derivatives.

Here, following previous notation of Eq. (9.121),

$$
g = \frac{1}{2} \sum_{l=1}^{N_j} \delta(t - t_l) \frac{\left(p_{wf,j}(t) - p_{wf,obs,j}(t)\right)^2}{\sigma_j^2}. \tag{9.138}
$$

To solve the adjoint equations for the total derivative of $\hat{g}$ with respect to model parameters, i.e. the sensitivity of $\hat{g}$ to model parameters, we need to compute the source terms on the right-hand sides of Eqs. (9.134) and (9.135). With the preceding function g, the sink term on the right-hand side of Eq. (9.134) is identically zero. The sink term on the right-hand side of Eq. (9.135) is zero except when $t = t_l$; see Eq. (9.138). If $t = t_l$,

$$
\nabla_{p_w(t_l)} g(t) = e_w \delta(t - t_l) \frac{\left(p_{wf,j}(t) - p_{wf,obs,j}(t)\right)}{\sigma_j^2}, \tag{9.139}
$$

where e_w is an N_w-dimensional vector with the entry corresponding to the location of p_{wf} in the vector p_w equal to one and all other entries zero.

9.6   Sensitivity calculation example

We consider a simple linear problem of the form

$$
\frac{du}{dt} + Au = q, \tag{9.140}
$$

$$
u(0) = u_0, \tag{9.141}
$$

where u_0 is a constant vector. Although writing down concise formulas will only be convenient when u and a are scalars, we start with the case where u is an N-dimensional column vector and A is a symmetric $N \times N$ constant matrix. We assume that the N-dimensional vector q is independent of time which allows for easy evaluation of some integrals. The single-phase flow problem considered previously is of this form in the two-dimensional case if the flow rate at each well is specified. (In the two-dimensional case, it is unnecessary to redistribute a total well rate among model layers.) Multiplying Eq. (9.140) by $\exp(At)$ gives

$$\exp(At)\left(\frac{du}{dt} + Au\right) = \frac{d}{dt}(\exp(At)u(t)) = \exp(At)q. \tag{9.142}$$

Integrating Eq. (9.142) from 0 to t and using the initial condition of Eq. (9.141) gives

$$\exp(At)u(t) - u_0 = \left(\int_0^t \exp(At)\,dt\right)q = A^{-1}(\exp(At) - I)q = (\exp(At) - I)A^{-1}q, \tag{9.143}$$

where here, I is the $N \times N$ identity matrix. From the last equation it follows that

$$u(t) = \exp(-At)u_0 + (I - \exp(-At))A^{-1}q. \tag{9.144}$$

9.6.1 Sensitivities from analytical solution

Although, the derivative of u at some time $\hat{t}$ with respect to a model parameter α (e.g. an entry of A) can be expressed formally, for simplicity, we now assume that $A = a$ is a scalar so that Eq. (9.144) reduces to

$$u(t) = \exp(-at)u_0 + \frac{q}{a}(1 - \exp(-at)). \tag{9.145}$$

Then,

$$\frac{\partial u(\hat{t})}{\partial a} = -\hat{t}\exp(-a\hat{t})u_0 + \hat{t}\exp(-a\hat{t})\frac{q}{a} - \frac{q}{a^2}(1 - \exp(-a\hat{t})), \tag{9.146}$$

gives the sensitivity of $u(\hat{t})$ to a. In the following, we will compute this sensitivity using the direct method and the adjoint method.

9.6.2 Sensitivities from the direct method

Taking the derivative of Eqs. (9.140) and (9.141) with respect to a model parameter α under the assumption that q does not change with α gives

$$\frac{d(\partial u/\partial \alpha)}{dt} + A\frac{\partial u}{\partial \alpha} + \frac{\partial A}{\partial \alpha}u = 0 \tag{9.147}$$

and

$$\frac{\partial u(0)}{\partial \alpha} = \frac{\partial u_0}{\partial \alpha} = 0, \tag{9.148}$$

where the last equality follows because u_0 is a constant vector. Letting

$$y(t) = \frac{\partial u(t)}{\partial \alpha} \tag{9.149}$$

for all $t > 0$, Eqs. (9.147) and (9.148), respectively, can be rewritten as

$$\frac{dy(t)}{dt} + Ay(t) = -\frac{\partial A}{\partial \alpha}u(t) \tag{9.150}$$

and

$$y(0) = \frac{\partial u_0}{\partial \alpha} = 0. \tag{9.151}$$

Letting

$$Q(t) = -\frac{\partial A}{\partial \alpha}u(t), \tag{9.152}$$

the analytical solution of Eqs. (9.150) and (9.151) is

$$y(t) = \exp(-At)\int_0^t \exp(At)Q(t)\,dt. \tag{9.153}$$

In the one-dimensional case, $A = a$, so $Q(t) = 0$ for $\alpha \neq a$ and for $\alpha = a$, $Q(t) = -u(t)$ so Eq. (9.153) gives

$$y(t) = -\exp(-at)\int_0^t \exp(at)u(t)\,dt. \tag{9.154}$$

Using the definition of $y(t)$ (Eq. 9.149) and Eq. (9.145), Eq. (9.154) becomes

$$\frac{\partial u(t)}{\partial a} = -\exp(-at)\int_0^t \exp(at)\left[\exp(-at)u_0 + \frac{q}{a}(1 - \exp(-at))\right]dt. \tag{9.155}$$

Performing the integration in the preceding equation gives

$$\begin{aligned}
\frac{\partial u(t)}{\partial a} &= -\exp(-at)\left[tu_0 + \frac{q}{a^2}(\exp(at) - 1) - t\frac{q}{a}\right] \\
&= -u_0 t \exp(-at) - \frac{q}{a^2}(1 - \exp(-at)) + t\exp(-at)\frac{q}{a}.
\end{aligned} \tag{9.156}$$

Note for $t = \hat{t}$, Eq. (9.156) is identical to Eq. (9.146).

9.6.3   Sensitivities from the adjoint method

Letting $\lambda(t)$ denote the vector of adjoint variables at time t, the analog of Eq. (9.134) for the case where the forward problem is given by Eqs. (9.140) and (9.141) is

$$-A\lambda(t) + \frac{d\lambda(t)}{dt} = -\left(\nabla_{u(t)}g(t)\right). \tag{9.157}$$

The preceding equation can be obtained by recognizing that except for the added source term q, Eq. (9.140) can be obtained from Eq. (9.114) by setting $A_{1,1} = -A$, $A_{1,2} = 0$, $V = I$, and $p = u$. If the matrix A is a function of the parameter vector k, then the analog of Eq. (9.129) is

$$\left(\frac{dJ}{dk}\right)^{\mathrm{T}} = (\nabla_k J)^{\mathrm{T}} = -\int_0^{t_f} \left[\lambda^{\mathrm{T}}(t)(\nabla_k [Au(t)]^{\mathrm{T}})^{\mathrm{T}} dt\right]. \tag{9.158}$$

In the one-dimensional case with $A = a$, $k = a$, and $g(t) = u(t)\delta(t - \hat{t})$, Eq. (9.157) becomes

$$\frac{d\lambda(t)}{dt} - a\lambda(t) = -\delta(t - \hat{t}). \tag{9.159}$$

Because the auxiliary equation for the adjoint problem is

$$\lambda(t_f) = 0, \tag{9.160}$$

integrating Eq. (9.159) from t to t_f gives

$$\lambda(t) = 0, \tag{9.161}$$

for $t > \hat{t}$. For any $t < \hat{t}$, we multiply Eq. (9.159) by $\exp(-at)$ and integrate to obtain

$$-\exp(-at)\lambda(t) = -\exp(-a\hat{t}), \tag{9.162}$$

i.e.

$$\lambda(t) = \exp(-a[\hat{t} - t]). \tag{9.163}$$

Using Eqs. (9.161) and (9.163) in Eq. (9.158) gives the sensitivity of $u(t)$ to a as

$$\frac{dJ}{da} = -\int_0^{t_f} \lambda(t)(\nabla_a[au(t)])dt = -\int_0^{\hat{t}} \exp(-a[\hat{t} - t])u(t)dt$$

$$= -\exp(-a\hat{t})\int_0^{\hat{t}} \exp(at)u(t)\,dt. \tag{9.164}$$

From Eq. (9.145), it follows that

$$\exp(at)u(t) = \exp(at)\left[\exp(-at)u_0 + \frac{q}{a}(1 - \exp(-at))\right] = u_0 + \frac{q}{a}(\exp(at) - 1). \tag{9.165}$$

Thus,

$$\int_0^{\hat{t}} \exp(at)u(t)dt = \left[u_0\hat{t} + \frac{q}{a^2}(\exp(a\hat{t}) - 1) - \frac{q}{a}\hat{t}\right]. \tag{9.166}$$

Using Eq. (9.166) in Eq. (9.164) gives

$$\frac{dJ}{da} = -\exp(-a\hat{t})\left[u_0\hat{t} + \frac{q}{a^2}(\exp(a\hat{t}) - 1) - \frac{q}{a}\hat{t}\right], \tag{9.167}$$

which is the same result obtained analytically (Eq. 9.146) and by the direct method (Eq. 9.156).

9.7   Adjoint method for multi-phase flow

The derivation of the adjoint method used to generate the sensitivity of dynamic data to reservoir model parameters is presented. Although the derivation is general, the adjoint procedure is easiest to implement when the discrete equations solved by the reservoir simulator are based on a fully implicit finite-difference formulation. For computation of the sensitivity of an objective function, or predicted data, to model parameters, the adjoint method relies on using adjoint variables to adjoin the finite-difference equations satisfied by the reservoir simulation to the objective function or predicted data. Thus, accurate formulation and solution of the adjoint requires specific knowledge of the discrete equations that are solved in the reservoir simulator.

9.7.1   The reservoir simulator

We assume that the reservoir occupies a three-dimensional region Ω and define a grid and gridblocks such that the gridblocks encompass Ω. We use an $x–y–z$ coordinate system. For simplicity, we suppose there are n_x, n_y, n_z gridblocks in the x, y, and z directions respectively. If a gridblock is outside of Ω, the cell is inactive and we set permeability equal to zero and porosity equal to a very small number in this gridblock. For concreteness, we assume the simulator used is based on a finite-difference formulation of the three-phase flow, black-oil equations expressed in an $x–y–z$ coordinate system, but there is nothing that precludes incorporating the adjoint method in a compositional simulator or one based on a finite volume method on an unstructured grid. We let $N = n_x n_y n_z$ denote the total number of gridblocks. At each gridblock, we have three finite-difference equations which represent the mass balances for each of the three components, oil, gas, and water. In addition, at each well at each time step, we assume that either an individual phase flow rate, the total flow rate or wellbore pressure is specified as a target condition with constraints specified as secondary conditions. For example, if a specific oil rate is specified as the target condition at a well, a secondary constraint is usually the minimum bottomhole pressure. If the well can not produce at the target rate with a wellbore pressure at above the minimum specified, the producing condition is specified as equal to the minimum bottomhole pressure. A well can be shut-in if the well or field producing WOR or producing GOR exceeds a specified limit. In general, any producing condition at a well can be incorporated; one simply needs to keep track of the well conditions used in the forward reservoir simulation run when formulating the adjoint problem.

Table 9.1. *Equations and unknowns solved for in the simulator.*

Phases	Equations	Unknowns	Auxiliary equations
$S_g > 0$	Sum, Oil, Gas	p, S_o, S_g	$S_w = 1 - S_o - S_g$; R_s from PVT table
$S_g = 0$	Sum, Oil, Gas	p, S_o, R_s	$S_g = 0$; $S_w = 1 - S_o$

In summary, at each time step, there are $3N$ mass balance equations and N_w equations representing well constraints. In the black-oil simulator that we use, the three mass balance equations are summed to obtain an overall mass balance equation. Thus, the three finite-difference equations that are solved in the simulator at each gridblock are this overall equation plus two finite-difference equations for two components. If capillary pressure is included, there are auxiliary equations for capillary pressure. Also the sum of the three-phase saturations is required to be equal to unity.

For gridblock i, the primary variables that are solved for in the reservoir simulator depend on whether we are below or above bubble point. Table 9.1 summarizes the two cases. Under the column "Equations," Sum denotes the total mass balance equation, Oil represents the finite-difference equation for the oil component mass balance equation, and Gas represents the finite-difference equation for the gas component mass balance equation. Here p refers to the oil phase pressure and R_s is the dissolved gas–oil ratio.

At each time step, we can output p, S_o, S_g, S_w, and R_s from the simulator and from these values we can use the PVT tables to construct derivatives required for forming the adjoint system. At each gridblock the three primary variables are the ones solved for by the simulator. In addition, the flowing wellbore pressure, $p_{wf,l}$ at the lth well at a specified datum depth is also a primary variable. We let y^n denote a column vector which contains the set of primary variables. The finite-difference equation for component u at gridblock i is then represented by

$$f_{u,i}(y^{n+1}, y^n, m) = 0, \qquad (9.168)$$

for $u = o, w, g$ and $i = 1, \ldots, N$. The well constraint at well l is represented by

$$f_{wf,l}(y^{n+1}, y^n, m) = 0, \qquad (9.169)$$

for $l = 1, 2, \ldots, N_w$. Here n refers to the time step so the preceding equations represent the equations that must be solved to obtain the primary variables, y^{n+1}, at the $n + 1$st time given the solution at the nth time step. The primary variables can change from time step to time step as noted in Table 9.1. In particular,

$$y^{n+1} = [p_1^{n+1}, S_{o,1}^{n+1}, x_1^{n+1}, p_2^{n+1}, \ldots, p_i^{n+1}, S_{o,i}^{n+1}, x_i^{n+1}, \ldots, x_N^{n+1}, p_{wf,1}^{n+1}, \ldots, p_{wf,N_w}^{n+1}]^{\mathsf{T}},$$
$$(9.170)$$

where

$$x_i^{n+1} = \begin{cases} S_{g,i}^{n+1} & \text{for} \quad S_{g,i}^{n+1} > 0 \\ R_{s,i}^{n+1} & \text{for} \quad S_{g,i}^{n+1} = 0. \end{cases} \tag{9.171}$$

For simplicity in notation, let

$$f_{u,i}^{n+1} = f_{u,i}(y^{n+1}, y^n, m), \tag{9.172}$$

$$f_{s,i}^{n+1} = f_{o,i}^{n+1} + f_{w,i}^{n+1} + f_{g,i}^{n+1}, \tag{9.173}$$

and

$$f_{wf,l}^{n+1} = f_{wf,l}(y^{n+1}, y^n, m). \tag{9.174}$$

The finite-difference equations that are solved for in the simulator are then represented by

$$f_{s,i}^{n+1} = 0 \tag{9.175}$$

$$f_{o,i}^{n+1} = 0 \tag{9.176}$$

$$f_{g,i}^{n+1} = 0 \tag{9.177}$$

for $i = 1, 2, \ldots, N$ and

$$f_{wf,l}^{n+1} = 0, \tag{9.178}$$

for $l = 1, 2, \ldots, N_w$. If the value of the flowing wellbore pressure at well l at the datum depth at time t^{n+1} is specified as equal to $p_{wf,l,0}^{n+1}$, then Eq. (9.178) becomes simply

$$f_{wf,l}^{n+1} = p_{wf,l}^{n+1} - p_{wf,l,0}^{n+1} = 0. \tag{9.179}$$

Eqs. (9.175)–(9.178) comprise a system of

$$N_e = 3N + N_w, \tag{9.180}$$

equations in N_e unknowns. This system is solved to obtain the values of the primary variables at time $t^{n+1} = t^n + \Delta t^n$. For wells at which the flowing bottomhole pressure is specified, phase flow rates at each well are computed by Peaceman's equation [11] in the particular simulator we use, but there is nothing in the adjoint formulation that precludes an alternate treatment of a well's producing equation. For example, for a vertical well, a radial grid could be used around the well in which case, the flow rate can be related to the flowing wellbore pressure and pressures in the adjacent blocks by direct differencing of the correct boundary condition.

The complete system of equations can be written as

$$
f^{n+1} = f(y^{n+1}, y^n, m) =
\begin{bmatrix}
f_{s,1}^{n+1} \\
f_{o,1}^{n+1} \\
f_{g,1}^{n+1} \\
f_{s,2}^{n+1} \\
f_{o,2}^{n+1} \\
f_{g,2}^{n+1} \\
\vdots \\
f_{g,N}^{n+1} \\
f_{wf,1}^{n+1} \\
\vdots \\
f_{wf,N_w}^{n+1}
\end{bmatrix}
= 0,
\tag{9.181}
$$

where the vector of model parameters is given by

$$
m = [m_1, m_2, \ldots, m_{N_m}]^{\mathrm{T}},
\tag{9.182}
$$

and y^{n+1} is given by Eq. (9.170). The model parameters can include the set of grid-block permeabilities and porosities, well skin factors, parameters that describe relative permeability curves [73, 101, 102], a multiple of fault transmissibility, or the location of boundaries between facies [103, 104].

As noted in the next subsection, if Eq. (9.181) is solved by the Newton–Raphson method [105], much of the information needed for the adjoint equations will be computed in the forward simulation run. The Newton–Raphson method can be written as

$$
J^{n+1,k} \delta y^{n+1,k+1} = -f^{n+1,k},
\tag{9.183}
$$

$$
y^{n+1,k+1} = y^{n+1,k} + \delta y^{n+1,k+1},
\tag{9.184}
$$

where k is the Newton–Raphson iteration index, n is the time step index and

$$
J^{n+1,k} = \left[\nabla_{y^{n+1}} (f^{n+1})^T \right]^T_{y^{n+1,k}},
\tag{9.185}
$$

is the Jacobian matrix evaluated at $y^{n+1,k}$, which represents the kth approximation for y^{n+1}. The initial guess for y^{n+1} is chosen as the solution at the previous time step, i.e.

$$
y^{n+1,0} = y^n.
\tag{9.186}
$$

9.7.2   Adjoint equations and sensitivities

We define a scalar function by

$$\beta = \beta(y^1, \ldots, y^L, m), \tag{9.187}$$

where L corresponds to the last time step t^L at which one wishes to compute sensitivities. In the LBGS algorithm, β will represent the data mismatch part of the objective function. When using the Gauss–Newton method, β will represent one of the predicted dynamic data which can include production data and/or seismic data. In any case, we wish to compute the sensitivity of β to the model m, i.e. $d\beta/dm$.

The development here follows that of Li [69] and Li $et\ al.$ [63]. We obtain an adjoint functional J by adjoining Eq. (9.181) to the function β to obtain

$$J = \beta + \sum_{n=0}^{L} (\lambda^{n+1})^T f^{n+1}, \tag{9.188}$$

where

$$\lambda^{n+1} = \left[\lambda_1^{n+1}, \lambda_2^{n+1}, \ldots, \lambda_{N_e}^{n+1} \right]^T, \tag{9.189}$$

is the vector of adjoint variables.

Taking the total differential of Eq. (9.188), changing the index of summation in terms involving dy^{n+1} and dm and rearranging slightly the resulting equation yields

$$dJ = d\beta + \sum_{n=0}^{L} \left\{ (\lambda^{n+1})^T [\nabla_{y^{n+1}}(f^{n+1})^T]^T dy^{n+1} + [\nabla_m (f^{n+1})^T]^T dm \right\}$$

$$+ \sum_{n=0}^{L} (\lambda^{n+1})^T [\nabla_{y^n}(f^{n+1})^T]^T dy^n \tag{9.190}$$

$$= d\beta + BT + \sum_{n=1}^{L} \left\{ [(\lambda^n)^T [\nabla_{y^n}(f^n)^T]^T \right.$$

$$\left. + (\lambda^{n+1})^T [\nabla_{y^n}(f^{n+1})^T]^T] dy^n + (\lambda^n)^T [\nabla_m (f^n)^T]^T dm \right\},$$

where the boundary terms are

$$BT = (\lambda^{L+1})^T \left\{ [\nabla_{y^{L+1}}(f^{L+1})^T]^T dy^{L+1} + [\nabla_m (f^{L+1})^T]^T dm \right\} + (\lambda^1)^T [\nabla_{y^0}(f^1)^T]^T dy^0. \tag{9.191}$$

The total differential of β can be written as

$$d\beta = \sum_{n=1}^{L} [\nabla_{y^n} \beta]^T dy^n + [\nabla_m \beta]^T dm. \tag{9.192}$$

The initial conditions are fixed, so

$$dy^0 = 0. \tag{9.193}$$

As in subsection 9.5, it is appropriate to set

$$\lambda^{L+1} = 0, \tag{9.194}$$

which implies that $BT = 0$. Using $BT = 0$, Eqs. (9.192) and (9.193) in Eq. (9.190) and doing some simple rearranging gives

$$
dJ = \sum_{n=1}^{L} \left[\left\{ (\lambda^n)^{\mathrm{T}} [\nabla_{y^n} (f^n)^{\mathrm{T}}]^{\mathrm{T}} + (\lambda^{n+1})^{\mathrm{T}} [\nabla_{y^n} (f^{n+1})^{\mathrm{T}}]^{\mathrm{T}} + [\nabla_{y^n} \beta]^{\mathrm{T}} \right\} dy^n \right]
$$
$$
+ \left\{ [\nabla_m \beta]^{\mathrm{T}} + \sum_{n=1}^{N} (\lambda^n)^{\mathrm{T}} [\nabla_m (f^n)^{\mathrm{T}}]^{\mathrm{T}} \right\} dm. \tag{9.195}
$$

To obtain the adjoint system, the coefficients multiplying dy^n in Eq. (9.195) are set equal to zero; i.e.

$$(\lambda^n)^{\mathrm{T}} [\nabla_{y^n} (f^n)^{\mathrm{T}}]^{\mathrm{T}} + (\lambda^{n+1})^{\mathrm{T}} [\nabla_{y^n} (f^{n+1})^{\mathrm{T}}]^{\mathrm{T}} + [\nabla_{y^n} \beta]^{\mathrm{T}} = 0. \tag{9.196}$$

Taking the transpose of the preceding equations gives the system of adjoint equations

$$[\nabla_{y^n} (f^n)^{\mathrm{T}}] \lambda^n = -[\nabla_{y^n} (f^{n+1})^{\mathrm{T}}] \lambda^{n+1} - \nabla_{y^n} \beta. \tag{9.197}$$

which we solve backward in time for $n = L, L-1, \ldots, 1$ where Eq. (9.194) gives the starting condition for the backward solution in time. In Eq. (9.197),

$$
\nabla_{y^n} [f^n]^{\mathrm{T}} =
\begin{bmatrix}
\dfrac{\partial f_{s,1}^n}{\partial p_1^n} & \dfrac{\partial f_{o,1}^n}{\partial p_1^n} & \cdots & \dfrac{\partial f_{g,N}^n}{\partial p_1^n} & \dfrac{\partial f_{wf,1}^n}{\partial p_1^n} & \cdots & \dfrac{\partial f_{wf,Nw}^n}{\partial p_1^n} \\[2ex]
\dfrac{\partial f_{s,1}^n}{\partial S_{o,1}^n} & \dfrac{\partial f_{o,1}^n}{\partial S_{o,1}^n} & \cdots & \dfrac{\partial f_{g,N}^n}{\partial S_{o,1}^n} & \dfrac{\partial f_{wf,1}^n}{\partial S_{o,1}^n} & \cdots & \dfrac{\partial f_{wf,Nw}^n}{\partial S_{o,1}^n} \\[2ex]
\dfrac{\partial f_{s,1}^n}{\partial x_1^n} & \dfrac{\partial f_{o,1}^n}{\partial x_1^n} & \cdots & \dfrac{\partial f_{g,N}^n}{\partial x_1^n} & \dfrac{\partial f_{wf,1}^n}{\partial x_1^n} & \cdots & \dfrac{\partial f_{wf,Nw}^n}{\partial x_1^n} \\[2ex]
\dfrac{\partial f_{s,1}^n}{\partial p_2^n} & \dfrac{\partial f_{o,1}^n}{\partial p_2^n} & \cdots & \dfrac{\partial f_{g,N}^n}{\partial p_2^n} & \dfrac{\partial f_{wf,1}^n}{\partial p_2^n} & \cdots & \dfrac{\partial f_{wf,Nw}^n}{\partial p_2^n} \\[2ex]
\vdots & \vdots & \cdots & \vdots & \vdots & \cdots & \vdots \\[2ex]
\dfrac{\partial f_{s,1}^n}{\partial x_N^n} & \dfrac{\partial f_{o,1}^n}{\partial x_N^n} & \cdots & \dfrac{\partial f_{g,N}^n}{\partial x_N^n} & \dfrac{\partial f_{wf,1}^n}{\partial x_N^n} & \cdots & \dfrac{\partial f_{wf,Nw}^n}{\partial x_N^n} \\[2ex]
\dfrac{\partial f_{s,1}^n}{\partial p_{wf,1}^n} & \dfrac{\partial f_{o,1}^n}{\partial p_{wf,1}^n} & \cdots & \dfrac{\partial f_{g,N}^n}{\partial p_{wf,1}^n} & \dfrac{\partial f_{wf,1}^n}{\partial p_{wf,1}^n} & \cdots & \dfrac{\partial f_{wf,Nw}^n}{\partial p_{wf,1}^n} \\[2ex]
\vdots & \vdots & \cdots & \vdots & \vdots & \cdots & \vdots \\[2ex]
\dfrac{\partial f_{s,1}^n}{\partial p_{wf,Nw}^n} & \dfrac{\partial f_{o,1}^n}{\partial p_{wf,Nw}^n} & \cdots & \dfrac{\partial f_{g,N}^n}{\partial p_{wf,Nw}^n} & \dfrac{\partial f_{wf,1}^n}{\partial p_{wf,Nw}^n} & \cdots & \dfrac{\partial f_{wf,Nw}^n}{\partial p_{wf,Nw}^n}
\end{bmatrix}, \tag{9.198}
$$

$$\nabla_{y^n}[f^{n+1}]^{\mathrm{T}} = \begin{bmatrix} \dfrac{\partial f_{s,1}^{n+1}}{\partial p_1^n} & \dfrac{\partial f_{o,1}^{n+1}}{\partial p_1^n} & \cdots & \dfrac{\partial f_{g,N}^{n+1}}{\partial p_1^n} & \dfrac{\partial f_{wf,1}^{n+1}}{\partial p_1^n} & \cdots & \dfrac{\partial f_{wf,N_w}^{n+1}}{\partial p_1^n} \\[2.2ex] \dfrac{\partial f_{s,1}^{n+1}}{\partial S_{o,1}^n} & \dfrac{\partial f_{o,1}^{n+1}}{\partial S_{o,1}^n} & \cdots & \dfrac{\partial f_{g,N}^{n+1}}{\partial S_{o,1}^n} & \dfrac{\partial f_{wf,1}^{n+1}}{\partial S_{o,1}^n} & \cdots & \dfrac{\partial f_{wf,N_w}^{n+1}}{\partial S_{o,1}^n} \\[2.2ex] \dfrac{\partial f_{s,1}^{n+1}}{\partial x_1^n} & \dfrac{\partial f_{o,1}^{n+1}}{\partial x_1^n} & \cdots & \dfrac{\partial f_{g,N}^{n+1}}{\partial x_1^n} & \dfrac{\partial f_{wf,1}^{n+1}}{\partial x_1^n} & \cdots & \dfrac{\partial f_{wf,N_w}^{n+1}}{\partial x_1^n} \\[2.2ex] \dfrac{\partial f_{s,1}^{n+1}}{\partial p_2^n} & \dfrac{\partial f_{o,1}^{n+1}}{\partial p_2^n} & \cdots & \dfrac{\partial f_{g,N}^{n+1}}{\partial p_2^n} & \dfrac{\partial f_{wf,1}^{n+1}}{\partial p_2^n} & \cdots & \dfrac{\partial f_{wf,N_w}^{n+1}}{\partial p_2^n} \\[2.2ex] \vdots & \vdots & \cdots & \vdots & \vdots & \cdots & \vdots \\[2.2ex] \dfrac{\partial f_{s,1}^{n+1}}{\partial x_N^n} & \dfrac{\partial f_{o,1}^{n+1}}{\partial x_N^n} & \cdots & \dfrac{\partial f_{g,N}^{n+1}}{\partial x_N^n} & \dfrac{\partial f_{wf,1}^{n+1}}{\partial x_N^n} & \cdots & \dfrac{\partial f_{wf,N_w}^{n+1}}{\partial x_N^n} \\[2.2ex] \dfrac{\partial f_{s,1}^{n+1}}{\partial p_{wf,1}^n} & \dfrac{\partial f_{o,1}^{n+1}}{\partial p_{wf,1}^n} & \cdots & \dfrac{\partial f_{g,N}^{n+1}}{\partial p_{wf,1}^n} & \dfrac{\partial f_{wf,1}^{n+1}}{\partial p_{wf,1}^n} & \cdots & \dfrac{\partial f_{wf,N_w}^{n+1}}{\partial p_{wf,1}^n} \\[2.2ex] \vdots & & \cdots & \vdots & \vdots & \cdots & \vdots \\[2.2ex] \dfrac{\partial f_{s,1}^{n+1}}{\partial p_{wf,N_w}^n} & \dfrac{\partial f_{o,1}^{n+1}}{\partial p_{wf,N_w}^n} & \cdots & \dfrac{\partial f_{g,N}^{n+1}}{\partial p_{wf,N_w}^n} & \dfrac{\partial f_{wf,1}^{n+1}}{\partial p_{wf,N_w}^n} & \cdots & \dfrac{\partial f_{wf,N_w}^{n+1}}{\partial p_{wf,N_w}^n} \end{bmatrix}, \tag{9.199}$$

and

$$\nabla_{y^n}\beta = \left[\frac{\partial \beta}{\partial p_1^n}, \frac{\partial \beta}{\partial S_{o,1}^n}, \frac{\partial \beta}{\partial x_1^n}, \frac{\partial \beta}{\partial p_2^n}, \dots, \frac{\partial \beta}{\partial x_N^n}, \frac{\partial \beta}{\partial p_{wf,1}^n}, \dots, \frac{\partial \beta}{\partial p_{wf,N_w}^n} \right]^{\mathrm{T}}. \tag{9.200}$$

If the simulation equations represent the purely implicit equations which are solved by the Newton–Raphson method, then the forward run requires on average a few Newton–Raphson iterations per time step. The simulation equations are nonlinear whereas the adjoint system is linear as the coefficient matrices in Eq. (9.197) are independent of λ. Thus the adjoint equation solution typically takes on the order of 1/4 the computation time required by a forward reservoir simulation run. Readers familiar with reservoir simulation will note that the coefficient matrix $(\nabla_{y^n}(f^n)^{\mathrm{T}})$ in Eq. (9.198) in the adjoint system is simply the transpose of the Jacobian matrix of Eq. (9.185) evaluated at y^n where the equations and primary variables used to construct the adjoint system are the same as used in the forward equations. This fact, which was first pointed out by Li *et al.* [63], makes the adjoint method particularly easy to attach to a simulator which solves fully implicit finite-difference equations using Newton–Raphson iteration because code for computing the Jacobian is already incorporated. Again assuming purely implicit finite-difference formulation, the submatrix of Eq. (9.199) formed by deleting the last N_w rows and N_w columns is a pentadiagonal matrix which is only related to the accumulation terms in the reservoir simulation equations and can easily be formulated.

It is important to note, however, that the system of adjoint equations is general as is the equation for calculating the sensitivities given by Eq. (9.202) below. The adjoint procedure does not rely on having a fully implicit-finite difference formulation or the use of Newton–Raphson iteration.

Formulation and solution of the adjoint problem requires that we save the information from the forward run that is necessary to compute the coefficient matrices involved in the adjoint system, Eq. (9.197). If Newton–Raphson iteration is used in the forward run then one could save the final Jacobians at each time step, but our preference is to save the values of all primary variables at each time step and then formulate the coefficient matrices for the adjoint solution as they are needed. In this approach, we write all the values of the primary variables to disk using unformatted direct access I/O which uses the binary format to store data record by record. We can also store adjoint variables this way, particularly if we are computing the sensitivities of a large number of data to model parameters.

When the λ^ns satisfy the adjoint system (Eq. 9.197), Eq. (9.195) simplifies to

$$dJ = \left\{ [\nabla_m \beta]^T + \sum_{n=1}^{N} (\lambda^n)^T [\nabla_m (f^n)^T]^T \right\} dm. \tag{9.201}$$

It follows that the total derivative of J with respect to m, i.e. the sensitivities are given by

$$\frac{dJ}{dm} = \begin{bmatrix} \frac{dJ}{dm_1} \\ \vdots \\ \frac{dJ}{dm_{N_m}} \end{bmatrix} = \nabla_m \beta + \sum_{n=1}^{L} \left[\nabla_m (f^n)^T \right] (\lambda^n), \tag{9.202}$$

where

$$\nabla_m [f^n]^T = \begin{bmatrix} \frac{\partial f_{s,1}^n}{\partial m_1} & \frac{\partial f_{o,1}^n}{\partial m_1} & \frac{\partial f_{g,1}^n}{\partial m_1} & \frac{\partial f_{s,2}^n}{\partial m_1} & \cdots & \frac{\partial f_{g,N}^n}{\partial m_1} & \frac{\partial f_{wf,1}^n}{\partial m_1} & \cdots & \frac{\partial f_{wf,N_w}^n}{\partial m_1} \\ \frac{\partial f_{s,1}^n}{\partial m_2} & \frac{\partial f_{o,1}^n}{\partial m_2} & \frac{\partial f_{g,1}^n}{\partial m_2} & \frac{\partial f_{s,2}^n}{\partial m_2} & \cdots & \frac{\partial f_{g,N}^n}{\partial m_2} & \frac{\partial f_{wf,1}^n}{\partial m_2} & \cdots & \frac{\partial f_{wf,N_w}^n}{\partial m_2} \\ \vdots & \vdots & \vdots & \vdots & \vdots & \vdots & \vdots & \vdots & \vdots \\ \frac{\partial f_{s,1}^n}{\partial m_{N_m}} & \frac{\partial f_{o,1}^n}{\partial m_{N_m}} & \frac{\partial f_{g,1}^n}{\partial m_{N_m}} & \frac{\partial f_{s,2}^n}{\partial m_{N_m}} & \cdots & \frac{\partial f_{g,N}^n}{\partial m_{N_m}} & \frac{\partial f_{wf,1}^n}{\partial m_{N_m}} & \cdots & \frac{\partial f_{wf,N_w}^n}{\partial m_{N_m}} \end{bmatrix}, \tag{9.203}$$

and

$$\nabla_m \beta = \left[\frac{\partial \beta}{\partial m_1}, \frac{\partial \beta}{\partial m_2}, \dots, \frac{\partial \beta}{\partial m_{N_m}} \right]^T. \tag{9.204}$$

It is important to note that in the preceding two equations, the partial derivative of a term with respect to a model parameter is zero unless the expression for the term explicitly involves that model parameter. Again, we note that it is common to write the left hand side of Eq. (9.202) as the gradient of J, or the gradient of β with respect to m. Similar to our comments on Eqs. (9.129) and (9.133), this is fine as long as the proper interpretation is given.

The matrix $\nabla_m[f^n]^T$ is an $N_m \times N_e$ sparse matrix and $\nabla_m\beta$ is an N_m-dimensional column vector. In Eq. (9.202), the gradient $\nabla_m\beta$ involves the partial derivatives of β with respect to the model parameters. If the jth model parameter does not explicitly appear in the expression for β, then $\partial\beta/\partial m_j = 0$. For example, if $\beta = p^L_{wf}$, then we set $\nabla_m\beta = 0$ in Eq. (9.202).

Note if we wish to compute the sensitivities of N_d different predicted, β_j, $j = 1$, $2, \ldots, N_d$, then we have to solve the adjoint system, Eq. (9.197), with N_d different βs; each β will have its own set of adjoint variables. Because the coefficient matrices in Eq. (9.197) are independent of β, one obtains the N_d sets of adjoint variables by solving the adjoint system with N_d right-hand sides backward in time.

To apply a nonlinear conjugate gradient [90] or quasi-Newton method such as LBFGS, we need only compute the gradient (total derivative) of the objective function and this can be done by setting β equal to the data mismatch part of the objective function in the adjoint procedure. In this case, one only needs to solve the adjoint system Eq. (9.197) once and substitute the resulting adjoint solutions into Eq. (9.202) to obtain the gradient.

To apply the adjoint method to calculate the sensitivity of the variable β to model parameters m, one needs to solve the adjoint system equation Eq. (9.197) to obtain the adjoint variables and then use Eq. (9.202) to calculate sensitivity coefficients. If the model m consists of only permeabilities (k_x, k_y, and k_z) and porosities (ϕ) in each individual gridblock, i.e.

$$m_{k_x} = k_x = [k_{x,1}, k_{x,2}, \ldots, k_{x,N}]^T, \tag{9.205}$$

$$m_{k_y} = k_y = [k_{y,1}, k_{y,2}, \ldots, k_{y,N}]^T, \tag{9.206}$$

$$m_{k_z} = k_z = [k_{z,1}, k_{z,2}, \ldots, k_{z,N}]^T, \tag{9.207}$$

and

$$m_\phi = \phi = [\phi_1, \phi_2, \ldots, \phi_N]^T, \tag{9.208}$$

then from Eq. (9.202), the equations to calculate the derivatives with respect to k_x, k_y, k_z, and ϕ are given by

$$\nabla_{k_x} J \equiv \frac{dJ}{dk_x} = \nabla_{k_x}\beta + \sum_{n=1}^{L} [\nabla_{k_x}(f^n)^T](\lambda^n), \tag{9.209}$$

$$\nabla_{k_y} J \equiv \frac{dJ}{dk_y} = \nabla_{k_y}\beta + \sum_{n=1}^{L} [\nabla_{k_y}(f^n)^T](\lambda^n), \tag{9.210}$$

$$\nabla_{k_z} J \equiv \frac{dJ}{dk_z} = \nabla_{k_z}\beta + \sum_{n=1}^{L} [\nabla_{k_z}(f^n)^T](\lambda^n), \tag{9.211}$$

and

$$\nabla_\phi J \equiv \frac{dJ}{d\phi} = \nabla_\phi \beta + \sum_{n=1}^{L} \left[\nabla_\phi (f^n)^\mathrm{T}\right](\lambda^n), \tag{9.212}$$

where β is pressure, GOR, WOR or a phase flow rate at some well at a specified time step L, the whole data mismatch part of the objective function $O_d(m)$ or any other term for which we wish to calculate sensitivities. Again, we have used gradient notation on the left-hand sides of the preceding four equations because it is fairly common to do so. However, these gradients really represent total derivatives (sensitivities). For example, the formal partial derivative of $J = p_{wf}$ with respect to the horizontal permeability in the gridblock penetrated by the well would be zero, but the total derivative is not zero if the well is operating under a nonzero rate constraint.

In order to calculate the gradient (total derivative) of the objective function, we consider β as the whole data mismatch part of the objective function, i.e.

$$\beta = O_d(m) = \frac{1}{2}(g(m) - d_{\mathrm{obs}})^\mathrm{T} C_D^{-1}(g(m) - d_{\mathrm{obs}}), \tag{9.213}$$

or in the case of stochastic simulation of m,

$$\beta = O_d(m) = \frac{1}{2}(g(m) - d_{\mathrm{uc}})^\mathrm{T} C_D^{-1}(g(m) - d_{\mathrm{uc}}). \tag{9.214}$$

To formulate the adjoint problem, Eq. (9.197), we need to calculate

$$\begin{aligned}
\nabla_{y^n} \beta &= \nabla_{y^n} \left[\frac{1}{2}(g(m) - d_{\mathrm{obs}})^\mathrm{T} C_D^{-1}(g(m) - d_{\mathrm{obs}})\right] \\
&= \left[\nabla_{y^n}(g(m) - d_{\mathrm{obs}})^\mathrm{T}\right] C_D^{-1}(g(m) - d_{\mathrm{obs}}) \\
&= \nabla_{y^n}[g(m)]^\mathrm{T} C_D^{-1}(g(m) - d_{\mathrm{obs}}).
\end{aligned} \tag{9.215}$$

In the case where β is given by Eq. (9.214), the d_{obs} in Eq. (9.215) should be replaced by d_{uc}. The matrix $\nabla_{y^n}[g(m)]^\mathrm{T}$ is an $N_e \times N_d$ matrix and is given by

$$\nabla_{y^n}[g(m)]^\mathrm{T} =
\begin{bmatrix}
\frac{\partial g_1}{\partial p_1^n} & \frac{\partial g_2}{\partial p_1^n} & \cdots & \frac{\partial g_{N_d}}{\partial p_1^n} \\
\frac{\partial g_1}{\partial S_{w,1}^n} & \frac{\partial g_2}{\partial S_{w,1}^n} & \cdots & \frac{\partial g_{N_d}}{\partial S_{w,1}^n} \\
\frac{\partial g_1}{\partial x_1^n} & \frac{\partial g_2}{\partial x_1^n} & \cdots & \frac{\partial g_{N_d}}{\partial x_1^n} \\
\frac{\partial g_1}{\partial p_2^n} & \frac{\partial g_2}{\partial p_2^n} & \cdots & \frac{\partial g_{N_d}}{\partial p_2^n} \\
\vdots & \vdots & \cdots & \vdots \\
\frac{\partial g_1}{\partial x_N^n} & \frac{\partial g_2}{\partial x_N^n} & \cdots & \frac{\partial g_{N_d}}{\partial x_N^n} \\
\frac{\partial g_1}{\partial p_{wf,1}^n} & \frac{\partial g_2}{\partial p_{wf,1}^n} & \cdots & \frac{\partial g_{N_d}}{\partial p_{wf,1}^n} \\
\vdots & \vdots & \cdots & \vdots \\
\frac{\partial g_1}{\partial p_{wf,N_w}^n} & \frac{\partial g_2}{\partial p_{wf,N_w}^n} & \cdots & \frac{\partial g_{N_d}}{\partial p_{wf,N_w}^n}
\end{bmatrix}
\tag{9.216}$$

The vector $g(m)$ contains all the dynamic data that we wish to history match. The implementation details are simplest if the observed data corresponds to a time t^n at which the simulator equations are solved, but this is seldom the case in practice. Otherwise, one must either interpolate the data to a simulation time step, which requires a dense data set for accuracy, interpolate primary variables to the observation times, or interpolate sensitivities. It is important to note that in Eq. (9.216), only columns corresponding to data that are measured at time n are nonzero. For example, if the N_d entries of g correspond to the flowing wellbore pressure at a single well (well 1) at N_d distinct times and $g_1 = p_{wf}^n$, then all entries of the matrix of Eq. (9.216) are zero except for $\partial g_1/\partial p_{wf,1}^n$ which is equal to unity. Details for calculating each entry of the matrix $\nabla_{y^n}[g(m)]^T$ can be found in Li [69]. After we evaluate nonzero entries of the matrix $\nabla_{y^n}[g(m)]^T$, we multiply $C_D^{-1}(g(m) - d_{obs})$ by this matrix to obtain $\nabla_{y^n}\beta$. Once we have $\nabla_{y^n}\beta$, we can apply Eq. (9.197) to compute the adjoint variables.

To apply Eq. (9.202) to compute the sensitivities, we need to evaluate $\nabla_m\beta$ first. The vector $\nabla_m\beta$ is given by

$$
\begin{aligned}
\nabla_m\beta &= \nabla_m O_d(m) \\
&= \nabla_m \left\{ \frac{1}{2}(g(m) - d_{obs})^T C_D^{-1}(g(m) - d_{obs}) \right\} \\
&= \left[\nabla_m(g(m) - d_{obs})^T\right] C_D^{-1}(g(m) - d_{obs}) \\
&= \nabla_m[g(m)]^T C_D^{-1}(g(m) - d_{obs}).
\end{aligned}
\tag{9.217}
$$

In the case where β is given by Eq. (9.214), the d_{obs} in Eq. (9.217) should be replaced by d_{uc}. The matrix $\nabla_m[g(m)]^T$ is an $N_m \times N_d$ matrix and defined as

$$
\nabla_m[g(m)]^T =
\begin{bmatrix}
\frac{\partial g_1}{\partial m_1} & \frac{\partial g_2}{\partial m_1} & \cdots & \frac{\partial g_{N_d}}{\partial m_1} \\
\frac{\partial g_1}{\partial m_2} & \frac{\partial g_2}{\partial m_2} & \cdots & \frac{\partial g_{N_d}}{\partial m_2} \\
\vdots & \vdots & \cdots & \vdots \\
\frac{\partial g_1}{\partial m_{N_m}} & \frac{\partial g_2}{\partial m_{N_m}} & \cdots & \frac{\partial g_{N_d}}{\partial m_{N_m}}
\end{bmatrix}.
\tag{9.218}
$$

Again if the formula for g_i does not explicitly involve m_j, then $\partial g_i/\partial m_j = 0$, but the total derivative dg_i/dm may not be zero. Detailed formulas for calculation of elements in the matrix $\nabla_m[g(m)]^T$ can be found in Li [69]. After we compute $\nabla_m\beta$, we can use Eq. (9.202) to compute the derivatives of the objective function with respect to model parameters, i.e., the sensitivity of the objective function which represents the gradient we need in quasi-Newton methods.

Some example formulas

The discussion here pertains only to the fully implicit finite-difference scheme. In this case, the gas material balance at gridblock ℓ of Eq. (9.168) in oil field units can be written in the form

$$f_{g,\ell}^{n+1} = f_{g,\ell}(y^{n+1}, y^n, m)$$

$$= \frac{V_\ell}{5.615\Delta t_n}\left(\left[\phi_\ell^{n+1}\left(\frac{S_{g,\ell}^{n+1}}{B_{g,\ell}^{n+1}} + \frac{R_{s,\ell}^{n+1}S_{o,\ell}^{n+1}}{B_{o,\ell}^{n+1}}\right)\right] - \left[\phi_\ell^n\left(\frac{S_{g,\ell}^n}{B_{g,\ell}^n} + \frac{R_{s,\ell}^n S_{o,\ell}^n}{B_{o,\ell}^n}\right)\right]\right) + \cdots = 0,$$

$$(9.219)$$

where V_l is the bulk volume of the gridblock, R_s denotes dissolved gas–oil ratio, ϕ is porosity, B_g and B_o, respectively denote the gas and oil formation volume factors, S_g is gas saturation and Δt_n is the length of the time step from time t_n to time t_{n+1}. The terms represented by $\cdots$ in Eq. (9.219) are evaluated at t_{n+1} so the gradient of the missing terms with respect to y^n is a zero vector. Assuming the primary variables for the gridblock at time t_n are pressure, p, gas saturation and oil saturation, then letting $a_\ell = V_\ell/(5.615\Delta t_n)$, the only nonzero entries of the gradient of this equation with respect to y^n are

$$\frac{\partial f_{g,\ell}^{n+1}}{\partial p_\ell^n} = -a_\ell\left[\frac{\partial \phi_\ell^n}{\partial p_\ell^n}\left(\frac{S_{g,\ell}^n}{B_{g,\ell}^n} + \frac{R_{s,\ell}^n S_{o,\ell}^n}{B_{o,\ell}^n}\right)\right.$$

$$\left. + \phi_\ell^n\left(S_{g,\ell}^n\frac{-1}{(B_{g,l}^n)^2}\frac{\partial B_{g,\ell}^n}{\partial p_\ell^n} + R_{s,\ell}^n S_{o,\ell}^n\frac{-1}{(B_{o,\ell}^n)^2}\frac{\partial B_{o,\ell}^n}{\partial p_\ell^n} + \frac{S_{o,\ell}^n}{B_{o,\ell}^n}\frac{\partial R_{s,\ell}^n}{\partial p_\ell^n}\right)\right], \quad (9.220)$$

$$\frac{\partial f_{g,\ell}^{n+1}}{\partial S_{o,\ell}^n} = -a_\ell\frac{\phi_\ell^n R_{s,\ell}^n}{B_{o,\ell}^n}, \quad (9.221)$$

and

$$\frac{\partial f_{g,\ell}^{n+1}}{\partial S_{g,\ell}^n} = -a_\ell\frac{\phi_\ell^n}{B_{g,\ell}^n}. \quad (9.222)$$

The derivatives of $f_{s,\ell}^{n+1}$ and $f_{o,\ell}^{n+1}$ with respect to these same three primary variables are also nonzero. The derivatives of $f_{g,\ell}^{n+1}$, $f_{s,\ell}^{n+1}$, and $f_{o,\ell}^{n+1}$ with respect to any primary variable in y^n other than p_ℓ^n, $S_{o,\ell}^n$, and $S_{g,\ell}^n$ is zero. Thus, it is easy to see that the upper left $3N \times 3N$ submatrix of Eq. (9.199) is pentadiagonal and can easily be calculated. Assuming a purely implicit formulation, the $f_{wf,j}^{n+1}$ terms do not involve any primary variables at time t_n and f^{n+1} does not involve wellbore pressures at time t_n; thus the last N_w rows and N_w columns of the matrix of Eq. (9.199) contain only zeros.

To illustrate how to compute derivatives of the well constraint equations needed in the matrix of Eq. (9.198), consider the case where the oil rate at well 1 is specified as the production condition at time t_n, i.e. a value is specified for $q_{o,1}^n$ which applies on the time step from t_{n-1} to t_n. Also assume this well produces only from gridblocks (i, j, k) and $(i, j, k + 1)$ which correspond, respectively, to gridblocks ℓ_1 and ℓ_2. Then using Peaceman's equation, the well equation has the form

$$f_{wf,1}^n = WI_{i,j,k} \frac{k_{ro,\ell_1}^n}{\mu_{o,\ell_1}^n B_{o,\ell_1}^n}(p_{\ell_1}^n - p_{wf,1}^n)$$

$$+ WI_{i,j,k+1} \frac{k_{ro,\ell_2}^n}{\mu_{o,\ell_2}^n B_{o,\ell_2}^n}(p_{\ell_2}^n - p_{wf,1}^n - \bar{\rho}(z_{k+1} - z_k)) - q_{o,1}^n = 0, \tag{9.223}$$

where we have used standard reservoir simulation notation, e.g. the WI denote the standard well index terms, k_{ro} represents relative permeability, the datum depth for wellbore pressure is z_k and $\bar{\rho}$ is an average wellbore density term that accounts for the difference in depth. Assuming the same primary variables as before and assuming three-phase flow then k_{ro} is a function of both oil and gas saturation. Thus, the derivatives of Eq. (9.223) with respect to $p_{\ell_1}^n$, $p_{\ell_2}^n$, $p_{wf,1}^n$, S_{o,ℓ_1}^n, S_{o,ℓ_2}^n, S_{g,ℓ_1}^n, S_{g,ℓ_2}^n may all be nonzero but the derivative of the equation with respect to any other component of y^{n+1} is zero.

9.7.3   Compact derivation of sensitivities

Here we provide an alternate way to derive the formula for the entries of the sensitivity matrix G. In the process, we show how to calculate the product of G times a vector and G^T times a vector without explicitly computing the entries of G, i.e. provide the necessary user input to implement a Paige–Saunders type algorithm [106]. Having a procedure to do this is advantageous if one wishes to solve the matrix problem

$$(C_D + GC_M G^T)x = b, \tag{9.224}$$

by the conjugate gradient method. Note this is the matrix problem that must be solved at each iteration when the Gauss–Newton method is formulated as in Eq. (8.48). Chu *et al.* [107] implemented a conjugate gradient algorithm for solving Eq. (9.224) that avoids explicit computation of G. They, however, consider only a single-phase flow problem and use somewhat complicated mathematical notation. The development given here essentially follows that of Rodrigues [108, 109] and Kraaijevanger *et al.* [110].

To begin we write the system of reservoir simulator equations given in Eq. (9.181) as

$$f^n = f(y^n, y^{n-1}, m) = 0, \text{ for } n = 1, 2, \ldots, L, \tag{9.225}$$

where y^0 denotes the initial conditions which we assume are fixed. Next, we combine vectors of primary variables at all time steps into one overall primary vector Y and combine the simulation equations at all time steps into one overall simulation equation

$F = F(Y, m)$, i.e. we define

$$Y = \begin{bmatrix} y^1 \\ y^2 \\ \vdots \\ y^L \end{bmatrix}, \tag{9.226}$$

$$F(Y, m) = \begin{bmatrix} f^1(y^1, y^0, m) \\ f^2(y^2, y^1, m) \\ \vdots \\ f^L(y^L, y^{L-1}, m) \end{bmatrix}. \tag{9.227}$$

From Eq. (9.225), it follows that the complete set of equations solved in the simulator are

$$F(Y, m) = 0. \tag{9.228}$$

This implies that the total differential of F is zero, from which it follows that

$$\left(\nabla_Y F^T\right)^T dY + \left(\nabla_m F^T\right)^T dm = 0. \tag{9.229}$$

Defining

$$F_Y = \left(\nabla_Y F^T\right)^T, \tag{9.230}$$

and

$$F_m = \left(\nabla_m F^T\right)^T, \tag{9.231}$$

it follows from Eq. (9.229) that

$$F_Y \frac{dY}{dm} = -F_m. \tag{9.232}$$

The matrix of total derivatives, dY/dm, contains sensitivity matrices for all time steps and is given by

$$\frac{dY}{dm} = \begin{bmatrix} dy^1/dm \\ dy^2/dm \\ \vdots \\ dy^L/dm \end{bmatrix} = \begin{bmatrix} [\nabla_m(y^1)^T]^T \\ [\nabla_m(y^2)^T]^T \\ \vdots \\ [\nabla_m(y^L)^T]^T \end{bmatrix} = \begin{bmatrix} \hat{G}_1 \\ \hat{G}_2 \\ \vdots \\ \hat{G}_L \end{bmatrix}. \tag{9.233}$$

Some comments on notation are in order. The second equality of Eq. (9.233) indicates that if we had an expression for each y^ℓ in terms of only m, then the total derivatives would be equivalent to partial derivatives, i.e. we could simply compute

$$\frac{dy^\ell}{dm} = [\nabla_m(y^\ell)^T]^T, \tag{9.234}$$

which is an $N_e \times N_m$ matrix with the entry in the ith row and jth column given by dy_i^ℓ/dm_j. As this matrix is a sensitivity matrix, we have denoted it by $\hat{G}_\ell$. Note this matrix contains the sensitivities of all primary variables to all model parameters. The sensitivities of predicted data can be constructed from these sensitivities. As shown more clearly later, solving Eq. (9.232) represents the direct (or gradient simulator) method.

The matrix F_Y is block bi-diagonal and given by

$$
F_Y = \left(\nabla_Y F^T\right)^T
$$

$$
= \begin{bmatrix}
[\nabla_{y^1}(f^1)^T]^T & O & O & \cdots & O \\
[\nabla_{y^1}(f^2)^T]^T & [\nabla_{y^2}(f^2)^T]^T & O & \cdots & O \\
O & [\nabla_{y^2}(f^3)^T]^T & [\nabla_{y^3}(f^3)^T]^T & \cdots & O \\
\vdots & \vdots & \vdots & \ddots & \vdots \\
O & O & \cdots & [\nabla_{y^{L-1}}(f^L)^T]^T & [\nabla_{y^L}(f^L)^T]^T
\end{bmatrix}.
$$

$$(9.235)$$

F_m is given by

$$
F_m = \left(\nabla_m F^T\right)^T = \begin{bmatrix}
[\nabla_m(f^1)^T]^T \\
[\nabla_m(f^2)^T]^T \\
\vdots \\
[\nabla_m(f^L)^T]^T
\end{bmatrix}.
$$

$$(9.236)$$

From Eqs. (9.233)–(9.236), it follows that the complete system of Eq. (9.232) can be written in the following recursive form:

$$
\left(\nabla_{y^{\ell-1}}[(f^\ell)^T]^T\right)\hat{G}_{\ell-1} + \left(\nabla_{y^\ell}[(f^\ell)^T]^T\right)\hat{G}_\ell = -[\nabla_m(f^\ell)^T]^T, \text{ for } \ell = 1, 2, \ldots, L,
$$

$$(9.237)$$

where G_0 is a null matrix because y^0 is a fixed vector and hence independent of m. The preceding equation represents the normal form of the direct (gradient simulator) method.

Now suppose we wish to compute the sensitivity (total derivative) of a real-valued function $\beta = \beta(Y, m)$. So the final formula will involve the same notation we used previously, we let $J = J(Y, m) = \beta$. In our notation, dJ/dm is a column vector but chain rules for total derivatives are more naturally written when the total derivative is expressed as a row vector as in [110]. So we can use the more natural formula, we first derive the formula for the transpose of dJ/dm. We have

$$
\left[\frac{dJ}{dm}\right]^T = (\nabla_m J)^T + (\nabla_Y J)^T \frac{dY}{dm}.
$$

$$(9.238)$$

Using the definition of F_Y and F_m (Eqs. 9.230 and 9.231) in Eq. (9.232), solving for dY/dm and using the result in Eq. (9.238) gives

$$\left[\frac{dJ}{dm}\right]^{\mathrm{T}} = (\nabla_m J)^{\mathrm{T}} - (\nabla_Y J)^{\mathrm{T}}(\nabla_Y F^{\mathrm{T}})^{-\mathrm{T}}(\nabla_m F^{\mathrm{T}})^{\mathrm{T}}, \tag{9.239}$$

where the superscript $-\mathrm{T}$ always denotes the dual operation of taking the transpose and then the inverse. If we define λ^{T} by

$$\lambda^{\mathrm{T}} = -(\nabla_Y J)^{\mathrm{T}}(\nabla_Y F^{\mathrm{T}})^{-\mathrm{T}}, \tag{9.240}$$

then Eq. (9.239) becomes

$$\left[\frac{dJ}{dm}\right]^{\mathrm{T}} = (\nabla_m J)^{\mathrm{T}} + \lambda^{\mathrm{T}}(\nabla_m F^{\mathrm{T}})^{\mathrm{T}}, \tag{9.241}$$

Using β in place of J on the right-hand side of Eq. (9.241) and taking the transpose of the equation gives

$$\frac{dJ}{dm} = \nabla_m \beta + (\nabla_m F^{\mathrm{T}})\lambda. \tag{9.242}$$

From Eq. (9.240), we see that λ can be obtained by solving

$$(\nabla_Y F^{\mathrm{T}})\lambda = -\nabla_Y \beta, \tag{9.243}$$

where again we have replaced J by β. Partition λ into L column subvectors of length N_e so that

$$\lambda = \begin{bmatrix} \lambda^1 \\ \lambda^2 \\ \vdots \\ \lambda^L \end{bmatrix}. \tag{9.244}$$

Then using Eq. (9.235) we can write Eq. (9.243) as the system

$$[\nabla_{y^L}(f^\ell)^{\mathrm{T}}]\lambda^L = -\nabla_{y^L}\beta, \tag{9.245}$$

$$[\nabla_{y^\ell}(f^\ell)^{\mathrm{T}}]\lambda^\ell = -[\nabla_{y^\ell}(f^{\ell+1})^{\mathrm{T}}]\lambda^{\ell+1} - \nabla_{y^\ell}\beta, \tag{9.246}$$

for $\ell = L-1, L-2, \ldots, 1$. Note this is the same adjoint system we derived previously in Eq. (9.197). Similarly, Eq. (9.242) is equivalent to Eq. (9.202).

9.7.4   G times a vector

Here, it is convenient to partition the sensitivity matrix G into a column of block submatrices where the ℓth block is denoted by G^ℓ and represents sensitivities of all predicted data at time t_ℓ to model parameters. Thus, if we can find a method for computing G^ℓ times a vector, we can compute G times a vector. We let $g^\ell = [g_1^\ell, g_2^\ell, \ldots, g_{N_\ell}^\ell]^{\mathrm{T}}$ be the

N_ℓ-dimensional column vector of predicted data corresponding to the set of observed data at time t_ℓ. Thus G^ℓ is an $N_\ell \times N_m$ matrix with the entry in the ith row and j column given by dg_i^ℓ / dm_j. Here, the entry in the ith row and j column of the $N_e \times N_m$ matrix $\hat{G}^\ell$ is dy_i^ℓ / dm_j. If we multiply Eq. (9.237) by any N_m-dimensional column vector v, the resulting system can be solved recursively to obtain

$$z^\ell = \hat{G}^\ell v, \tag{9.247}$$

for $\ell = 1, 2, \ldots, L$. The kth component of z^ℓ is

$$z_k^\ell = \sum_{j=1}^{N_m} \frac{dy_k^\ell}{dm_j} v_j. \tag{9.248}$$

The ith component of the vector $G^\ell v$ is given by

$$x_i = \sum_{j=1}^{N_m} \frac{dg_i^\ell}{dm_j} v_j = \sum_{j=1}^{N_m} \sum_{k=1}^{N_e} \frac{\partial g_i^\ell}{\partial y_k^\ell} \frac{dy_k^\ell}{dm_j} v_j + \sum_{j=1}^{N_m} \frac{\partial g_i^\ell}{\partial m_j} v_j. \tag{9.249}$$

Using Eq. (9.248), Eq. (9.249) can be written as

$$x_i = \sum_{k=1}^{N_e} \frac{\partial g_i^\ell}{\partial y_k^\ell} z_k^\ell + \sum_{j=1}^{N_m} \frac{\partial g_i^\ell}{\partial m_j} v_j. \tag{9.250}$$

The partial derivatives on the right-hand side of Eq. (9.250) are computed directly from the expression for the predicted data.

9.7.5 G^T times a vector

The product of G^T times v can be obtained by an analogous procedure except here we solve for $\tilde{G}^T v$ using a recursive formula applied backward in time; see Rodrigues [108, 109]. Here, we give an alternate approach. If g is the vector of all predicted data corresponding to all observed data at all times, then for any fixed vector v

$$\beta = g^T v, \tag{9.251}$$

has as its sensitivity vector

$$\frac{d\beta}{dm} = G^T v. \tag{9.252}$$

Thus to compute $G^T v$, we simply apply the adjoint method to compute $d\beta/dm$.

9.7.6   Comments on sensitivities

As the equations we have presented indicate, it is most straightforward to compute the sensitivities of predicted data to variables that appear in simulation equations, i.e.

sensitivity to a gridblock saturation, pressure, permeability, or porosity. But sensitivities to other variables that depend on these variables can often be computed from these sensitivities using the chain rule. In this way, one can for example, compute the sensitivities of parameters (width, aspect ratio, sinuosity) describing a channel [103] or to time-lapse seismic data [111].

9.8   Reparameterization

For history matching by minimizing an objective function, our optimization algorithm of choice for large-scale problems has been LBFGS. With this approach, one only needs to compute the gradient of the objective function at each iteration of LBFGS. As shown the computation of this gradient simply requires one adjoint solution regardless of the number of model parameters.

However, there exists a large body of literature on trying to reduce the number of model parameters, and historically in manual history matching, engineers have tried to choose a small set of critical parameters to adjust to obtain a history match where the choice of parameters is based on physical intuition and experience.

Zonation [97, 112] represents the earliest work on a reduced parameterization. In this approach, one divides the reservoir into a small number of subregions and assumes each subregion as homogeneous. Then one adjusts the permeabilities and porosity on each subregion by history matching dynamic data. The "gradzone" procedure [72] which simply uses a single multiplier in each subregion to adjust all transmissibilities in the subregion by history matching is an outgrowth of zonation, although Bissell provides some guidelines for choosing the grad zones based on approximate sensitivities. Adaptive multi-scale methods [113, 114] essentially represent a modern extension of zonation. They start with a small number of zones and gradually increase the number of zones until they are able to achieve a history match of data. Each zone has constant permeability and porosity. At each refinement step, they add additional zones where the choice of which zones to add is based on the choice that minimizes a predicted value of the objective function which is estimated using sensitivities.

Parameterization in terms of the principal orthonormal eigenvectors of $G^{T}G$ where G is the sensitivity matrix for a least-squares problems was considered by Shah *et al.* [115]. As the eigenvalues of $G^{T}G$ decay rapidly to zero, one does not need to keep many eigenvectors in the parameterization to obtain a good history match and good estimate of model parameters. One can extend the procedure to parameterization in terms of the principal eigenvalues of any real-symmetric positive-definite Hessian matrix. One should always bear in mind, however, that when one deletes eigenvectors corresponding to small eigenvalues and expands the model (vector of model parameters) as a linear combination of the remaining eigenvectors, one is deleting high frequency components from the solution. While this is not necessarily bad, it can lead to a model substantially smoother than the truth. The work of Rodrigues [108, 109] is reminiscent of the early work of Shah *et al.* [115] except Rodrigues parameterizes in terms of the principal right and left singular vectors of the sensitivity matrix G where these vectors are obtained by singular value decomposition. By this, we mean keeping only singular vectors corresponding to the largest singular values. The vector of model parameters m is then expanded as a linear combination of the right singular vector retained and the coefficients in this expansion are obtained by minimizing the appropriate objective function. The predicted data is a linear combination of the left singular vectors so to match observed data well, it must lie in the subspace spanned by the left singular vectors retained in the parameterization.

In terms of the Bayesian formulation used here where the objective function has the form of Eq. (8.23), Rodrigues suggests a parameterization based on the principal left and right singular vectors of the matrix

$$S_D = C_D^{-1/2} G C_M^{1/2}, \tag{9.253}$$

where G is the sensitivity matrix, $C_D^{-1/2}$ is the square root of C_D^{-1} and $C_M^{1/2}$ is a square root of the prior covariance matrix C_M. Rodrigues uses $C_M^{1/2} = L$ where $C_M = LL^T$ is the Cholesky decomposition of C_M. The matrix in Eq. (9.253) is actually the dimensionless sensitivity coefficient matrix defined in Zhang *et al.* [103] as the matrix that directly indicates how conditioning to data reduces the uncertainty in the model parameters. Thus, a parameterization based on the dimensionless sensitivity matrix may prove to be the best parameterization, assuming that its computation is feasible. Parameterization based on the principal eigenvectors of the covariance matrix have also been used with some success by many authors [116, 117].

The geophysics literature contains substantial discussion of the subspace method for parameterization [118–120]. In this approach, over each iteration of a minimization algorithm, one expands the change in model parameters in terms of subspace vectors. There are a wide variety of ways to choose the subspace vectors including partitioning the data mismatch of the objective function into subobjective functions and using the gradients of subobjective functions as subspace vectors. With this choice, one must also add to this space some subspace vectors from the gradient of the model mismatch part of the objective function. In a history-matching context with a Gauss–Newton type method, one would compute directly the sensitivity of data to subspace vectors and this effectively requires computation of the normal sensitivity matrix times subspace vectors which can be accomplished using the gradient simulator method in the form given by Rodrigues [109] discussed earlier in this chapter, see Abacioglu [65], Abacioglu *et al.* [121], and Reynolds *et al.* [117] for additional details. The iterative subspace method of Vogel and Wade [122] represents a somewhat similar approach to those given above that has yet to be applied in a history-matching context.

Although it is not our preferred method of parameterization, we discuss the pilot-point method in a separate subsection because it has been used extensively in the ground water and petroleum engineering literature.

9.8.1   Pilot- and master-point methods

The "pilot-point method" of de Marsily *et al.* [123] refers to a method for limiting the number of parameters needed for reservoir characterization by using kriging to interpolate permeability values between pilot-point locations at which the permeability is determined by an inversion procedure. The original application was to the interpretation of interference test data and the result was a type of best estimate, that is a mostly smooth surface. Bréfort and Pelcé [124] applied nearly the same method to a problem involving a highly compressible gas field and LaVenue and Pickens [125] added a feature in which pilot-point locations were calculated automatically. Because the results depend on the locations of the pilot points, it is probably accurate to say that it generated the maximum a posteriori surface conditional to the data (measured permeability and head), the variogram, and the location of the pilot points.

The method was later modified by RamaRao *et al.* [126] to generate "equally likely" realizations of the permeability field. The procedure for doing this was to first generate a permeability field that was conditional to measurements of permeability, then use the pilot-point method to generate a "smooth" correction to the permeability field so that it is also honors the pressure data. They also proposed a method for automatically placing the pilot points at locations which would be most beneficial for reducing the misfit in the pressure data. LaVenue *et al.* [127] applied this approach to the assessment of the probability of containment of radioactive waste at the Waste Isolation Pilot Plant (WIPP) site in New Mexico.

The pilot-point method can be thought of as a subspace method, since it effectively expands the change in the property field, δm over one iteration of an optimization algorithm as a sum of subspace vectors where the subspace vectors are certain columns of the prior covariance matrix. Letting $C_{i,j}$ denote the element in the ith row and jth column of C_M, the expansion is given by

$$\delta m = \begin{bmatrix} C_{r_1,1} & C_{r_2,1} & \cdots & C_{r_{n_p},1} \\ C_{r_1,2} & C_{r_2,2} & & C_{r_{n_p},2} \\ \vdots & & \ddots & \vdots \\ C_{r_1,N_m} & C_{r_2,N_m} & \cdots & C_{r_{n_p},N_m} \end{bmatrix} \begin{bmatrix} \alpha_1 \\ \alpha_2 \\ \vdots \\ \alpha_{n_p} \end{bmatrix} = A\alpha = \sum_{j=1}^{n_p} \alpha_j a_j. \tag{9.254}$$

Note that column j of A is the transpose of row r_j of C_M, but because C_M is a real-symmetric matrix, column j of A is identical to column r_j of C_M. Here a_j denote the jth column of A.

If the objective is to obtain a MAP estimate, then we wish to minimize the objective function of Eq. (8.2). If this is done with the Gauss–Newton method, then at each iteration the change

$$(C_M^{-1} + G^T C_D^{-1} G)\, \delta m = -[C_M^{-1}(m - m_{\text{prior}}) + G^T C_D^{-1}(g(m) - d_{\text{obs}})]. \tag{9.255}$$

Multiplying both sides by A^T and replacing δm by $A\alpha$ results in the following equation for α:

$$[A^T C_M^{-1} A + (GA)^T C_D^{-1}(GA)]\alpha = -A^T C_M^{-1}(m - m_{\text{prior}}) - (GA)^T C_D^{-1}(g(m) - d_{\text{obs}}) \tag{9.256}$$

Note that the coefficient matrix on the left-hand side of Eq. (9.256) is $n_p \times n_p$, i.e. the matrix problem we solve for α is of much lower dimension than the $N_m \times N_m$ matrix problem we must solve for δm if Eq. (9.255) is solved directly. In addition, we show that many of the matrix products in Eq. (9.256) are particularly simple to evaluate. Because A can be written as the product

$$A = C_M \left[e_{r_1} \cdots e_{r_{n_p}} \right], \tag{9.257}$$

where e_i is the unit vector with all zeros except for a 1 in the r_i location, it is easy to establish that

$$A^T C_M^{-1} A = \begin{bmatrix} C_{r_1,r_1} & C_{r_2,r_1} & \cdots & C_{r_{n_p},r_1} \\ C_{r_1,r_2} & C_{r_2,r_2} & & C_{r_{n_p},r_2} \\ \vdots & & \ddots & \vdots \\ C_{r_1,r_{n_p}} & C_{r_2,r_{n_p}} & \cdots & C_{r_{n_p},r_{n_p}} \end{bmatrix} \tag{9.258}$$

and

$$A^T C_M^{-1}(m - m_{\text{prior}}) = \begin{bmatrix} (m - m_{\text{prior}})_{r_1} \\ (m - m_{\text{prior}})_{r_2} \\ \vdots \\ (m - m_{\text{prior}})_{r_{n_p}} \end{bmatrix}. \tag{9.259}$$

The product GA can be calculated efficiently using the "gradient simulator" method discussed in Section 9.7.3.

Remark 9.3. If a pilot point is placed at every gridblock, then $A = C_M$ and the pilot point method is equivalent to the standard Gauss–Newton. It is common to use pilot points without the regularization provided by the prior model. In that case, the method becomes ill posed as the number of pilot points increases and it is necessary to use an ad hoc procedure to limit the changes in the value of the model parameter at pilot-point locations.

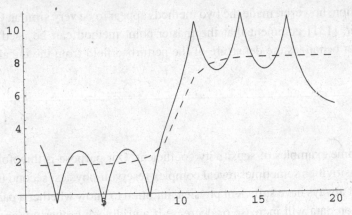

Figure 9.2. A comparison of the "smooth" pilot-point solution (solid line) with correct solution (dashed) for information on averages over the intervals $x \le 10$ and $x \ge 10$.

Assessment of the pilot-point method

Several problems occur in the usage of the pilot-point method. The first is that it does not necessarily produce a "smooth" correction because kriged surfaces are not necessarily smooth. Figure 9.2 shows an example of a "smooth" surface generated by a pilot-point method versus the proper smooth surface generated by a linear inversion technique using the same exponential covariance model. It would give the correct result for a linear problem, however, if a pilot point were to be placed at every node of the grid.

The main problem with the pilot-point method, then, is that the solution is constructed as a linear combination of localized covariance functions (localized at the pilot points) while the correct solutions should be constructed as linear combinations of sensitivity functions convolved with the covariance [see 128]. If the sensitivity functions for dynamic data were isolated delta functions the pilot-point method would work very well. We have a great deal of information showing that sensitivity functions for pressure are diffuse averages over large areas [16, 17, 27, 53, 55, 129], and that sensitivity functions for tracer concentration data, for example, are nearly linear along streamlines [39, 130]. Linear combinations of delta functions at pilot-point locations provide a poor approximation of these sensitivities.

9.8.2   The master-point method

Gómez-Hernández *et al.* [131] proposed the "master-point method" as a method for generating "equally likely" realizations of the permeability field. At the time of their development of this method for conditional simulation they differentiated it from the method of pilot points which had been only used for estimation. The papers by RamaRao *et al.* [126] and LaVenue *et al.* [127] in which the pilot-point method was used for simulation, however, made the two methods appear to be very similar. Gómez-Hernández *et al.* [131] comment that the master-point method can be used with a penalty term that penalizes the departure of the perturbed field from the uncalibrated realization.

9.9   Examples

We first show some examples of sensitivity coefficient. Our purpose is threefold, first to show how sensitivities sometimes reveal complex reservoir physics, second to show that one does not always have sufficient physical intuition to know whether a particular type of production data will increase or decrease if a gridblock property is increased or decreased, and third to provide examples that could be used to check adjoint code.

After illustrating computation of sensitivity in an example, we show some results from history matching using LBFGS.

9.9.1   Sensitivity examples

The first example pertains to vertical flow in an x–z cross section with 15 gridblocks in the horizontal direction and eight gridblocks in the vertical direction. The gridblocks in the z direction are ordered from top to bottom, that is, the bottom row of gridblocks has indices $(i, 8)$, $i = 1, 2, \ldots, 15$. A well is produced at a constant oil rate from gridblock $(8, 8)$. The initial pressure is above bubble-point pressure in all gridblocks and drops below bubble-point pressure fairly soon after the beginning of production. Water saturation is fixed at irreducible so only oil and gas can flow. Critical gas saturation is given by $S_{g,c} = 0.05$. After 30 days of production, free gas has evolved but gas saturation is below critical gas saturation so free gas is not flowing. After 400 days of production, gas saturation is above 0.25 in all simulator gridblocks. Some free gas has been produced, but much of the free gas has migrated to the top of the structure to form a gas cap, Fig. 9.3.

Figure 9.3. Gas saturation distribution after 400 days of production. Figure from Li *et al.* [42] is being used with permission from the Petroleum Society of CIM, who retains the copyright.

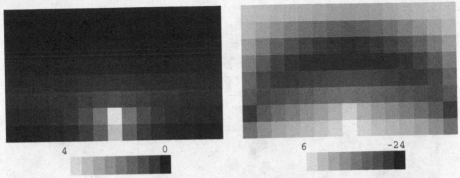

Figure 9.4. Sensitivity of producing GOR to vertical permeability field at 30 days (left) and 400 days (right). Figure from Li *et al.* [42] is being used with permission from the Petroleum Society of CIM, who retains the copyright.

The producing GOR is given by

$$GOR = R_s + \frac{k_{rg}}{\mu_g B_g} \frac{\mu_o B_o}{k_{ro}}. \tag{9.260}$$

If we increased pressure in the well gridblock, we would force additional gas into solution so the dissolved gas–oil ratio (R_s) would increase, but the free gas component would decrease. Thus, whether the producing GOR would increase or decrease due to an increase in pressure depends on the relative changes in the two components of the producing GOR. Which is dominant is difficult to predict in general and is even more complicated in this simple example because of counter current flow of oil and gas. The sensitivities were computed with the adjoint procedure and verified by comparison with sensitivities computed with the finite-difference method.

To illustrate the complexity, we map the sensitivity of the producing GOR to each gridblock vertical permeability as shown in Fig. 9.4. The figure on the left-hand side shows the map of these sensitivities at 30 days. Here, $\partial GOR / \partial k_{z,i,k}$ is nonnegative at all gridblocks. This is the expected result. Gas saturation is below critical so increasing k_z should result in lower pressure gradients (higher pressures) so R_s and hence the producing GOR should increase resulting in a positive sensitivity as shown in the left-hand plot of Fig. 9.4. Note the sensitivity to vertical permeability in gridblock $(8, 7)$ is higher than the sensitivity to the vertical permeability in the well gridblock $(8, 8)$. This is because increasing the vertical permeability in gridblock $(8, 7)$ increases two vertical transmissibilities. On the other hand, at 400 days, the free gas saturation component of the producing GOR dominates. At this time, only gridblocks near the well gridblock have positive sensitivities and the most negative ones are near the center of the cross section (right-hand plot of Fig. 9.4). This largely reflects that increasing vertical permeability in this part of the reservoir causes more gas to migrate to the top and thus less flows to the well resulting in a decrease in the GOR and hence a negative

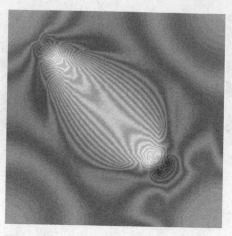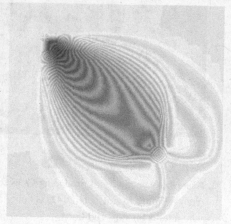

Figure 9.5. Sensitivity of producing WOR to log-horizontal permeability (left) and porosity (right). (Adapted from Wu *et al.* [41], copyright SPE.)

sensitivity. Many interesting examples of sensitivity coefficients can be found in Li [69] and Wu [132].

The second example of sensitivities is more straightforward and is taken from Wu [132]. It pertains to two-dimensional two-phase (oil–water) flow. The finite-difference simulator equations are solved on a $25 \times 25 \times 1$ grid system. There is one producing well and one water-injection well. The injection rate is constant and the producer is constrained to produce at a constant liquid rate. The rates are such that there is little change in the pressure during the run. Letting (i, j) denote the indices of the areal gridblocks, the producer is located at gridblock $(5, 5)$ and the injector at gridblock $(18, 18)$ in the results shown. The permeability and porosity fields are homogeneous. Again, sensitivities are computed from the adjoint method, but verified by comparison to the same sensitivities computed from the finite-difference method. Figure 9.5 shows the sensitivity of the producing WOR ratio to the log-permeability field (left) and the porosity field (right) at a time soon after breakthrough. Note, that even though we are computing the sensitivities of WOR to gridblock porosities and log-permeabilities, the pattern of the sensitivity plots resemble streamlines and reflects the contribution of water flow along each streamline to water production. Thus, we will speak of sensitivity along streamlines. Note that when the sensitivity of WOR to porosity along streamlines is nonzero, it is negative because increasing the porosity along a streamline will decrease water production. Because viscosities are constant and the change in formation volume factors is quite small, the change in water production to a fixed perturbation to porosity will be greatest when the change in porosity results in the greatest change in saturation at the producer, or more accurately, the change in the relative permeability ratio. For the results in Fig. 9.5, this occurs along the shortest streamlines between injectors and producers.

Similarly, the sensitivity of WOR to $\ln(k)$ reflects flow along streamlines. Increasing log-permeability along a streamline results in a higher flow velocity along the streamline and hence a higher water saturation and higher producing WOR. Thus, the sensitivity to permeability along a streamline is positive. One interesting feature that is apparent in the left-hand plot of Fig. 9.5 is that the sensitivity of WOR to log-permeability in gridblocks adjacent to the injector, but not between injector and producer, is approximately equal to -0.1. These sensitivities are negative because increasing permeability in these blocks causes part of the injected water to flow in the direction of gridblock (25, 25) instead of towards the producing well. The reader is referred to Wu [132] for a lengthy discussion of the sensitivity of WOR to rock properties at different times as well as results on the sensitivities of pressures to the rock property fields.

9.9.2   History-matching example

Here we consider an example from Gao [73]. The example pertains to the well known PUNQ-S3 model [71, 133] which was formulated to test methods for the quantification of uncertainty. The simulation model of the dome-shaped reservoir consists of a $20 \times 30 \times 5$ grid with about 1800 active gridblocks. Note we have more gridblocks than for the original PUNQ-S3 study because we use a numerical aquifer. Although the stochastic model from which the true reservoir property fields were generated was not exactly a Gaussian (http://www.nitg.tno.nl/punq/), we approximate it by a Gaussian based on the prior PDF estimated by Gao [73]. The model parameters are of gridblock porosities, horizontal log-permeabilities and vertical log-permeabilities. All other rock properties and fluid properties are known. The true horizontal log-permeability field is shown in Fig. 9.6. Most gridblocks colored black in this and similar figures correspond to inactive cells although a few correspond to the numerical aquifer.

The reservoir has a small gas cap and is connected to an aquifer. Six producing wells are completed in the oil column. There are no injection wells. Ten realizations are generated by RML by minimizing the objective function of Eq. (8.23) for ten different realizations of (m_{uc}, d_{uc}). Minimization is done with LBFGS. Data matched correspond to 84 wellbore pressures, 25 producing GOR data and only eight producing water–oil ratio data. Thus, in accordance with the original PUNQ-S3 study, we have a

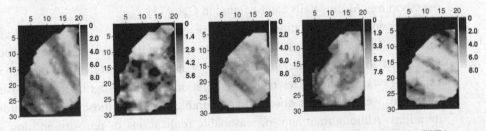

Figure 9.6. True horizontal log-permeability. (Adapted from Gao *et al.* [78], copyright SPE.)

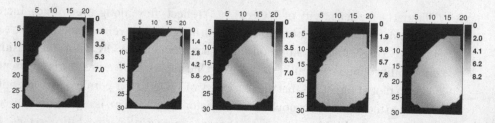

Figure 9.7. MAP estimate of $\ln(k)$.

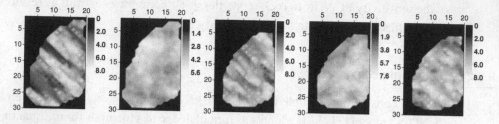

Figure 9.8. Conditional realization of $\ln(k)$. (Adapted from Gao *et al.* [78], copyright SPE.)

small number of data. Moreover the data contain a high degree of noise. Although the original study provided hard data at the wells, in the example presented here, we treat the more difficult problem where we do not use hard data. In minimizing Eq. (8.23) it is natural to use m_{uc} as the initial guess of the conditional realization. Then we can essentially think of the minimization process as finding the closest model to m_{uc} that matches the data d_{uc}. This is definitely the procedure we advocate. Here, however, we will start with an initial guess based on the MAP estimate which is shown in Fig. 9.7. If all ten conditional realizations used the MAP estimate as the initial guess, it is doubtful that one would obtain conditional realizations corresponding to several modes of a multi-modal distribution. Even in the minimization with C_M providing regularization, it is possible to obtain unrealistically high (overshooting) or low (undershooting) values of gridblock rock property fields, if one does not provide some form of damping at early iterations or limit the changes in the model at early iterations. In the results presented here, we simply use the log-transform [78] to eliminate undershooting and overshooting. More details can be found in Gao [73].

The conditional realization of the log-permeability field is shown in Fig. 9.8. In this case, LBFGS required 93 iterations to converge. At convergence, all data were well matched and the criteria of Eq. (8.24) were satisfied. In fact, the upper bound of Eq. (8.24) was satisfied in fewer than 40 iterations so with a slightly looser convergence tolerance, we could have observed comparable results to those shown in around 50 iterations. Although not shown, reasonable realizations of porosity and log vertical permeability were also obtained.

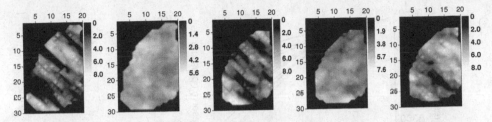

Figure 9.9. Unconditional realization of ln (*k*). (Adapted from Gao *et al.* [78], copyright SPE.)

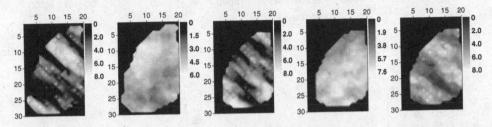

Figure 9.10. Conditional realization of ln (*k*). (Adapted from Gao *et al.* [78], copyright SPE.)

The results of Fig. 9.8 are not atypical. Some conditional realizations resemble the truth less than this one does, but all conditional realizations of $(\ln(k), \ln(k_z), \phi)$ show basic features from the geological model because the prior information is incorporated into the PDF, and all honor the production data. Figure 9.9 shows the unconditional realization that led to a conditional realization whose log-horizontal permeability field is shown in Fig. 9.10. The history matching of production data results in significant changes near the center of the layers which is where the wells are located.

The production data matched correspond to times up to 2920 days. After generating ten realizations of the conditional PDF by RML, we predicted reservoir performance for a total time of about 6000 days. As all wells operated at a fixed oil rate as long as the bottomhole pressure did not become too low, there is little uncertainty in the predicted cumulative oil rate.

Predictions of the cumulative gas and water rate and the average pressure are more interesting. In these plots, we compare predictions from the set of unconditional realizations (left-hand graph) with those from the conditional realizations (right-hand graph). In Figs. 9.11–9.13 the thick black curve always corresponds to a prediction based on the true model and the gray curves correspond to predictions from unconditional (left) or conditional (right) realizations. Figure 9.11 shows results on the predicted average pressure. In the truth case, the wells are produced at a constant oil rate. However, with the property fields of most of the unconditional realizations, some wells can not meet the target oil rate and are switched to bottomhole pressure control. As less fluid is then produced, the unconditional realizations give a biased (high) prediction of average pressure. However, after conditioning to production data, the predictions of average

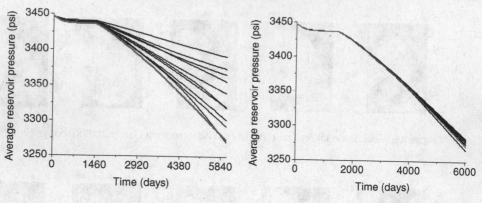

Figure 9.11. Predicted average pressure from unconditional (left) and conditional (right) realizations.

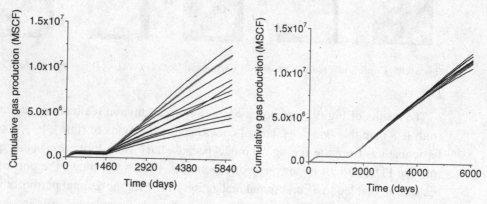

Figure 9.12. Predicted cumulative gas production from unconditional (left) and conditional (right) realizations.

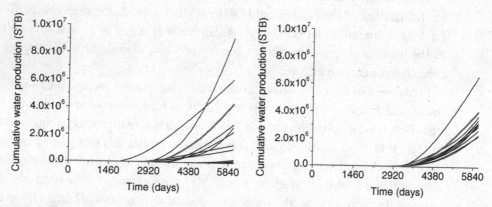

Figure 9.13. Predicted cumulative water production from unconditional (left) and conditional (right) realizations.

pressure encompass the predicted average pressure from the true model and appear to give a reasonable characterization of the uncertainty in predictions. Of course, a set of ten realizations is far too few to give an accurate mean and variance. Moreover, RML can be shown to sample correctly only in the case where data are linearly related to the model. Results similar to those obtained for average pressure are also obtained for cumulative gas production and cumulative water production. Most unconditional realizations of the model result in a later water breakthrough time than the true, whereas the conditional realizations give much more reasonable breakthrough times in general but tend to slightly underestimate cumulative water production. As there are only ten conditional realizations and the set of predictions of cumulative water predictions from them bound the prediction from the truth, one can not conclude that our predictions are biased. Note we only matched eight water-cut data which may not be sufficient to resolve the geological features that control future water production.

9.10 Evaluation of uncertainty with a posteriori covariance matrix

9.10.1 Approximate a posteriori covariance matrix

Let m_∞, denote the MAP estimate obtained by minimizing the objective function

$$O(m) = \frac{1}{2}(m - m_{\text{prior}})^{\text{T}} C_M^{-1}(m - m_{\text{prior}}) + \frac{1}{2}(g(m) - d_{\text{obs}})^{\text{T}} C_D^{-1}(g(m) - d_{\text{obs}}),$$

(9.261)

by applying some optimization algorithm. As m_∞ corresponds to a minimum of $O(m)$,

$$\nabla O(m_\infty) = C_M^{-1}(m_\infty - m_{\text{prior}}) + G_\infty^{\text{T}} C_D^{-1}(g(m_\infty) - d_{\text{obs}}) = 0,$$

(9.262)

where G_∞ denotes the sensitivity matrix evaluated at m_∞. Solving for m_∞ gives

$$\begin{aligned} m_\infty &= m_{\text{prior}} - (G_\infty^{\text{T}} C_D^{-1} G_\infty + C_M^{-1})^{-1} G_\infty^{\text{T}} C_D^{-1}[g(m_\infty) - d_{\text{obs}} - G_\infty(m_\infty - m_{\text{prior}})] \\ &= m_{\text{prior}} - C_M G_\infty^{\text{T}} (C_D + G_\infty C_M G_\infty^{\text{T}})^{-1}[g(m_\infty) - d_{\text{obs}} - G_\infty(m_\infty - m_{\text{prior}})], \end{aligned}$$

(9.263)

where the second equality of Eq. (9.263) follows from the matrix inversion lemma, Eq. (7.44). In some neighborhood of m_∞, $g(m)$ can be well approximated by the first order Taylor series:

$$g(m) = g(m_\infty) + G_\infty(m - m_\infty).$$

(9.264)

When Eqs. (9.262) and (9.264) hold, the following second order Taylor's series expansion of $O(m)$ about m_∞ is exact:

$$O(m) = O(m_\infty) + \frac{1}{2}(m - m_\infty)^{\text{T}} H_\infty(m - m_\infty).$$

(9.265)

Using the last result, we can approximate the a posteriori PDF of Eq. (8.1) as

$$f(m|d_{\text{obs}}) = \hat{a} \exp\left[-\frac{1}{2}(m - m_\infty)^{\text{T}} H_\infty (m - m_\infty)\right], \tag{9.266}$$

where $\hat{a}$ is equal to the original normalizing constant multiplied by $\exp\left(-O(m_\infty)/2\right)$ and the Hessian H_∞ is given by

$$H_\infty = C_M^{-1} + G_\infty^{\text{T}} C_D^{-1} G_\infty. \tag{9.267}$$

Note Eq. (9.266) is in the form of a Gaussian PDF with mean m_∞ and covariance matrix given by

$$C_{M'} = H_\infty^{-1} = (G_\infty^{\text{T}} C_D^{-1} G_\infty + C_M^{-1})^{-1}. \tag{9.268}$$

$C_{M'}$ is referred to as the posterior or a posteriori covariance matrix and is often denoted by C_{MAP}. If the linearized approximation of Eq. (9.264) is exact for all m, Eq. (9.266) holds for all m so the posterior PDF is Gaussian.

9.10.2 Normalized variance

Using the matrix inversion lemma of Eq. (7.49), Eq. (9.268) can be rewritten as

$$C_{M'} = C_M - C_M G_\infty^{\text{T}} (G_\infty C_D^{-1} G_\infty^{\text{T}} + C_D)^{-1} G_\infty C_M. \tag{9.269}$$

Here, the diagonal entries of C_M and $C_{M'}$ are denoted, respectively, by $c_{i,i}$ and $c'_{i,i}$ and represent the prior and posterior variances of m_i, for $i = 1, \ldots, N_m$. As the matrix subtracted from C_M is real symmetric positive semidefinite, its diagonal entries, denoted by $r_{i,i}$, $i = 1, 2, \ldots, N_m$, must be nonnegative. The normalized posterior variance of each model parameter, m_i, is defined by

$$\text{var}_{n,i} \equiv \frac{c'_{i,i}}{c_{i,i}} = 1 - \frac{r_{i,i}}{c_{i,i}}. \tag{9.270}$$

Note the normalized covariance gives a measure of the reduction in the variance obtained by integrating the observed data. If, for example, the normalized variance of m_i is 0.6, then we have obtained a forty percent reduction in the variance obtained by conditioning the prior model to the observed data d_{obs}. Although it is convenient to identify a reduction in variance to a reduction in uncertainty, considering only variances ignores the change in uncertainty due to the change in the covariances. Nevertheless various authors, e.g. Li [69], He *et al.* [100], Gavalas *et al.* [116], Li *et al.* [134] have compared priori and posteriori variances (or their sums) to obtain a measure of the reduction in uncertainty obtained by conditioning to different types of data.

9.10.3   Dimensionless sensitivity matrix

To obtain a better measure of the change in uncertainty, it is useful to work in terms of the dimensionless sensitivity matrix G_D defined by

$$G_D = C_D^{-1/2} G C_M^{1/2}, \tag{9.271}$$

where G is the sensitivity coefficient evaluated at some particular m, $C_M^{1/2}$ denotes the square root of the covariance matrix and $C_D^{-1/2}$ denotes the inverse of the square root of C_D. For our purposes, these square roots can be generated by either the Schur decomposition or the Cholesky decomposition. Although the two methods give different dimensionless sensitivity matrices, for the purposes of characterizing the reduction in uncertainty, both methods give the same result. Here, we use the general notation

$$C_M = C_M^{1/2} C_M^{T/2}, \tag{9.272}$$

where $C_M^{T/2}$ denotes the transpose of $C_M^{1/2}$. If $C_M = LL^T$ is the Cholesky decomposition of C_M where L is lower triangular, then $C_M^{1/2} = L$. From Eq. (9.272), it follows that

$$C_M^{-1} = C_M^{-T/2} C_M^{-1/2}, \tag{9.273}$$

where $C_M^{-T/2}$ denotes the inverse of $C_M^{T/2}$. Similarly,

$$C_D = C_D^{1/2} C_D^{T/2}, \tag{9.274}$$

and

$$C_D^{-1} = C_D^{-T/2} C_D^{-1/2}. \tag{9.275}$$

Introducing the square roots of the covariance matrices and letting G and G_D represent the sensitivity and dimensionless sensitivity matrices evaluated at the MAP estimate, and letting I_{N_m} and I_{N_d}, respectively, denote the $N_m \times N_m$ and $N_d \times N_d$ identity matrices, we can rewrite Eq. (9.269) as

$$\begin{aligned}
C_{M'} &= C_M^{1/2} \big(I_{N_m} - C_M^{T/2} G^T C_D^{-T/2} \big[C_D^{-1/2} G C_M^{-1} G^T C_D^{-T/2} + I_{N_d} \big]^{-1} C_D^{-1/2} G C_M^{1/2} \big) C_M^{T/2} \\
&= C_M^{1/2} \big(I_{N_m} - G_D^T \big[G_D G_D^T + I_{N_d} \big]^{-1} G_D \big) C_M^{T/2}.
\end{aligned} \tag{9.276}$$

Assuming the a posteriori PDF is Gaussian (Eq. 9.266) with covariance matrix $C_{M'}$, a surface of the form

$$(m - m_\infty)^T C_{M'}^{-1} (m - m_\infty) = r^2 \tag{9.277}$$

is a surface of constant probability density and the interior of this ellipsoid represents a confidence region [135]. Smaller values of r^2 correspond to lower confidence percentages and consequently smaller regions. Thus, the volume of the ellipsoid reflects the uncertainty in m. This volume is given by

$$V' = \frac{\sqrt{r^2 \pi}^{N_m}}{\Gamma(1 + (N_m/2))} \sqrt{\det C_{M'}}, \tag{9.278}$$

where Γ is the Gamma (generalized factorial) function. For the same value of r^2, the corresponding volume of the ellipsoid

$$(m - m_{\text{prior}})^{\text{T}} C_{M'}^{-1} (m - m_{\text{prior}}) = r^2 \tag{9.279}$$

is given by

$$V = \frac{\sqrt{r^2 \pi}^{N_m}}{\Gamma(1 + (N_m/2))} \sqrt{\det C_M}. \tag{9.280}$$

The ratio of V' to V represents the reduction in uncertainty obtained by integrating d_{obs} into the PDF. From Eqs. (9.279), (9.280), and (9.276), it follows that

$$\frac{V'}{V} = \frac{\sqrt{\det C_{M'}}}{\sqrt{\det C_M}} = \left[\det \left(I_{N_m} - G_D^{\text{T}} [G_D G_D^{\text{T}} + I_{N_d}]^{-1} G_D \right) \right]^{1/2}. \tag{9.281}$$

It is straightforward to show that

$$\frac{V'}{V} = \frac{\sqrt{\det C_{M'}}}{\sqrt{\det C_M}} = \left[\Pi_{j=1}^{p} \frac{1}{1 + \lambda_j^2} \right]^{1/2}, \tag{9.282}$$

where λ_j, $j = 1, 2, \ldots, p$ are the p nonzero singular values of the dimensionless sensitivity matrix. Thus, small singular values (say $\lambda_j < 0.01$) have a completely negligible influence on the reduction in uncertainty. Although Eq. (9.282) gives an accurate characterization of the relative uncertainty in m based on the prior and posterior probability density functions, if one wishes to measure the uncertainty in an individual model parameter, the normalized variance defined in Eq. (9.270) may still be useful. For large problems, however, the normalized variance for most of the parameters may be very close to unity [100].

9.10.4 Example

The example considered here was originally presented by Li [69]. It pertains to a three-phase flow problem where a MAP estimate of the log-horizontal permeability field was generated by matching data for flowing bottomhole pressure, gas–oil ratio (GOR) and water–oil ratio, or various combinations of these data. All reservoir variables except $\ln(k)$ were known exactly. It is a two-dimensional simulation example on a $21 \times 21 \times 1$ reservoir simulation grid with all gridblocks having equal volume. The reservoir contains four producing wells, one near each corner with each well produced at a constant total rate. Water is injected at a constant rate. Producing wellbore pressures fall below bubble-point pressures soon after production begins. By the end of a 240 day producing period, the producing GOR at three out of the four producing wells (wells 1, 3, and 4) is 150–250 scf/STB higher than the initial dissolved gas–oil ratio which is equal to 1400 scf/STB. The other producing well (well 2) is located in a high permeability region and there, at the end of the 240 day producing period, the producing GOR is only slightly higher than the initial dissolved GOR. On the other hand, because of a high permeability connection between well 2 and the injection well, water breaks through first at well 2; breakthrough time is equal to 140 days. The latest water breakthrough time is 190 days. Production data were generated from a reservoir simulation run with the observed data to be history matched selected from the output file from the simulation run. Specifically, the vector of observed data consisted of flowing bottomhole pressure, producing GOR, and producing WOR at each of the four producing wells "measured" at 30 day intervals with the first "measurement" occurring at 30 days. Thus we had eight "measurements" for each type of data at each well. No noise was added to the data generated from the simulator so the data are very accurate even though we specified the standard deviation of the pressure measurement error as 1 psi and the standard deviation of the GOR measurement error as 10 scf/STB. Because the observed data used did not actually include noise, increasing the standard deviation of the GOR measurement error to 100 scf/STB did not appreciably affect the results. The standard deviation of the water measurement error was variable but was in general about one percent of the observed WOR. The prior model for $\ln(k)$ was based on a spherical covariance function with correlation length equal to four gridblocks and variance equal to 0.5.

To estimate the value of each type of data, Li [69] history matched each type of data individually as well as various combinations of the data types. The results in Fig. 9.14 show the normalized a posteriori variances of $\ln(k)$ at each gridblock computed from Eq. (9.270). Figure 9.14(a) shows the normalized variances obtained when the MAP estimate of $\ln(k)$ was generated by matching only WOR data at the four producing

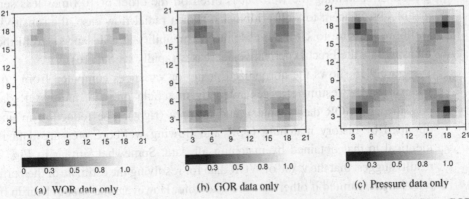

(a) WOR data only   (h) GOR data only   (c) Pressure data only

Figure 9.14. Normalized posterior variance for $\ln(k)$ field, conditioning to only WOR (left) or GOR (center) or pressure (right) [69].

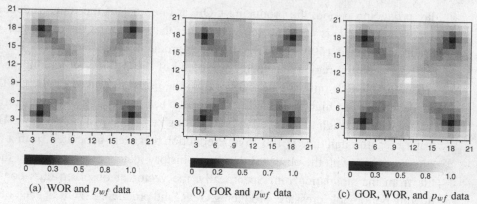

(a) WOR and p_{wf} data   (b) GOR and p_{wf} data   (c) GOR, WOR, and p_{wf} data

Figure 9.15. Normalized posterior variance for $\ln(k)$ field, conditioning to two or three types of data [69].

wells, Fig. 9.14(b) pertains to the case where only producing GOR data were matched, and Fig. 9.14(c) represents normalized a posteriori variance when the MAP estimate was generated by matching only flowing bottomhole pressure at the producing wells. Note that for most gridblocks, the normalized variances are equal to 1.0 when only WOR data is history matched, i.e. at these gridblocks the reduction in uncertainty in $\ln(k)$ due to integrating WOR data is negligible. On the other hand, when only the GOR data is history matched, at most gridblocks, the normalized variance is reduced to between 0.8 and 0.5, a 20–50 per cent reduction in uncertainty. The normalized variances obtained by conditioning only to pressure data (right-hand plot in Fig. 9.14) are similar to those obtained by conditioning only to GOR. The biggest difference is that in the gridblocks containing wells, the normalized variance is less than 0.1, i.e. the uncertainty in the absolute permeability in the well gridblocks is very small. This is a typical result in that the flowing wellbore pressure is extremely sensitive to the gridblock containing the well but is often on the order of 10 times less sensitive to gridblocks adjacent to well gridblocks. (For a radial flow case, the sensitivity of the wellbore pressure to $k(r)$, the absolute permeability at a distance r from the well as measured by the Fréchet derivative decreases rapidly with r [16].)

Figure 9.15 shows the normalized posterior variances computed from Eq. (9.270) when the MAP estimate is generated by matching pressure and WOR data (left), pressure and GOR data (center), and all data (right). The reduction in uncertainty obtained by history matching GOR and flowing wellbore pressure data is almost identical to that obtained by matching all data. Somewhat similar to Fig. 9.14, the results suggest that the value of WOR data for resolving the estimate of the permeability field is quite limited if other data are available. However, one should bear in mind that this is a two-dimensional example, data have no noise and that gas and water flow along similar paths to the producing wells. If one injects water near the bottom of a true three-dimensional reservoir, producing WOR data may be useful for resolving the permeability fields along the streamlines from an injector to a producer.

In the final example of this chapter, we compare sensitivity coefficients with dimensionless sensitivity coefficients. The results are from an example considered by Zhang et al. [103] who considered the problem of generating realizations of the parameters which determine the size, shape, and position of a channel within a reservoir together with the permeability and porosity in the channel facies and the permeability and porosity in the nonchannel facies. They assumed uniform permeability and porosity within each of the two facies. For our illustration, we use a diagonal prior covariance matrix, C_M, and a diagonal data covariance matrix C_D with all diagonal entries of C_D equal to 1 psi^2. With C_D and C_M diagonal, the element in the ith row and jth column of the dimensionless sensitivity matrix (Eq. 9.271) is identical to the dimensionless sensitivity of the ith data, g_i, to the jth model parameter, m_j which is defined by

$$s_{i,j} = \frac{\partial g_i}{\partial m_j} \frac{\sigma_{m_j}}{\sigma_{d,i}}, \qquad (9.283)$$

where $\sigma_{m,j}^2$ is the prior variance of m_i and $\sigma_{d,i}^2$ is the variance of the ith measurement error. Here we consider the case where $g_i = g_i(m)$ represents predicted data corresponding to the ith observation of the flowing wellbore pressure at a well in the channel. We consider the case where m_i is equal to either channel permeability k_c or channel porosity, ϕ_c. The left-hand plot in Fig. 9.16 shows the sensitivity of flowing wellbore pressure to ϕ_c and k_c as a function of time. Note wellbore pressure is much more sensitive to porosity than to permeability which might lead one to believe that pressure data would resolve porosity better than permeability. However, from basic well-testing theory, we know that, during the infinite acting radial flow period, permeability can be estimated accurately from flowing wellbore pressure data, but porosity can not. This apparent discrepancy can be resolved by considering dimensionless

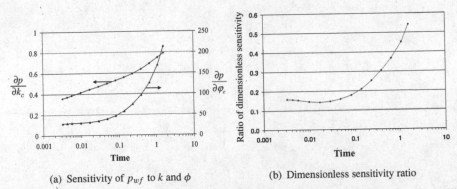

(a) Sensitivity of p_{wf} to k and ϕ (b) Dimensionless sensitivity ratio

Figure 9.16. Sensitivities (left), and ratio of dimensionless sensitivities (right), of flowing wellbore pressure to porosity and permeability. (Reproduced from Zhang et al. [103] with kind permission from Springer Science and Business Media.)

sensitivities. Here, we use 25 md as the prior standard deviation for k_c and 0.05 as the prior variance for ϕ_c. Note in a relative sense, the prior standard deviation for k_c is much smaller that the prior standard deviation for ϕ_c. The right-hand plot in Fig. 9.16 shows the ratio of the dimensionless sensitivity of pressure data to porosity to the dimensionless sensitivity to permeability. The results indicate that the dimensionless sensitivity of pressure data to porosity is much less than the dimensionless sensitivity to permeability particularly when $t \leq 0.02$ days which corresponds to the time when the pressure data exhibit infinite acting radial flow. During later times ($t > 0.4$) days the pressure approximately exhibits linear flow. During linear flow, the wellbore pressure drop is dominated by a term which is inversely proportional to $t/\sqrt{k\phi}$, and then it is not surprising that dimensionless sensitivities become more nearly equal. In fact, during pseudosteady-state flow, the dimensionless porosity sensitivity would be much larger than the dimensionless permeability sensitivity.

10 Quantifying uncertainty

Reservoir engineers are not typically concerned with the value of permeability and its uncertainty at some particular location in the reservoir. More commonly, they need to quantify uncertainty in some function of all the reservoir variables. They might, for example, require an estimate of time until water-cut reaches 30%, or the oil production rate five years in the future, and the uncertainty in those estimates. In Chapter 8, we discussed computation of the mean and the covariance for linear inverse problems whose posterior probability distributions were multi-normal. We also discussed computation of the mode and a local approximation to the covariance when the problem was nonlinear. In most history-matching applications, the covariance is a relatively poor measure of uncertainty, in which case Monte Carlo methods may be required. By simulating the future performance of many realizations that have been sampled according to their probability of being correct, it is possible to characterize the uncertainty in reservoir production. For Monte Carlo methods, the key to quantifying uncertainty is to generate the samples correctly and efficiently.

In this chapter, we introduce methods of generating realizations from a PDF either unconditional or conditional to observations. We first present methods for unconditional simulation of Gaussian and truncated Gaussian random fields because this is the initial step for many methods of generating conditional realizations. The three methods introduced in this category are: Cholesky method, moving average method, and truncated Gaussian method. For multi-variate models, the unconditional realizations honor the spatial correlation among model parameters through the covariance modeling. The basis of these methods of unconditional simulation is filtering a field of Gaussian white noise and subsequent adjustment of the expectation of the realizations.

Following a description of unconditional simulation, the rest of this chapter is devoted to methods of generating realizations conditional to data. For problems having a linear relationship between the model variables and the data, the a posteriori PDF of the model variables is a Gaussian distribution, for which the problem of sampling is trivial. Unfortunately, the relation between reservoir model variables (such as permeability and porosity) and the data (such as measured water-cut or pressure) is often highly nonlinear, in which case it can be very difficult to characterize the a posteriori PDF. For problems such as this, approximate methods of uncertainty evaluation often provide better solutions than exact methods considering the efficiency of generating realizations and the quality of the approximation to the correct a posteriori PDF. Among the conditional simulation methods, we first introduce the rejection method as it is relatively straightforward to implement and easy to understand. We then introduce the LU or Cholesky method for generating conditional realizations for linear problems (with Gaussian PDF) and the Linearization about the Maximum A Posteriori model (LMAP) method for nonlinear problem sampling. In both methods, roughness is added to the relatively smooth maximum a posteriori model. We then introduce the Randomized Maximum Likelihood (RML) method, in which conditional realizations are generated by calibration of unconditional realizations. Several other approximate methods, such as the pilot-point method and the gradual deformation method, are described briefly. The Markov chain Monte Carlo (McMC) method is an exact method for sampling. Although the McMC is computationally expensive and impractical for uncertainty evaluation of reservoir performance, it can be used to validate the performance of approximate methods in cases where it is possible to generate a large number of realizations.

Finally each of the conditional simulation methods introduced in this chapter is applied to a synthetic highly nonlinear flow problem to evaluate the viability of uncertainty quantification in a nonlinear problem for each of the approximate methods. The distribution of realizations from the Markov chain Monte Carlo method is used as a standard for comparison.

Remark 10.1. It has been common in the geostatistics literature to stress the importance of generating "equally probable realizations." In reality, this is neither necessary nor desirable. Ironically, it does turn out that realizations drawn from the a posteriori PDF of a large (many variable) model are often approximately "equally probable" in some sense. Consider, for example, the distribution of realizations of the objective function in a large multi-Gaussian model. As the number of variables gets large, the mean value of $S(m) = (m - \mu)^T C_M^{-1}(m - \mu)$ is k where k is the dimension of the model m. The standard deviation of the realizations of the objective function approaches $\sqrt{2k}$ for large k. Thus the distribution of the $S(m)$ becomes narrower as the size of the model increases, and for very large models the realizations are approximately equiprobable. To simply state that all realizations are equally probable is a dangerous fallacy, however, that makes it difficult to justify rejection of a clearly implausible realization from a bad algorithm.

10.1 Introduction to Monte Carlo methods

The uncertainty in a linear functional u of model parameters m that are multi-normal, can be characterized by the covariance, which is easily computed,

$$C_U = AC_M A^T,$$

(10.1)

where $u = Am$. Because the relationship is linear, the PDF for the random vector u is also Gaussian, so its PDF is completely defined by its mean and covariance. Examples of this type of relationship include average porosity for a reservoir or the pressure drop at an unmeasured location in steady flow.

Suppose, however, that we wanted to characterize the uncertainty in some *nonlinear* functional of the model parameters, say $u = f(m)$ for a fairly general function f. By definition, the probability that u is within a region B is

$$Pr(u \in B) = Pr(m : f(m) \in B) = \int_A p_M(m) \, dm, \qquad (10.2)$$

where A is the region of space for which $f(m) \in B$. Assuming (1) that the transformation is one-to-one and the inverse $g(u) = m$ exists, and (2) that the derivative of g exists and is continuous, then, by change of variables in Eq. (10.2), we obtain

$$\int_A p_M(m) \, dm = \int_B p_M(g(u)) |g'(u)| \, du. \qquad (10.3)$$

Clearly, the integrand on the right-hand side of Eq. (10.3) must be the PDF of u. Unfortunately, because of the dimension of the problem and the complexity of the relationship between model parameters and predictions, use of this relationship for reservoir characterization problems is impossible. In most cases the only practical alternative is the use of Monte Carlo methods.

10.1.1 Calculating expectations

Let m be a vector of M random variables, with probability density $\pi(\cdot)$. In reservoir characterization applications, m will usually comprise reservoir model parameters such as permeability and porosity. The task is to evaluate the expectation

$$E[f(m)] = \frac{\int f(m) \pi(m) \, dm}{\int \pi(m) \, dm} \qquad (10.4)$$

for some function of interest $f(\cdot)$. Here we allow for the possibility that $\pi(m)$ is known only up to a constant of normalization. That is, $\int \pi(m) \, dm$ is unknown. This is a common situation in practice, for example in Bayesian inference we know $P(m|d_{\mathrm{obs}}) \propto P(m)P(d_{\mathrm{obs}}|m)$, but the normalization constant $\int P(m)P(d_{\mathrm{obs}}|m) \, dm$ is not easy to estimate.

Monte Carlo integration evaluates $E[f(m)]$ by drawing samples $\{m_i, i = 1, \ldots, n\}$ from $\pi(\cdot)$ and then approximating the integral,

$$E[f(m)] \approx \frac{1}{n} \sum_{i=1}^{n} f(m_i). \qquad (10.5)$$

So the population mean of $f(m)$ is estimated by a sample mean. When the samples $\{m_i\}$ are independent, laws of large numbers ensure that the approximation can be made as accurate as desired by increasing the sample size n. Note that here n is under our control: we can generate as many samples (plausible reservoirs) as desired or computationally feasible. It is important, however, that the $\{m_i\}$ be generated by a process which draws samples from throughout the support of $\pi(\cdot)$ in the correct proportions.

Remark 10.2. It is sometimes difficult for students to rectify the difference between Eqs. (10.4) and (10.5). The probability density explicitly appears in Eq. (10.4), but not in Eq. (10.5). Because of this, students sometimes believe that the realizations must be equally probable in order to use this approach. In fact, what is necessary is that the realizations be sampled from $\pi(\cdot)$. We illustrate this with a simple example. Consider the sum of the values on the upper faces of two fair dice. The possible outcomes and their probabilities are as follows.

Throw	2	3	4	5	6	7	8	9	10	11	12
Probability	$\frac{1}{36}$	$\frac{2}{36}$	$\frac{3}{36}$	$\frac{4}{36}$	$\frac{5}{36}$	$\frac{6}{36}$	$\frac{5}{36}$	$\frac{4}{36}$	$\frac{3}{36}$	$\frac{2}{36}$	$\frac{1}{36}$

Computing the expectation of the random sum using Eq. (10.4), we obtain

$$E[X] = 2 \times \frac{1}{36} + 3 \times \frac{2}{36} + 4 \times \frac{3}{36} + \cdots + 11 \times \frac{2}{36} + 12 \times \frac{1}{36} = 7.$$

On the other hand, we could use the Monte Carlo approach, in which we use the results of a large number of throws of the dice to estimate the expectation. Suppose we throw the dice 100 times with the following results: (8, 6, 7, 11, 5, 4, 8, 10, 10, 7, 6, 5, 5, 7, 6, 9, 4, 7, 5, 5, 9, 10, 8, 6, 3, 5, 6, 7, 9, 6, 3, 8, 3, 10, 9, 6, 6, 10, 4, 8, 7, 10, 7, 4, 7, 2, 7, 2, 9, 6, 7, 5, 7, 8, 5, 11, 4, 7, 8, 5, 9, 7, 6, 9, 7, 9, 7, 11, 7, 5, 6, 3, 3, 9, 10, 8, 12, 11, 9, 5, 5, 6, 10, 9, 9, 10, 3, 5, 3, 6, 6, 9, 5, 9, 6, 2, 6, 5, 7, 7). Computing the expectation of the random sum using Eq. (10.5), we obtain

$$E[X] = (8 + 6 + 7 + 11 + 5 + 4 + 8 + 10 + 10 + \cdots + 7 + 7)/100 = 6.8.$$

Note that in this case, some of the outcomes occur more than once. In fact, the frequency of occurrence will be proportional to the probability for that outcome. If we wished, we could count that the number 2 occured three times, the number 3 occured seven times, the number 4 occured five times, and so on, then write our Monte Carlo estimate as

$$E[X] = 3 \times 2 + 7 \times 3 + 5 \times 4 + 15 \times 5 + \cdots + 4 \times 11 + 1 \times 12)/100 = 6.8.$$

Only by grouping all identical realizations together do we get an equation for approximating the expectation that appears to contain the probability.

Simple example

Consider the application of these ideas to a highly nonlinear problem in reservoir engineering and suppose that we have been able to estimate fracture density within a

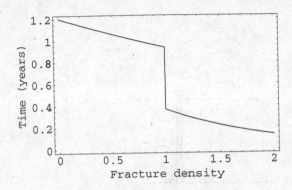

Figure 10.1. The relationship between breakthrough time (vertical axis) and fracture density (horizontal axis) for a fractured reservoir.

very low permeability matrix from core observations. Suppose, also, that the estimate of fracture density is 1.1 fracture centers per cubic meter. The estimate of the variance of the fracture density is 0.01 (in the same system of units). We want to estimate the time it will take for water to travel from an injector to a producer, when both are maintained at constant pressure. For low fracture densities, the flow is controlled by the matrix permeability, and the breakthrough time is large. For high fracture densities, it is very likely that a connected fracture path will exist from one well to the other and breakthrough will be very fast. This is a type of percolation problem, for which the change from long breakthrough time to short breakthrough time is quite abrupt (see Fig. 10.1). Suppose that the critical density for this particular reservoir is 1.0 fracture centers per cubic meter and that the goal is not to estimate the time it will take, but to characterize the level of uncertainty in the time it will take for water to break through.

Linearized Estimate

We might first try to characterize the uncertainty using the relationship appropriate for linear transformations. To do this, we would expand t_{bt} in a Taylor series about the point $f_D = 1.1$,

$$t_{bt}(f_D) = t_{bt}(1.1) + \left.\frac{dt_{bt}}{df_D}\right|_{1.1}(f_D - 1.1) + \cdots. \tag{10.6}$$

The linearized estimate of the mean of t_{bt} is just $t_{bt}(1.1) = 0.333$, and the linearized covariance estimate (from Eq. 10.1) is

$$\sigma^2_{t_{bt}} = \left.\frac{dt_{bt}}{df_D}\right|_{1.1} \sigma^2_{f_D} \left.\frac{dt_{bt}}{df_D}\right|_{1.1} \tag{10.7}$$

$$= 0.001\,11,$$

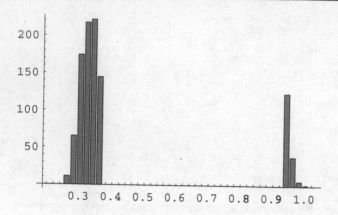

Figure 10.2. The histogram of one thousand realizations of the breakthrough time. Clearly, the PDF is not Gaussian, and the covariance would not be a particularly good measure of the uncertainty.

which is equivalent to a standard deviation of 0.0333. Clearly, this is too small, primarily because the derivative of t_{bt} with respect to fracture density at $f_D = 1.1$ is a very poor measure of the relationship between t_{bt} and f_D over the interval $0.9 \leq f_D \leq 1.3$.

Monte Carlo approach

In the Monte Carlo approach to the problem of estimating uncertainty in breakthrough time, we generate a large number of realizations of fracture density, f_D. Recall that we do not know its value exactly, and the PDF represents the uncertainty in our knowledge of f_D, so it is important to generate the realizations from the correct PDF. In this case we have assumed that the PDF for f_D is Gaussian so it is relatively easy to generate a large number of realizations. For each realization of f_D, we calculate the breakthrough time. These realizations of breakthrough time characterize the uncertainty in breakthrough time that results from uncertainty in fracture density.

Figure 10.2 shows the distribution of 1000 realizations of the breakthrough time. For this problem, we are able to adequately characterize the entire PDF by sampling. We could also use Eq. (10.5) to calculate functions of the distribution, such as the mean, $\langle t_{bt} \rangle = 0.43$, or the standard deviation, $\sigma_{t_{bt}} = 0.23$. Note that the standard deviation estimated from the Monte Carlo method is almost seven times as large as the standard deviation estimated from a linearized method.

10.2 Sampling based on experimental design

The methods of experimental design are strategies for setting up experiments such that the information needed for a desired statistical analysis is collected in an efficient and precise manner. The relationship between the perturbations applied to the system, and the system response is approximated by a simple linear or quadratic mathematical model based on a limited number of experiments. Two keys to this approach to uncertainty quantification are the efficiency of the experiments in exploring the sampling space, and the precision of the approximation of the response surface to the real system response at any point of the sampling space.

Experimental design methods are widely used in reservoir management for various purposes, such as probabilistic production forecasting, optimization of operational decisions, and history matching. Depending on the objective of the problem, we may choose a screening design for identifying variables with significant impact, or a response surface modeling design, or a combination of both types of designs. The screening designs are normally linear designs to filter and rank the uncertainty variables in scope, based on their effect on certain decision making metrics. The standard texts, such as Plackett and Burman [136], Winer [137], Box *et al.* [138], and McKay *et al.* [139], primarily discuss designs with unbiased main effect estimates from relatively small numbers of observations. For production forecasting, the metrics for decision making could be the recovery factor (RF) and time-discounted cumulative production. For history-matching problems, the "objective function" should be one of the key metrics. The "heavy hitter" variables are selected after filtering and ranking.

The response surface modeling designs are often quadratic designs for capturing non linear relationships between the variables and the metrics. An analytical function (often polynomial up to the second order) is fit to the experiment observations in the sampling space to form a "response surface." Box and Draper [140] and Myers and Montgomery [141] introduced interpolation and emulation algorithms for generating response surfaces with observation output from a given set of input settings. The analytical function describing the response surface is used in place of the real simulation model for estimation of the model values at any variable combinations, and for uncertainty quantification. Typically the robustness of the response surface is tested by making a number of independent experimental runs in the design space, and checking if the simulation results fall on the response surface. When the response surface is proven valid within the parameter distribution ranges, it may be used for production forecasting and uncertainty assessment. The response surface is used in place of the reservoir simulator to efficiently generate a large number of Monte Carlo predictions in the sampling space. The distribution of the key metrics from the group of Monte Carlo runs is often the basis for field development decision making.

The focus of this section is on the application of experimental design methods for estimation and uncertainty quantification of subsurface reservoir parameters. The objective function for minimization is approximated by a continuous response surface in the parameter space. When solving history-matching problems, experimental design requires a large number of simulation runs to build an accurate response surface for problems with highly nonlinear objective function. While gradient based methods are capable of solving history-matching problems with a large number of model parameters (i.e. thousands to millions), it is impractical to apply the experimental design approach for the same objective. Experimental design is, however, gradually gaining popularity for assisted history matching, because of the simplicity in algorithm and application. There are a good number of publications in recent years on experimental design for subsurface uncertainty and sensitivity analysis using the response surface approach. Eide *et al.* [142] introduced an iterative work flow for estimating reservoir parameters by approximating the reservoir response to changing of the variables as a response surface, and searching for the prediction from the response surface that is closest to the production history. White *et al.* [143] applied experimental design to the problem of recognizing and estimating significant geological parameters from the response surface. There are also many field application case studies on history matching and prediction, such as Kabir and Young [144] on Meren field, Alessio *et al.* [145] on a Luconia carbonate field, Peake *et al.* [146] on Minagish Oolite reservoir, and King *et al.* [147] on Nemba field.

The following sections introduce the experimental design approach for history matching around two aspects: screening designs for reducing the number of variables, and response surface designs for estimating model response.

10.2.1 Screening designs

In reservoir multi-phase flow systems, there are many possible variables impacting production/injection profiles, pressure measurements, and recovery. The parameter screening step defines the scope of uncertainty variables for history matching. The range of uncertainty for each of the n_m candidate variables forms the n_m-dimension sampling space to evaluate the objective function. The design of experiments is specified by a sample of n_d points in this sampling space. The objective function value at each of the n_d points requires a simulation run with a set of input value combinations of the n_m variables. The set of sampling points is denoted by the $n_d \times n_m$ matrix X given by

$$X = \begin{pmatrix} X_1 \\ \vdots \\ X_{n_d} \end{pmatrix} = \begin{pmatrix} x_{11} & x_{12} & \cdots & x_{1n_m} \\ \vdots & \vdots & \vdots & \vdots \\ x_{n_d1} & x_{n_d2} & \cdots & x_{n_dn_m} \end{pmatrix}. \tag{10.8}$$

X_i refers to the combination of model parameters for the ith measurement. If the operator $O(\cdot)$ represents the simulation and the objective function calculation process, the n_d sampling points in space X yield an array of n_d objective function values D through this process:

$$O(X) = D = \begin{pmatrix} d_1 \\ d_2 \\ \vdots \\ d_{n_d} \end{pmatrix}. \tag{10.9}$$

The relationship between parameter values X and the objective function $O(X)$ can be approximated by a quadratic matrix polynomial:

$$O(X) = A + XB + XCX^{\mathrm{T}} + \varepsilon_{ED}, \tag{10.10}$$

where A, B, and $O(X)$ are $n_d \times 1$ vectors, C is an $n_m \times n_m$ matrix, ε_{ED} is the error term for the polynomial approximation. The vector B and the diagonal elements of C represent the main effect of parameters X to the objective function, and the nondiagonal elements of C represent the two-way interaction effect between any two parameters. The three-way and higher level interaction effects are left in the error term and often omitted. For experimental design, the parameters are sampled at a certain number of values within the range of sampling, also referred to as "levels." In Eq 10.8, the columns x_{ij} for $i = 1$ to n_d vary at the value levels. The problem of history matching becomes one of solving for the coefficient matrices A, B, and C from n_d simulation runs that cover all levels of each parameter. As the number of simulations n_d needed for solving the second-order coefficients in C is in the order of $n_m \times n_m$, the number of parameters n_m is limited by computation resources.

Screening designs are utilized to reduce the list of parameters to the ones that significantly impact the objective function. Factorial and Plackett Burman designs [136] are two common methods for parameter screening purposes. Each parameter takes a prime number of levels (2, 3, 5, 7, ...), but generally two ("max" and "min") or three levels ("max," "min," and "mid"), and the levels are often represented by "1," "-1," and "0," or actual values of the variables. There are L^{n_m} possible combinations of the n_m parameters when all variables take L levels. For instance, reservoir pore volume can be calculated based on reservoir average thickness, equivalent diameter, and average porosity. If each of the three parameters take two levels "min" and "max," there are in total $2^3 = 8$ combinations. If three levels are taken for each variable, the full factorial design will need $3^3 = 27$ runs. The number of full factorial runs L^{n_m} quickly exceeds computational capacity with increased number of variables n_m and levels L.

Types of screening designs

Fractional factorial designs are subsets of the two-level full factorial design, with the number of experiments: $2^{n_m - p}$, where p is an integer indicating reduction of the design in size. The number of experiments should be always greater than the number of variables: $2^{n_m - p} > n_m$. The experiments for fractional factorial design should be properly selected, such that the sampling distribution of the variables are balanced and orthogonal. For a balanced design, each level of any certain variable is used an equal number of times in the set of experiments. Balanced designs are convenient for identifying and quantifying effects of model parameter variations. An experimental design is orthogonal if the inner product of any two columns in the design table is zero, i.e. $\sum_{i=1}^{n_d} x_{ij} x_{ik} = 0$ for any $j \neq k$. Orthogonal designs eliminate correlation between the main effects and interaction terms, so a minimal number of experiments are required for estimating the main effects. Detailed descriptions on how to generate fractional factorial designs are covered in many text books [148–150].

Plackett–Burman designs are subsets of the two-level full factorial designs with the maximum orthogonality among design columns. The number of experiments needs to be at least $n_m + 1$, and in multiples of 4 (e.g. 8, 12, 16, and so on). For instance, when $n_m = 7$, the number of experiments needed is 8; when $n_m = 16$, the number of experiments is 20. Eq. (10.11) represents the approximation of the Plackett–Burman design to the objective function at $X_i = x_{i1}, x_{i2}, \ldots, x_{in_m}$. To solve for the $n_m + 1$ unknowns $(a_0, b_1, \ldots, b_{n_m})$, at least $n_m + 1$ experiments are needed. When the number of experiments is exactly $n_m + 1$, there is a unique solution for the set of coefficients. When the number of experiments $n_d > n_m + 1$, the problem becomes overdetermined, and the solution from the least-square estimation is considered the best estimate for the coefficients.

$$O(x_i) = a_0 + b_1 x_{i1} + b_2 x_{i2} + \cdots + b_{n_m} x_{in_m}. \tag{10.11}$$

The theorems and procedures for generating Plackett–Burman design tables are fairly complex. Plackett and Burman [136] provided the first row of the design matrix for all $n_d < 100$ (except $n_d = 92$). A few examples with $n_d = 8, 12, 16$, and 20 are shown in Table 10.1. Each row has $n_d - 1$ components representing the maximum number of variables to which the row can be applied. All other rows or columns of the design are built by shifting the previous row one position at a time cyclically. The n_dth row is added to the design table consisting of the "-1" values of all variables.

Table 10.2 shows a seven-variable PB design with the first row the same as recommended in Table 10.1. Rows 2–7 are obtained by shifting the previous row by one position. There are 4 "-1"s and 4 "$+1$"s in each column, so the design is balanced. The inner product of any two different columns is 0, so the design is also orthogonal. In the case where there are $n_m < 7$ variables, only n_m columns in the table are required, which any n_m columns in Table 10.2 can be selected to form the new PB design table.

The applicability of the two-level screening designs is based on the linearity assumption of the dependent variable within the range of the sampling space. The two-level designs measuring the dependent variable at the high and low settings of the sampling

Table 10.1. *Root design for Plackett–Burman with number of runs* $p = 8, 12, 16,$ *and* 20.

	var 1	var 2	var 3	...															
$n_d = 8$	+	+	+	−	+	−	−												
$n_d = 12$	+	+	−	+	+	+	−	−	−	+	−								
$n_d = 16$	+	+	+	+	−	+	−	+	+	−	−	+	−	−	−				
$n_d = 20$	+	+	−	−	+	+	+	+	−	+	−	+	−	−	−	−	+	+	−

Table 10.2. *An $n_d = 8$ PB design table with seven variables.*

	var 1	var 2	var 3	var 4	var 5	var 6	var 7
1	+1	+1	+1	−1	+1	−1	−1
2	−1	+1	+1	+1	−1	+1	−1
3	−1	−1	+1	+1	+1	−1	+1
4	+1	−1	−1	+1	+1	+1	−1
5	−1	+1	−1	−1	+1	+1	+1
6	+1	−1	+1	−1	−1	+1	+1
7	+1	+1	−1	+1	−1	−1	+1
8	−1	−1	−1	−1	−1	−1	−1

Table 10.3. *A five-variable Plackett–Burman design with center run.*

	var 1	var 2	var 3	var 4	var 5
Run 1	+1	+1	+1	−1	+1
Run 2	−1	+1	+1	+1	−1
Run 3	−1	−1	+1	+1	+1
Run 4	+1	−1	−1	+1	+1
Run 5	−1	+1	−1	−1	+1
Run 6	+1	−1	+1	−1	−1
Run 7	+1	+1	−1	+1	−1
Run 8	−1	−1	−1	−1	−1
CP-Run	0	0	0	0	0

range can not detect any curvature or nonlinear relationship between the variables and dependent objective function. For instance, the magnitude of the aquifer strength can have nonlinear effect on recovery. In a fractured reservoir under primary depletion with fixed well bottomhole pressure, when the volume of aquifer connected to the reservoir is small, the field production rate will decline quickly with weak pressure support. Yet overly strong aquifer support may lead to early water breakthrough due to coning and fingering. So the low sweep efficiency also results in nonoptimal recovery. The main effect of aquifer strength on recovery would be underestimated if only accounting for the aquifer strength at the two extremes. As a result, the variable of aquifer strength could be screened out from history matching as insignificant to recovery. A center-point run with aquifer strength close to its best estimated value based on available information will help to identify and quantify the nonlinear relationship between aquifer strength and recovery.

A center-point run is often included in a two-level screening design, with all variables taking their center value. Table 10.3 shows an example five-variable Plackett–Burman design with a center run. The design table is constructed with the first five columns of Table 10.2, and a center run as the ninth row.

Table 10.4. *Folded design for the original five-variable Plackett–Burman design.*

	var 1	var 2	var 3	var 4	var 5
Run 9	−1	−1	−1	+1	−1
Run 10	+1	−1	−1	−1	+1
Run 11	+1	+1	−1	−1	−1
Run 12	−1	+1	+1	−1	−1
Run 13	+1	−1	+1	+1	−1
Run 14	−1	+1	−1	+1	+1
Run 15	−1	−1	+1	−1	+1
Run 16	+1	+1	+1	+1	+1

There are improved screening designs for fractional factorial and Plackett–Burman designs, including fold-over designs and Box–Behnken designs [138]. By making a number of additional experiments, the main effects confounded with interaction terms can be unmasked with the improved resolution. The mirror image fold-over designs are obtained by reversing the signs of all the entries in the base factorial or Plackett–Burman design table, and appending the new table to the base design. If the base case is a Plackett–Burman design with eight experimental runs, the resulted mirror image fold-over design will contain 16 runs. Table 10.4 shows the eight mirror image runs for the original PB design. Alternative fold-over designs may be used to augment selected main effects, sometimes with fewer additional runs. More details on fold-over designs can be found in Mee and Peralta [151], and John [152]. Box–Behnken designs are three-level factorial designs, and the number of experiments is $3^{n_m - p}$. If the center points are properly selected, this type of design is suitable for variable screening in nonlinear problems.

Analyzing screening designs

The primary goal of screening designs is to rank the variables by their main effects to the objective function, and to eliminate statistically insignificant variables from investigation. When there is no difference (or lack of evidence to prove the difference) in the mean and the variance of any subgroups of experimental results (null hypothesis), no statistical significance conclusion can be drawn from the test. To decide on the validity of a statistical significance, the variation of the test results should be greater than the variation of the measurement error carried in the experiment. An F-test [148] is used to evaluate and compare the magnitude of these two variations. For the "analysis of the variation" (ANOVA) test, the F-ratio is defined as the ratio of the "mean square treatment (MST)" to the "mean square error (MSE):"

$$F = \frac{\text{MST}}{\text{MSE}},$$

(10.12)

where the "treatment" is equivalent to "level" for computing a single variable F-ratio; the MST is the total variance of averaged data among all level groups normalized by its degree of freedom $L - 1$,

$$\text{MST} = \frac{\sum_{i=1}^{L} \frac{n_d}{L}(\bar{d}_i - \bar{d})^2}{L - 1};$$

(10.13)

and the MSE is the total variance of data within each level group normalized by the degree of freedom $n_d - L$,

$$\text{MSE} = \frac{\sum_{i=1}^{L} \sum_{j=1}^{n_d/L}(d_{ij} - \bar{d}_i)^2}{n_d - L}.$$

(10.14)

$\bar{d}$ is the average of all data from the experiments, and $\bar{d}_i$ is the average data from the subgroup of experiments with the variable of interest at its ith level. For the null hypothesis, there should be little difference between variation within level groups and variation among level groups, so the F-ratio is close to 1. The ANOVA test is based on the F-test. When the ANOVA test implies statistical significance in the experimental data, it becomes valid with further analysis of the variable rankings by their mean effect.

The Pareto chart is an easy and effective tool for ranking the main effect of the variables and the interaction effects. An example standard Pareto chart is shown in Fig. 10.3, where the variable with the most significant effect is ranked on the top, and the least significant variable at the bottom. In this Plackett–Burman design, the interaction effects are all lumped in the curvature term. The magnitude of each effect in the Pareto chart t_i is the standardized first-order coefficient of the variable i calculated

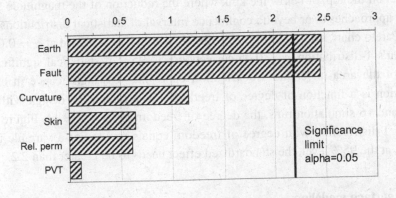

Figure 10.3. Pareto chart of standardized effects to discounted total oil production from a five-factor Plackett–Burman screening design.

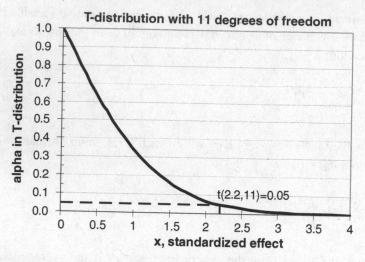

Figure 10.4. T-distribution with degree of freedom equals to 11.

from a least-square regression to the observed data:

$$t_i = \frac{b_i}{\sqrt{\sigma^2 \cdot C_{ii}}},$$

where b_i is the ith element of the first-order coefficient vector B in Eq. (10.10), and σ^2 is the variance of prediction error representing quality of the regression. The matrix C is the normalized square of the design matrix, also referred to as "Fisher information matrix." For two-level Plackett–Burman designs, the ith diagonal element of C is $C_{ii} = 1/n_d$ where n_d is the number of experiments.

The threshold line in a Pareto chart divides the variables into two groups: significant and insignificant, based on certain cut-off criteria, such as less than 10% of the maximum effect on the top, or below the kink where the reduction of the magnitude slows down and approaches 0, or beyond confidence interval of statistical distributions. In a standard Pareto chart, such as in Fig. 10.3, the threshold line is placed at $\alpha = 0.05$. In the Student's T-distribution, $\alpha = 0.05$ corresponds to 95% of statistical significance, i.e. 95% of the area under the T-distribution curve has less significance in effect. T-distribution is a function of degree of freedom. For a folded PB design with five variables and 16 simulation runs, the degree of freedom is $16 - 5 = 11$. Figure 10.4 shows the T-distribution with degree of freedom equals to 11. For a variable to be significant at the 95% level, the standardized effect needs to be greater than 2.2.

10.2.2 Response surface modeling

By regression on the experimental data, a response surface function is obtained representing the relationship between the variation of parameters and the objective function.

The response can be predicted continuously throughout the ranges of variables based on the information carried in the design. A robust response surface can be used for quantification of uncertainty in production forecasts through Monte Carlo sampling within the variation range of each parameter. The response surface is in general represented by polynomials of a given order. It is almost always an approximation to the actual model, so it is also called "proxy." A quadratic polynomial is often sufficient for describing the main effects and the second-order interactions. Based on Eq. (10.10) the second-order response surface model with n variables $\{x_1, \ldots, x_n\}$ is written as:

$$o(x) \approx a_0 + \sum_{i=1}^{n} b_i x_i + \sum_{i,j=1}^{n} c_{(i,j \geq i)} x_i x_j, \tag{10.15}$$

where $o(x)$ represents the objective function at any point in the sample space, $x = x_1, x_2, \ldots, x_n$. The elements in the coefficient terms B and C are regression results from a least-square calculation for matching data at experimental design points.

The three-level full factorial and the D-optimal are two most commonly used designs for selecting experimental design points for response surface modeling. For the size of a normal history-matching problem, a full factorial design is practical (number of runs $n_d \leq 27$) when the number of significant variables from the Plackett–Burman screening process is less than or equal to 3, such as the example shown in Fig. 10.3. The D-optimal design is applied for cases with more than 3 significant variables, as this design aims at maximizing the orthogonality of a set of variable combinations given a limited number of runs [141, 148].

Consider a simple history-matching problem with six variables, for matching the cumulative oil production at year 16. The variables are listed in the first row of Table 10.5. D-optimal design is used for selecting sampling points in the range of the parameter space. The measurements in this case are the difference between the recovery values from simulation at sample locations and the actual cumulative production:

$$O(X) = d_{\text{sim}}(X) - d_{\text{true}}.$$

The optimal locations for the solution would be in response surface regions with values close to 0. For the three-level design, the high $(+1)$, mid (0), low (-1) values of each variable are required as input. Although the distribution type of the variables can be triangular, Gaussian, or uniform, providing a median (P50) value as the best guess to the optimal value is recommended for a faster convergence to the solution.

The number of runs for a D-optimal design is largely decided by the user. The methodologies for generating D-optimal design tables are complex, and various optimal design algorithms are developed for improving the design efficiency, so the D-optimal design tables are software dependent. D-optimal design seeks to minimize the variance (or maximize the accuracy) of the coefficient estimates. The design matrices are not necessarily orthogonal or balanced. The quality of a D-optimal design is quantified

Table 10.5. *D-optimal design table and "measurement data" from simulation.*

	Skin Factors	Earth Model	Fault Sealing	Aquifer Strength	Perm Multipliers	Relative Perm.	Error (MMstb)
RUN 1	−1	−1	−1	−1	1	1	−49
RUN 2	−1	−1	0	−1	−1	0	−41
RUN 3	−1	−1	0	1	1	1	−20
RUN 4	−1	−1	1	−1	1	−1	−27
RUN 5	−1	−1	1	1	−1	−1	−26
RUN 6	−1	0	−1	0	−1	−1	−27
RUN 7	−1	0	1	1	0	1	−4
RUN 8	−1	1	−1	−1	1	−1	−43
RUN 9	−1	1	−1	1	−1	1	−23
RUN 10	−1	1	1	−1	−1	−1	−26
RUN 11	−1	1	1	−1	1	1	−6
RUN 12	−1	1	1	1	1	−1	4
RUN 13	0	−1	−1	1	0	−1	−19
RUN 14	0	−1	1	0	−1	1	−12
RUN 15	0	0	−1	1	1	0	4
RUN 16	0	1	−1	−1	−1	1	−25
RUN 17	0	1	0	1	−1	−1	6
RUN 18	1	−1	−1	−1	1	−1	−17
RUN 19	1	−1	−1	1	−1	1	−18
RUN 20	1	−1	1	−1	−1	−1	−14
RUN 21	1	−1	1	−1	1	1	−5
RUN 22	1	−1	1	1	1	−1	−3
RUN 23	1	0	0	−1	0	1	8
RUN 24	1	1	−1	−1	−1	−1	−6
RUN 25	1	1	−1	0	1	1	15
RUN 26	1	1	−1	1	1	−1	13
RUN 27	1	1	1	−1	−1	1	13
RUN 28	1	1	1	−1	1	−1	17
RUN 29	1	1	1	0	0	0	22
RUN 30	1	1	1	1	1	1	27
Ctr Run	0	0	0	0	0	0	7

by the D-efficiency criteria. The D-efficiency is a percentage number and its value depends on the number of runs, the number of independent variables in the design, and the composition of the design table.

$$\text{D-efficiency} = 100\frac{1}{n_d}|X'X|^{1/n_m},$$

where n_d is the number of experiments, n_m is the number of effects in the model including the interference terms. The best design is the one with the highest D-efficiency

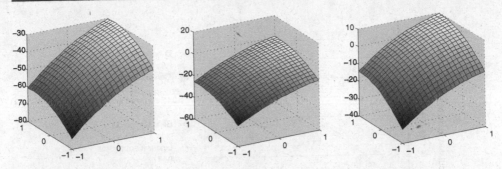

Figure 10.5. Response surface projection to a 2D variable range for well skins and fault sealing.

(100%). To maximize D-efficiency, the key is to maximize the determinant of the information matrix $|X^{\mathrm{T}}X|$. For our example, D-optimal design leads to 30 simulation runs for the six variables with acceptable D-efficiency (>40%). The list of variable combinations and measurements are shown in Table 10.5.

In this example, the response surface is nonlinear to the model variables and the correlation terms between the variables. Two variables, skin factors and fault sealing, are picked to form a 2D plane, and Fig. 10.5 shows the response surfaces projected on the plane defined by the ranges of the two variables. The amplitude of the 3D plots are the mismatch error, and the two horizontal axes are the normalized skin factors and the fault sealing factors. At the projection, all other parameter values are fixed at their low (left-hand plot), middle (middle plot), and high values (right-hand plot). The three plots show the same curvatures on the projection plane, as the second-order terms for defining the curvatures of the projected contours are the same terms.

When the variables carry large uncertainty, or the response is highly nonlinear in the range of the variables, the quadratic surface from regression can carry large error for predictions. Figure 10.6 shows a cross plot of the "measured data" versus response surface values for the example problem. The points should all line up alone the 45° line for a good proxy. At locations without data, "blind test" can be used for evaluating the approximation of a quadratic response surface to the real response. "Blind test" randomly proposes a few additional simulation runs that were not used in the response surface regression. In Fig. 10.6, these additional data points are plotted against response surface values as dots on the cross plot. When the response surface is not satisfactory in matching data, an iterative local refinement process is often suggested for searching optimum location in a reduced range. If the reservoir simulations were run in regions close to the optimal values, the relation between data mismatch and model parameters can be better linearized and it will be easier to find local minimum of the objective function. We may also select one or several optimal regions and model the response surfaces by separate refined experimental design process in each region. To improve the estimate of the response surface near a predicted optimum, a new design can be carried out close to this optimum [142].

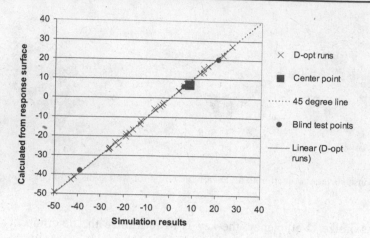

Figure 10.6. Cross-plot of experimental data vs predicted data from the response surface for validation of the response surface model. Dots are blind test points.

By approximation of a response surface to the real model response, it becomes possible to randomly sample a large number of locations from the range of the model and easily make model predictions from the response surface. This allows us to run a large number of Monte Carlo simulations for predicting the distribution of model response within the sampling range. In reserves booking, the key features of the model distribution would be the P10, P50, and P90 recovery values. The predicted distributions depend on selection of variable range and type of distribution. Uniform distribution of model variables will in general yield broader model distribution than truncated Gaussian distribution for a variable in the same range of sampling.

For history-matching problems, the error function based response surface is a reasonable tool for screening initial models, as running the Monte Carlo simulation on proxies is much faster than running full simulations. The key to this approach is to limit the number of model variables (normally less than 10) for feasibility in simulation time for the three-level experimental design runs. Obtaining a proxy in good match to the model is another challenge, which would require a set of blind tests for validation, or even breaking up the sampling range into multiple regions for refined experimental design.

10.3 Gaussian simulation

Reservoir model parameters that are typically estimated using inverse procedures include gridblock permeabilities, gridblock porosities, well skin factors, and relative permeabilities. Before conditioning to production data, the parameter variables are often assumed to be correlated Gaussian variables with specified prior mean and covariance. The prior PDF for model parameters m is given by

$$\pi_p(m) = c \exp\left[-\frac{1}{2}(m - m_{\text{prior}})^{\mathrm{T}} C_M^{-1}(m - m_{\text{prior}})\right], \tag{10.16}$$

where c is the normalizing constant. The term *prior* is used to differentiate from the distribution and parameter estimates after conditioning to observations. C_M is the prior covariance matrix of the model variables. It defines the relationship among the model variables and can be constructed from the prior information, mainly the variogram model. Eq. (10.17) shows the structure of C_M for a model with random permeability and porosity fields.

$$C_M = \begin{bmatrix} C_\phi & C_{\phi,k} \\ C_{k,\phi} & C_k \end{bmatrix}. \tag{10.17}$$

Matrices C_ϕ and C_k are the porosity covariance and permeability covariance among gridblocks respectively. $C_{\phi,k}$ is the cross covariance between the porosity and the permeability fields.

From Eq. (10.16), m_{prior} is not only the mean of the distribution, but also the model with the highest probability based on static data. Unconditional realizations of a reservoir model are generated from the prior PDF (Eq. 10.16) knowing the prior mean and the prior covariance. In the following sections we introduce two methods for generating unconditional realizations of Gaussian random fields and a method for generating reservoir facies. We begin, however, by discussing the generation of independent normal random deviates.

10.3.1 Generating (pseudo) random numbers

Before proceeding to the problem of generating large random vectors (reservoir models) from the a posteriori PDF, we briefly consider the related problem of generating random numbers from a univariate PDF. If a computer system has only one random number generator (called RANDOM or RAN), it probably generates pseudorandom numbers from a uniform distribution between 0 and 1. The discussion in Press *et al.* [8] on this type of random number generator is quite good, and will not be repeated here. The important points are that the distribution will probably be uniform, and that the system-supplied routine may not generate sufficiently large numbers of pseudorandom values before it begins to repeat. If that is the case, it may be necessary to replace the system routine with another routine that is more suitable for generating large numbers of pseudorandom numbers.

Random deviates with normal (Gaussian) probability distribution can be generated using many methods. One common approach is the Box–Muller method. From two uniform deviates x_1, x_2 on the interval $(0, 1)$, two deviates from the standard normal distribution, $N(0, 1)$, can be simulated as follows:

$$y_1 = \sqrt{-2 \ln x_1} \cos 2\pi x_2,$$
$$y_2 = \sqrt{-2 \ln x_1} \sin 2\pi x_2,$$

and y_1 and y_2 are independently distributed. Note that in addition to the cost of generating the uniform deviates, a logarithm, a square root, and a sine or cosine must be computed. Large vectors of identically distributed independent Gaussian deviates can be formed in this way from independent uniform deviates.

10.3.2 Cholesky or square-root method

The Cholesky method is a standard method generating realizations from a Gaussian PDF [153–155]. The core of this approach is the decomposition of the covariance matrix into the product of a matrix and its transpose:

$$C_M = LL^{\mathrm{T}}. \tag{10.18}$$

The decomposition of a symmetric positive-definite covariance matrix can be performed several ways; two conceptually straightforward methods are Cholesky decomposition [9, p. 309] and eigenvalue–eigenvector decomposition.

The Cholesky decomposition method solves for a lower triangular matrix L, which satisfies Eq. (10.18). This L matrix is often referred to as the square root of C_M. The elements of L are denoted as $l_{i,j}$, where i is the row index and j is the column index. The elements of C_M are denoted $c_{i,j}$. The Cholesky decomposition can be obtained by the following iterative procedure:

For $j = 1$ to N,

1. Compute the jth diagonal element of the L matrix

$$l_{j,j} = \left(c_{j,j} - \sum_{k=1}^{j-1} l_{j,k}^2 \right)^{1/2}. \tag{10.19}$$

2. Compute the remaining elements of the jth column, $l_{i,j}$ with $i = j+1, j+2, \ldots, N$. As L is lower triangular, only $l_{i,j}$ for $i \geq j$ are unknown.

$$l_{i,j} = \frac{1}{l_{j,j}} \left(c_{j,i} - \sum_{k=1}^{j-1} l_{i,k} l_{j,k} \right). \tag{10.20}$$

One difficulty of this method is that the accumulated roundoff errors may result in the right-hand side of Eq. (10.19) taking imaginary values for poorly conditioned matrices C. Another way to compute a square root of a covariance matrix is to start with the eigenvalue–eigenvector decomposition:

$$CU = U\Lambda,$$

where Λ is a diagonal matrix of the eigenvalues and the columns of U are eigenvectors of C. Then,

$$
\begin{aligned}
C &= U\Lambda U^{\mathrm{T}} \\
&= U\Lambda^{1/2}\Lambda^{1/2}U^{\mathrm{T}} \\
&= U\Lambda^{1/2}U^{\mathrm{T}}U\Lambda^{1/2}U^{\mathrm{T}} \\
&= (U\Lambda^{1/2}U^{\mathrm{T}})(U\Lambda^{1/2}U^{\mathrm{T}})^{\mathrm{T}} \\
&= RR^{\mathrm{T}}.
\end{aligned}
$$

Another, nonsymmetric, square root is $R' = U\Lambda^{1/2}$.

Unconditional realizations are generated from the prior PDF $N(m_{\mathrm{prior}}, C_M)$ by

$$
m_i = m_{\mathrm{prior}} + LZ_i, \tag{10.21}
$$

where m_i is the ith unconditional realization corresponding to the vector of independent normal deviates Z_i. The expectation and the covariance of the realizations from the Cholesky method can be shown to be correct on average, since

$$
\begin{aligned}
E[m_i] &= E[m_{\mathrm{prior}} + LZ_i] \\
&= m_{\mathrm{prior}} + LE[Z_i] \\
&= m_{\mathrm{prior}}
\end{aligned} \tag{10.22}
$$

and

$$
\begin{aligned}
E[(m_i - m_{\mathrm{prior}})(m_i - m_{\mathrm{prior}})^{\mathrm{T}}] &= L\,E[Z_iZ_i^{\mathrm{T}}]L^{\mathrm{T}} \\
&= LL^{\mathrm{T}} \\
&= C_M.
\end{aligned} \tag{10.23}
$$

Dietrich and Newsam [156] proposed an alternative method for generating the product LZ by using the first few terms in a Taylor series expansion of the square root of the covariance matrix. The method has been applied to simulation of reservoir properties [107].

Numerical examples

As an example of a matrix with a simple square root, let

$$
C = \begin{bmatrix} 4 & -1 & 1 \\ -1 & 4.25 & 2.75 \\ 1 & 2.75 & 3.5 \end{bmatrix},
$$

in which case the Cholesky decomposition of C is

$$L = \begin{bmatrix} 2 & 0 & 0 \\ -0.5 & 2 & 0 \\ 0.5 & 1.5 & 1 \end{bmatrix}$$

and it is easy to verify that $LL^{\mathrm{T}} = C$, so L is a type of matrix square root of C.

The symmetric square-root decomposition computed from the eigensystem of C is

$$R = U\Lambda^{1/2}U^{\mathrm{T}} = \begin{bmatrix} 1.938\,24 & -0.340\,223 & 0.357\,025 \\ -0.340\,223 & 1.860\,87 & 0.819\,394 \\ 0.357\,025 & 0.819\,394 & 1.643\,51 \end{bmatrix} \quad \text{and again} \quad RR^{\mathrm{T}} = C.$$

Another, nonsymmetric, square root is

$$R' = U\Lambda^{1/2} = \begin{bmatrix} -0.107\,486 & -1.978\,35 & 0.273\,076 \\ 1.959\,55 & 0.460\,45 & 0.445\,142 \\ 1.674\,23 & -0.665\,929 & -0.503\,469 \end{bmatrix}.$$

Figure 10.7 shows simulation results from two different square roots.

10.3.3 Moving average

The moving average method for generating realizations of a random field is closely related to the Cholesky method but has some advantages for stationary fields [157, 158]. The Cholesky method is not particularly useful for practical problems because of the expense of computing the decomposition of the covariance matrix for problems with large numbers of variables. When the size of a reservoir is large compared to the range of the correlation for properties, it is possible to obtain an approximation to the square root by consideration of the limit of the lattice or the dimension of the vector m_i becoming infinite.

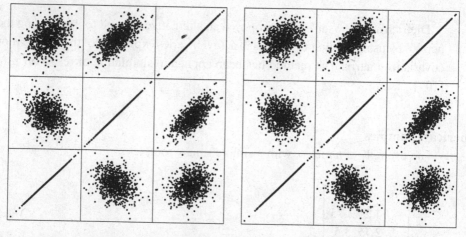

Figure 10.7. Cross-plots of variables generated using the square-root method, for Cholesky (left) and symmetric eigensystem decomposition (right).

Consider the following example of a Cholesky decomposition of a 1D exponential covariance matrix. For this example, the practical range of the exponential covariance is three grid cells. (We only show parts of several rows near the middle of the matrix.)

$$
L = \begin{bmatrix}
\ddots & \ddots & \ddots & \ddots & \ddots & \ddots & \ddots & \ddots & \ddots & \ddots & \ddots \\
0.02 & 0.05 & 0.13 & 0.34 & 0.93 & 0 & 0 & 0 & 0 & 0 & 0 \\
0.01 & 0.02 & 0.05 & 0.13 & 0.34 & 0.93 & 0 & 0 & 0 & 0 & 0 \\
0 & 0.01 & 0.02 & 0.05 & 0.13 & 0.34 & 0.93 & 0 & 0 & 0 & 0 \\
0 & 0 & 0.01 & 0.02 & 0.05 & 0.13 & 0.34 & 0.93 & 0 & 0 & 0 \\
0 & 0 & 0 & 0.01 & 0.02 & 0.05 & 0.13 & 0.34 & 0.93 & 0 & 0 \\
0 & 0 & 0 & 0 & 0.01 & 0.02 & 0.05 & 0.13 & 0.34 & 0.93 & 0 \\
0 & 0 & 0 & 0 & 0 & 0.01 & 0.02 & 0.05 & 0.13 & 0.34 & 0.93 \\
\ddots & \ddots & \ddots & \ddots & \ddots & \ddots & \ddots & \ddots & \ddots & \ddots & \ddots
\end{bmatrix}.
$$

$$(10.24)$$

Although the first few and last few rows might be different, all rows in the interior of the matrix appear to be identical. In this case a single row characterizes the covariance matrix and its Cholesky decomposition. Multiplication of matrices LL^T to compute the covariance is replaced by the convolution of one row of L with a column of L^T:

$$
[LL^T]_{i,i-3} =
\begin{array}{ccccccccc}
0.01 & 0.02 & 0.05 & 0.13 & 0.34 & 0.93 & 0 & 0 & 0 \\
\times & \times & \times & \times & \times & \times & \times & \times & \times \\
0 & 0 & 0 & 0.01 & 0.02 & 0.05 & 0.13 & 0.34 & 0.93 \\
\hline
0 & 0 & 0 & 0 & 0.01 & 0.05 & 0 & 0 & 0
\end{array} = 0.06,
$$

$$
[LL^T]_{i,i-2} =
\begin{array}{ccccccccc}
0.01 & 0.02 & 0.05 & 0.13 & 0.34 & 0.93 & 0 & 0 & 0 \\
\times & \times & \times & \times & \times & \times & \times & \times & \times \\
0 & 0 & 0.01 & 0.02 & 0.05 & 0.13 & 0.34 & 0.93 & 0 \\
\hline
0 & 0 & 0 & 0 & 0.02 & 0.12 & 0 & 0 & 0
\end{array} = 0.14,
$$

$$
[LL^T]_{i,i-1} =
\begin{array}{ccccccccc}
0.01 & 0.02 & 0.05 & 0.13 & 0.34 & 0.93 & 0 & 0 & 0 \\
\times & \times & \times & \times & \times & \times & \times & \times & \times \\
0 & 0.01 & 0.02 & 0.05 & 0.13 & 0.34 & 0.93 & 0 & 0 \\
\hline
0 & 0 & 0 & 0.01 & 0.04 & 0.32 & 0 & 0 & 0
\end{array} = 0.37,
$$

$$
[LL^T]_{i,i} =
\begin{array}{ccccccccc}
0.01 & 0.02 & 0.05 & 0.13 & 0.34 & 0.93 & 0 & 0 & 0 \\
\times & \times & \times & \times & \times & \times & \times & \times & \times \\
0.01 & 0.02 & 0.05 & 0.13 & 0.34 & 0.93 & 0 & 0 & 0 \\
\hline
0 & 0 & 0 & 0.02 & 0.12 & 0.86 & 0 & 0 & 0
\end{array} = 1.00,
$$

$$
[LL^{\mathrm{T}}]_{i,i+1} = \begin{matrix} 0.01 & 0.02 & 0.05 & 0.13 & 0.34 & 0.93 & 0 & 0 & 0 \\ \times & \times & \times & \times & \times & \times & \times & \times & \times \\ 0.02 & 0.05 & 0.13 & 0.34 & 0.93 & 0 & 0 & 0 & 0 \\ \hline 0 & 0 & 0.01 & 0.04 & 0.32 & 0 & 0 & 0 & 0 \end{matrix} = 0.37.
$$

In general, we could compute the (i, j) entry of the covariance matrix as follows

$$
\begin{aligned}
c_{i,j} &= \sum_k \ell_{i,k}\ell_{k,j}^{\mathrm{T}} \\
&= \sum_k \ell_{i,k}\ell_{j,k} \\
&= \sum_k f(k-i)f(k-j) \\
&= \sum_m f(m)f(m-(j-i))
\end{aligned}
$$

or

$$
C(h) = \sum_m f(m)f(m-h).
$$

In order for the sums to be equal, the limits of summation must extend beyond the range of the nonzero elements. Therefore this method is appropriate for a large grid, and for points away from the boundary.

Unconditional Gaussian fields are generated by convolution of the uncorrelated random normal deviates Z_i with covariance templates, i.e.

$$
m_i = m_{\mathrm{prior}} + L * Z_i, \tag{10.25}
$$

where $*$ is the convolution operator. For simulating a 2D Gaussian random field with exponential type covariance,

$$
C(r) = \sigma^2 e^{-r/a}, \tag{10.26}
$$

the kernel operator of its square root can be derived from the 2D Fourier transform:

$$
f(r) = \sigma K_{1/4}\left(\frac{r}{a}\right)(2\pi^2 a^3 r)^{-1/4}\Gamma\left(\frac{3}{4}\right)^{-1}. \tag{10.27}
$$

The formulation of Gaussian type covariance and its kernel in 2D are:

$$
C(r) = \sigma^2 e^{-r^2/a^2}, \tag{10.28}
$$
$$
f(r) = \sigma(4/a^2\pi)^{\frac{1}{2}}\exp(-2r^2/a^2). \tag{10.29}
$$

And a 2D Gaussian random field is generated using

$$
m(x, y) = m_{\mathrm{prior}}(x, y) + \int\!\!\!\int_{-\infty}^{\infty} f(x-s, y-t)Z(s, t)\,ds\,dt \tag{10.30}
$$

or its discrete equivalent.

Figure 10.8. The 2D uncorrelated Gaussian random field $Z \in N(\mathbf{0}, I)$.

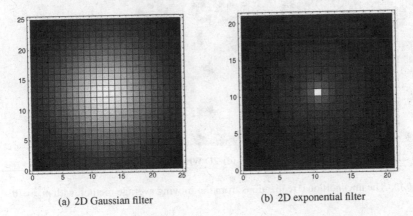

(a) 2D Gaussian filter   (b) 2D exponential filter

Figure 10.9. The moving average filters for square root of 2D Gaussian and exponential covariances.

There are two major advantages with the application of the moving average method for simulation of Gaussian random fields. First, the kernel is calculated explicitly, so it is not necessary to construct and store the covariance matrix. Second, the size of a kernel is generally determined by the correlation range in the field, and it is much smaller than the covariance matrix.

In order to demonstrate the simplicity of the method with fairly large grids, the moving average method is applied to a 256×256 grid of independent normal deviates Z as shown in Fig. 10.8. Two-dimensional filtering arrays L (25×25) for the square root of the 2D Gaussian and the exponential covariance are plotted in Fig. 10.9. Here assume $m_{\text{prior}} = \mathbf{0}$, then the random fields are $m = L * Z$ with different covariance types represented by the filtering matrix L. Figure 10.10 shows the 2D random field realizations with Gaussian, exponential, spherical, and Whittle covariance models.

The frames of uncorrelated cells surrounding the correlated images are a result of the fact that smoothing only is done for pixels that are surrounded by a large enough region for the filter to be applied. The practical consequence is that one would either

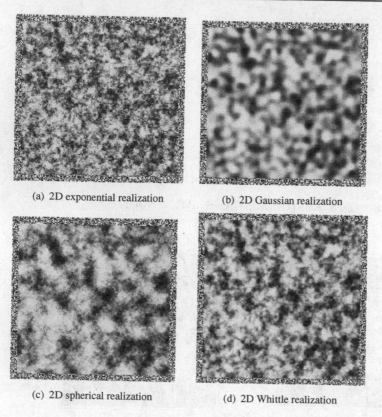

(a) 2D exponential realization

(b) 2D Gaussian realization

(c) 2D spherical realization

(d) 2D Whittle realization

Figure 10.10. The unconditional realizations from the moving average method with $m_{pr} = 0$.

have to generate a slightly larger grid than was truly necessary (and then discard the unwanted part after smoothing) or assume a periodic structure to the reservoir.

Example: generating a Gaussian random field using moving average

This example is dedicated to illustrate the algorithm of the moving average method. The porosity field on a uniform lattice of 128×128 gridlocks is assumed to be multi-normal with mean being 0.2 and variance 0.09. The principle correlation is along $-60°$ direction and equivalent to the length of 30 gridblocks. The shortest correlation range in the perpendicular direction has the length of 15 gridblocks. The covariance matrix in this case has the size $128^2 \times 128^2$, which is large for Cholesky decomposition. An unconditional realization of the porosity field can be generated by the moving average method following the steps below.

1. Rotate the coordinate system so that the longest correlation direction is along a coordinate axis, then stretch an axis to make the spatial correlation isotropic in the new coordinate system. This procedure has been explained in Section 5.4.4. The

coordinate transformation can be written as:

$$\begin{bmatrix} x'' \\ y'' \end{bmatrix} = \begin{bmatrix} 1 & 0 \\ 0 & 2 \end{bmatrix} \begin{bmatrix} \cos 120° & \sin 120° \\ -\sin 120° & \cos 120° \end{bmatrix} \begin{bmatrix} x \\ y \end{bmatrix}$$

in which case the distance measure is

$$r^2 = x''^2 + y''^2 = \begin{bmatrix} x & y \end{bmatrix} \begin{bmatrix} 3.249\,82 & 1.299 \\ 1.299 & 1.749\,96 \end{bmatrix} \begin{bmatrix} x \\ y \end{bmatrix}.$$

2. The filtering array has the largest value at the center where $r = 0$ and vanishes at the margins. We determine the necessary size of the filtering array and cut off the negligible margin values. The dimensions of the filtering array in the coordinate system of (x, y) are decided in the procedures below.
   - Let s be a cut-off criterion, $s \ll 1.0$.
   - Evaluate the kernel grid at the distance $r = 0$, i.e. $f(r = 0)$.
   - Increase r until $\frac{f(r=r_{max})}{f(0)} < s$, this r_{max} represents the distance in the coordinate system (x'', y'') within which $f(r)$ can not be cut off.

$$f(r) = 0.3 \left(\frac{12}{30^2 \pi} \right)^{1/2} \exp(-6r^2/30^2)$$

   - As r is decided by (x, y), the cut-off distance in the true coordinate system can be solved from:

$$r_{max}^2 = \begin{bmatrix} x & y \end{bmatrix} \begin{bmatrix} 3.249\,82 & 1.299 \\ 1.299 & 1.749\,96 \end{bmatrix} \begin{bmatrix} x \\ y \end{bmatrix},$$

   and

$$y = \tan(-60°)x.$$

   If $s = 0.02$, we obtain $x = 13$ and $y = 22.5$. The necessary dimensions of the kernel are $(27, 47)$ after rounding off the coordinates for the number of grids.

3. Compute the filtering array L in the necessary dimensions. The L filter is plotted in Fig. 10.11(a).

4. Generate an uncorrelated Gaussian random field $Z \in N(0, I)$. The dimensions of Z are: $(128 + 27 - 1, 128 + 47 - 1) = (154, 174)$, such that the convolution of L and Z yields a smoothed Gaussian field with 128×128 gridblocks.

5. Compute the unconditional realization by Eq. (10.25). An unconditional realization of the porosity field is shown in Fig. 10.11(b).

10.3.4 Truncated Gaussian simulation

In geostatistical simulation of permeability and porosity distributions, the assumption is almost always made that the rock properties are distributed randomly and that the

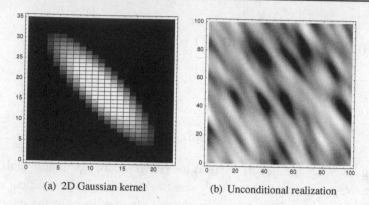

(a) 2D Gaussian kernel   (b) Unconditional realization

Figure 10.11. An unconditional realization from the moving average method.

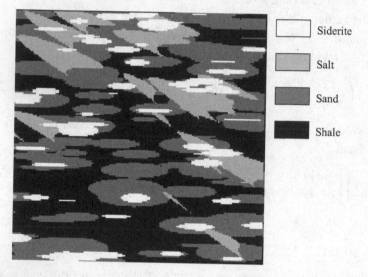

☐ Siderite

▨ Salt

▨ Sand

■ Shale

Figure 10.12. An example of sedimentary facies distribution in a formation.

randomness can be adequately described by the mean and the spatial covariance of the property fields. If there is more than one type of rock, region or facies, the assumption is usually made that the location of the boundaries of these regions is known. We define a facies to be a region of relatively uniform properties, i.e. the variation of properties within the facies is much less than the variation in properties between facies. Figure 10.12 shows a facies distribution map of the type that geologists might present. The truncated Gaussian simulation method simulates the randomness of the facies distributions by truncating a Gaussian random field into distinct regions, with each region representing a facies type.

The Gaussian random fields for generating facies distribution maps can be generated using any standard method. The lithotype in each grid is then decided by the value of

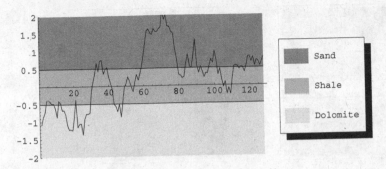

Figure 10.13. A schematic example shows the truncation of a one-dimensional random Gaussian field with exponential covariance. (Adapted from Liu and Oliver [104], copyright SPE.)

the random Gaussian field at this grid and the thresholds:

$$F(x) = \begin{cases} F_1 & \text{if } Y(x) \leq y_1, \\ F_2 & \text{if } y_1 < Y(x) \leq y_2, \\ \vdots & \vdots \\ F_N & \text{if } y_{N-1} < Y(x), \end{cases}$$

where x is the grid location, and y_i for $i = 1, \ldots, N - 1$ are the value of thresholds.

Figure 10.13 is a schematic example showing truncation of a one-dimensional random Gaussian field. In this one-dimensional reservoir model of 128 gridblocks, each gridblock is assigned a random Gaussian variable. Truncation thresholds are set at 0.5 and −0.5. For gridblocks in which the value of the Gaussian variable is greater than 0.5, they are assigned the sandstone facies type. For those with their Gaussian variables between 0.5 and −0.5, they are assigned as shale. The gridblocks with Gaussian variables below −0.5 are assigned as dolomite. Obviously, a slight perturbation of the threshold values will first change the facies type of the grids at the boundary of facies regions.

A more general application of the truncated Gaussian method is to simulate facies distribution in a 2D plane. The truncation of a 2D Gaussian random field using the same two thresholds is shown in Fig. 10.14 to illustrate the resulting facies distribution in a 2D plane. The figure on the left-hand side shows the random Gaussian function distributed on the field. The brighter the color, the greater the value of the Gaussian random variables in that area. The simulated facies map of the field is shown on the right. Three colors represent three different lithotypes. Note that the black and the white areas can not be directly adjacent to each other. This reflects one of the major drawbacks of the truncated "single" Gaussian method; the facies represented by nonadjacent scales of Gaussian values can never be in direct contact with each other in the facies field map

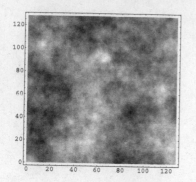 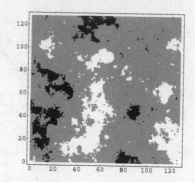

Figure 10.14. Truncation of a 2D random Gaussian field to simulate the distribution of three lithofacies.

unless the same facies is provided with separate threshold regions, which is unrealistic for most of the reservoir formations.

As an extension of the truncated Gaussian simulation method, the pluri-Gaussian simulation allows us to reproduce complex arrangements of several lithofacies. A second Gaussian random field is assigned to each of the gridblocks in the field and the two Gaussian fields together simulate the distribution of the lithofacies in the field. The two random functions can be either correlated or not. For the truncated pluri-Gaussian method, choosing a threshold scheme is no longer straightforward and the efficiency and flexibility of the truncated pluri-Gaussian method is largely ruled by the threshold scheme.

Le Loc'h *et al.* [159] have applied rectangular thresholds in truncated Gaussian simulation. In their approach, lithofacies F_i is modeled by

$$F_i = \{x \in R^3; S^i_{j-1} \le Y_j(x) < S^i_j, j = 1, \dots, p\}, \tag{10.31}$$

where $Y_j(x)$, for $j = 1, \dots, p$, are Gaussian random functions which can be independent or dependent. In practice, two Gaussian random functions are frequently used to determine lithotypes, i.e. $p = 2$. Figure 10.15 shows two anisotropic filtering stencils L_1 and L_2, which were used to generate random fields Y_1 and Y_2 shown in Figs. 10.16(a) and (b).

Now define a set of simple truncation thresholds:

$$F_i = \begin{cases} F_1 & \text{if } Y_1 > -4 \quad \text{and} \quad Y_2 \le -2, \\ F_2 & \text{if } Y_1 > -4 \quad \text{and} \quad Y_2 > -2, \\ F_3 & \text{if } Y_1 < -4, \end{cases}$$

where F_1, F_2, and F_3 are different lithotypes. The two random Gaussian fields, the threshold map, and the unconditional facies map realization are shown in Fig. 10.16. The black area in the facies map indicates facies F_1, grey for F_2, and white for F_3.

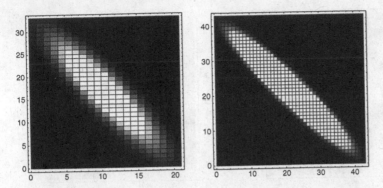

Figure 10.15. The Gaussian filters L_1 and L_2 for generating Gaussian random fields.

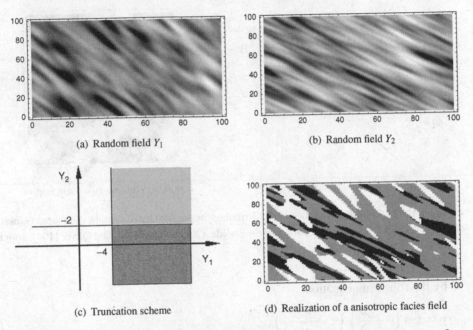

(a) Random field Y_1

(b) Random field Y_2

(c) Truncation scheme

(d) Realization of a anisotropic facies field

Figure 10.16. The unconditional realizations from the moving average method with $m_{pr} = 0$.

The rectangular thresholds approach has been used for geological simulation problems when conditioning to lithotype proportion data, however, it does not seem easy to apply to the problem of optimization of lithotype grouping in automatic history-matching problems. Liu and Oliver [160] introduced truncated Gaussian simulation using three intersecting lines. Three randomly generated lines intersecting each other without all passing through the same point divide the plane into seven regions. A facies type can be attributed to each region, so up to seven different lithotypes can be included in the same plane with appropriate relative percentage. This number of facies is normally enough for geology maps in petroleum reservoir study, but if not, another line could be added. Given an angle θ and a distance r, a threshold line can be described

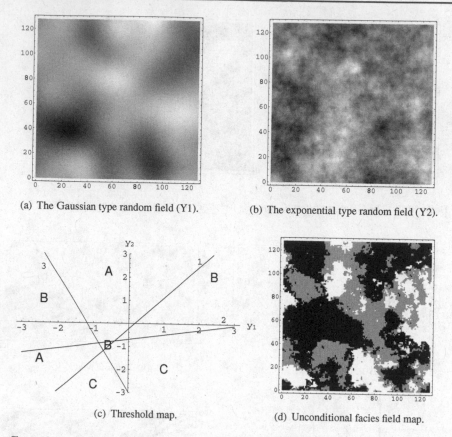

(a) The Gaussian type random field (Y1).

(b) The exponential type random field (Y2).

(c) Threshold map.

(d) Unconditional facies field map.

Figure 10.17. Simulation of lithofacies distribution in the field by truncation of random Gaussian fields Y_1 and Y_2 using intersecting line thresholds. (Adapted from Liu and Oliver [104], copyright SPE.)

by the following equation:

$$y = \tan\left(\theta - \frac{\pi}{2}\right)\left(x - \frac{r}{\cos\theta}\right),$$

(10.32)

i.e. the threshold line is perpendicular to the line passing through the origin with the slope θ and intersects the line at a distance r.

Figure 10.17 is an example illustrating the truncation scheme of intersecting threshold lines. The Gaussian random field Y_1 has Gaussian type covariance and Y_2 has exponential type covariance. The coordinates of the threshold map (Fig. 10.17c) are Y_1 and Y_2 respectively. Three kinds of lithotypes, A, B, and C are assigned to the seven regions in the threshold map. Facies type at any gridblock in the field is decided by taking the Y_1 and Y_2 value of that gridblock to the threshold map. For instance, the gridblock (20, 40) has low values for both its Y_1 and Y_2 variables. (They both are in areas with dark shade.) So it corresponds to the area in threshold map assigned facies A (see Fig. 10.17d).

10.4   General sampling algorithms

Suppose we have a vector of inaccurate data, d_{obs}, that are related to the model parameters by the relationship

$$d_{obs} = g(m) + \epsilon, \tag{10.33}$$

where the errors are assumed to be additive and Gaussian with mean 0 and covariance C_D. When the relation between the model parameters and the data is linear, $g(m) = Gm$. For Gaussian variables, the prior information on the model consists of an expectation vector, m_{prior}, and a model covariance matrix, C_M, that describes the uncertainty in the parameters. The algorithm for generating realizations that are conditional to the data will be introduced in the following sections.

Calculation of the most probable model is really only a part of the reservoir characterization problem, although it is the point at which most investigations stop. An equally important problem is to determine the uniqueness of the solution, i.e. how many other models will match the data and the a priori information adequately? In fact, for some types of prediction (e.g. EOR processes), it may be more important to generate a model that more closely matches the known variability than to generate the most probable model, which is a smoothed version of the true model.

10.4.1   Rejection method

The main idea in rejection sampling is to propose a sample from some relatively simple distribution, then apply a test to decide whether or not to accept it. The test does not depend on the most recent sample, so all accepted samples are truly independent. A general background on the rejection method for conditional simulation can be obtained from Ripley [161], Rubinstein [162], Kalos and Whitlock [163], Ripley [164].

We begin by examining two problems for which a method of sampling is intuitively obvious. We will then return to these examples and show the connection between our intuitive approach and the formal algorithm.

Example 1. Uniform distribution in irregular geometry

Suppose we need to sample points uniformly within the shaded region of Fig. 10.18. One obvious way to do this is to propose points from the uniform distribution over the rectangular region that bounds the shaded region, then check to see if the proposed point is located inside, which is very straightforward knowing the boundary of the shaded region. If it is inside, then accept it, otherwise reject. Hence the name.

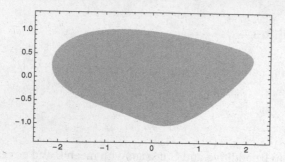

Figure 10.18. Uniformly distributed samples from the shaded region.

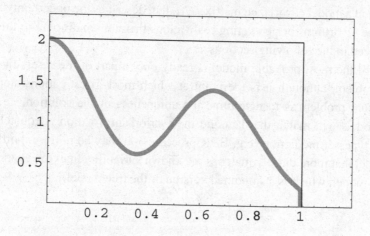

Figure 10.19. Sampling from a complicated PDF.

Example 2. Complicated single variable PDF

The second example is more closely related to our problem of sampling from a complicated PDF. Suppose we need to sample from the PDF represented by the thick curve in Fig. 10.19. It might be difficult to write an algorithm to sample this directly, but the same method that we used in the previous problem can be used here.

We can think of two equivalent ways of sampling in this problem. One is to propose points uniformly over the rectangle $(0, 1) \times (0, \max(f_X(x)))$, then only accept those that are beneath the curve $f_X(x)$. The other method would be to propose a value of X uniformly from the interval $(0, 1)$, then accept it with probability $f_X(X)/\max(f_X(x))$. The second method is equivalent to the first one with normalized scale of the PDF.

The rejection algorithm

To generate samples (realizations) from the target probability density $\pi(m) = f(m)/K$ where $f(m)$ is the un-normalized density (i.e. $\int f(m)\,dm \neq 1$) and K is the unknown normalizing constant, let $h(m)$ be a probability density that can be easily sampled (e.g. multi-variate Gaussian) and suppose that there is some constant $c \geq 1$ such that $\pi(m) \leq c\, h(m)$ for all m. Random samples from $\pi(m)$ can be obtained in the following procedure:

1. Generate a candidate sample m^* from the PDF $h(\cdot)$.
2. Generate a decision variable u from the uniform distribution on $(0, 1)$.
3. If $u \leq \pi(m^*)/[c\, h(m^*)]$ then accept the proposal and return $m = m^*$, else reject m^*.
4. Return to step 1 for the next sample.

We now show that the samples generated from this algorithm are in the correct PDF $\pi(m)$. There are two discrete events in the rejection algorithm: m^* is accepted (when $u \leq \pi(m^*)/[c\, h(m^*)]$), or m^* is rejected. The probability that the randomly proposed variate m^* is less than some value y is

$$P\{m^* < y\} = \int_{-\infty}^{y} h(m)\, dm \qquad (10.34)$$

because $h(m)$ is the PDF for y. Furthermore, the joint probability that $m^* < y$ and $\pi(m^*)/[c\, h(m^*)] \geq u$ is

$$P\left\{m^* < y \quad \text{and} \quad \frac{\pi(m^*)}{c\, h(m^*)} \geq u\right\} = \int_{-\infty}^{y} \frac{\pi(m)}{c\, h(m)} h(m)\, dm, \qquad (10.35)$$

where $h(m)\, dm$ is the probability of proposing a variate in the interval dm, and $\pi(m)/[c\, h(m)]$ is the probability of acceptance given m. Alternately, using Bayes' rule, we can write the joint probability of Eq. (10.35) as the product of a marginal probability for accepting the proposal and the conditional probability that m is less than some value y, given that the proposal is accepted,

$$P\left\{m^* < y \quad \text{and} \quad \frac{\pi(m^*)}{c\, h(m^*)} \geq u\right\} = P\left\{\frac{\pi(m)}{c\, h(m)} \geq u\right\} P\left\{m < y \,\middle|\, \frac{\pi(m)}{c\, h(m)} \geq u\right\}. \qquad (10.36)$$

The efficiency of the algorithm is simply the marginal probability that u is less than $\pi(m^*)/[c\, h(m^*)]$. It is found by integrating Eq. (10.35) over the range of all possible values of m, i.e.

$$P\left\{\frac{\pi(m^*)}{c\, h(m^*)} \geq u\right\} = \int_{-\infty}^{\infty} \pi(m)/c\, dm \qquad (10.37)$$

$$= \frac{1}{c}. \qquad (10.38)$$

The marginal distribution can not be greater than 1, that is why we specified earlier that $c \geq 1$.

The last term on the right-hand side of Eq. (10.36) is the probability distribution of m that results from the rejection algorithm. Hence,

$$P\left\{m < y \;\middle|\; \frac{\pi(m)}{c\,h(m)} \geq u\right\} = \frac{\int_{-\infty}^{y} \pi(m)\,dm}{\int_{-\infty}^{\infty} \pi(m)\,dm} \tag{10.39}$$

$$= \int_{-\infty}^{y} \pi(m)\,dm. \tag{10.40}$$

Example 1 revisited

Referring back to the first example of Section 10.4.1, we see that $h(x)$ is the uniform distribution on the rectangular region, and $\pi(x)/[c\,h(x)]$ is the function equal to 1 inside the shaded region and equal to 0 outside. The "efficiency" is just the area of the shaded region divided by the area of the rectangle from which the proposals are made.

Sampling efficiency with rejection algorithm

The key to the efficiency of the rejection algorithm is selecting a proposal density $g(x)$ that is a close approximation to the target density, in which case the acceptance rate is close to one. Unfortunately, it can be very difficult to find a simple distribution for the proposal of trial realizations that leads to an efficient algorithm, especially when the number of model variables is large.

Consider a very simple reservoir model in which the only unknown parameters are the porosities of three layers. We assume that the prior state of knowledge of porosity for each layer is represented by the prior mean

$$m_{\text{prior}} = \begin{bmatrix} 0.28 & 0.30 & 0.32 \end{bmatrix}^{\text{T}} \tag{10.41}$$

and the standard deviation of the prior estimate for each layer

$$\sigma_{\text{M}} = 0.1. \tag{10.42}$$

We assume that a measurement of the average reservoir porosity is available ($d_{obs} = \bar{\phi} = 0.25$) and that the uncertainty in the measurement is given by a standard deviation of the data ($\sigma_D = 0.01$). We represent the data relationship as

$$d = Gm = \begin{bmatrix} \frac{1}{3} & \frac{1}{3} & \frac{1}{3} \end{bmatrix} \begin{bmatrix} \phi_1 \\ \phi_2 \\ \phi_3 \end{bmatrix} = 0.25. \tag{10.43}$$

The (un-normalized) a posteriori PDF for the porosities is

$$f(m) = \exp\left[-\frac{1}{2\sigma_{\text{M}}^2}(m - m_{\text{prior}})^{\text{T}}(m - m_{\text{prior}}) - \frac{1}{2\sigma_D^2}(Gm - d_{\text{obs}})^2\right]. \tag{10.44}$$

In this case, the a posteriori PDF is Gaussian and sampling from this PDF is relatively simple. Although many other methods will do a good job in sampling the Gaussian

PDF, the rejection algorithm is applied here for demonstration purpose. The simplest possible method is one in which we propose porosities independently from the uniform distribution, i.e. $m_{\text{trial}} \sim [U[0, 1]\, U[0, 1]\, U[0, 1]]^{\text{T}}$. In order to use the rejection test, we need an upper bound for $f(m)$. Because this problem is linear, we know that the maximum value of $f(m)$ occurs at the maximum a posteriori estimate:

$$m_{\text{map}} = m_{\text{prior}} + C_{\text{M}} G^{\text{T}} (G C_{\text{M}} G^{\text{T}} + C_D)^{-1} (d_{\text{obs}} - G m_{\text{prior}}) = (0.231, 0.251, 0.271).$$
(10.45)

Substitution of m_{map} into $f(m)$ gives $f_{\text{max}} = 0.695$, hence the algorithm for sampling is as follows:

1. generate m_{trial} from PDF $[U[0, 1]\, U[0, 1]\, U[0, 1]]^{\text{T}}$;
2. generate a decision variable u from $\mathcal{U}[0, 1]$;
3. if $u \le f(m_{\text{trial}})/f_{\text{max}}$ return $m = m_{\text{trial}}$, else return to step 1.

As an indication of the performance, the first iteration of the algorithm went as follows:

1. $m_{\text{trial}} = (0.027, 0.252, 0.545)$ which gives $f(m_{\text{trial}})/f_{\text{max}} = 0.000\,21$;
2. $u = 0.852$;
3. since $u \nleq f(m_{\text{trial}})/f_{\text{max}}$ we reject m_{trial} and return to step 1.

The proposed porosity vector from the second iteration of the algorithm was much worse:

1. $m_{\text{trial}} = (0.357, 0.732, 0.917)$ which gives $f(m_{\text{trial}})/f_{\text{max}} = 1.1 \times 10^{-393}$;
2. $u = 0.316$;
3. since $u \nleq f(m_{\text{trial}})/f_{\text{max}}$ we reject m_{trial} and return to step 1.

The proposed distribution was not a particularly good one for this problem. Of 40 000 proposed vectors of porosity, only 99 were accepted. We see from Fig. 10.20 that the distribution of realizations of ϕ_3 is approximately Gaussian and centered on the MAP value of 0.27. The distribution of average porosity is more interesting. It seems to be a Gaussian centered on a value close to 0.25 with a relatively small variance. This is the distribution we expect from the data. Note that it would be very easy to improve

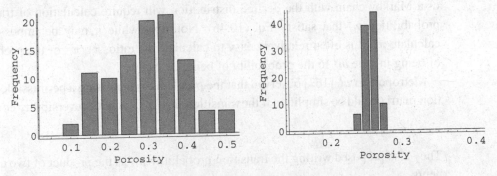

Figure 10.20. The sampled distribution of porosities for layer 3 (on the left) and the sampled distribution of average porosities (on the right) from the rejection algorithm.

the sampling efficiency for this problem, for example by proposing porosities from the interval $(0, 0.6)$ instead of $(0, 1)$. When the shape of the PDF is unknown, however, and the dimension of the problem is large, finding a PDF that tightly bounds the a posteriori PDF and is easy to sample from is a difficult problem.

10.4.2 Markov chain Monte Carlo

In the context of reservoir characterization, an ensemble of possible states or realizations of the reservoir permeability distribution can be thought of as a Markov chain if the probability of generating some particular new realization depends only on the preceding realization in the sequence. For simplicity in notation, assume that there are a countable number of possible realizations and that the possible outcomes are denoted by m^i. Each realization, m^i, will have a probability density, π_i, associated with it. In our context π_i is the probability density of the ith realization being the correct vector of reservoir properties. Recall that in the Monte Carlo approach, the realizations need to be sampled with the correct probability density. Unfortunately, in most real applications it is difficult to characterize the probability of the ith realization, but the relative probabilities are often straightforward to compute, so we generally use a method that relies only on the relative probabilities to generate realizations. This can be done using Markov chains if we are careful in the specification of the conditional probability, $p_{ij} = p(m^j|m^i)$, of transition to realization m^j from realization m^i.

If transition probabilities p_{ij} are specified such that it is possible to get from any state m^i to state m^j in a finite number of transitions and such that the probability of state m^j is the sum of the probabilities of being in a state m^i times the probability of transition from m^i to m^j, i.e.

$$\pi_j = \sum_i \pi_i p_{ij} \tag{10.46}$$

then the Markov chain will be stationary and ergodic (independent of initial conditions) and π_j will be the probability distribution for the realizations [165, 166]. Generation of a Markov chain with the desired distribution will require calculation of transition probabilities p_{ij} that satisfy Eq. (10.46). Note that while it may be impossible to calculate π_j, it is often relatively easy to calculate the ratio, π_j/π_i, of the probability of being in state m^j to the probability of being in state m^i.

Metropolis *et al.* [167] observed that the problem of calculating a permissible transition matrix could be simplified if the transition matrix satisfied a reversibility condition

$$\pi_i p_{ij} = \pi_j p_{ji}. \tag{10.47}$$

They also proposed writing the transition probability p_{ij} as the product of two components,

$$p_{ij} = \alpha_{ij} q_{ij} \tag{10.48}$$

where q_{ij} is the probability of proposing a transition from state m^i to state m^j and α_{ij} is the probability of accepting the proposed transition. The q_{ij} can be chosen somewhat arbitrarily, subject to the restrictions that it must be possible to propose a transition to any state in a finite number of steps. It should be clear, however, that some choices for q_{ij} will result in high probabilities of acceptance of transition and thus substantially reduce the computation time. The α_{ij} are not completely determined by Eqs. (10.47) and (10.48). Hastings [168], therefore, proposed using

$$\alpha_{ij} = s_{ij} \left[1 + \frac{\pi_i q_{ij}}{\pi_j q_{ji}} \right]^{-1}. \tag{10.49}$$

where s_{ij} is symmetric, i.e. $s_{ij} = s_{ji}$ and is chosen to ensure that $0 \le \alpha_{ij} \le 1$. From Eq. (10.49) and the constraints on α_{ij} we obtain

$$0 < s_{ij} \le 1 + \frac{\pi_i q_{ij}}{\pi_j q_{ji}}. \tag{10.50}$$

Since s_{ij} is symmetric, then from Eq. (10.50) we also obtain

$$0 < s_{ij} \le 1 + \frac{\pi_j q_{ji}}{\pi_i q_{ij}}.$$

Based on these constraints, Hastings' [168] choice for s_{ij} is

$$s_{ij} = \min \left[1 + \frac{\pi_i q_{ij}}{\pi_j q_{ji}}, 1 + \frac{\pi_j q_{ji}}{\pi_i q_{ij}} \right],$$

which gives the following acceptance probability for state j with respect to state i:

$$\alpha_{ij} = \min \left[1, \frac{\pi_j q_{ji}}{\pi_i q_{ij}} \right]. \tag{10.51}$$

Note that if the proposed transitions are symmetric, i.e. if $q_{ij} = q_{ji}$, then the calculation of whether or not to accept a transition is based only on the ratio of the probability of being in the two states. If the proposed transition is rejected, the old state is repeated in the chain. Swapping permeability values in two randomly chosen gridblocks is an example of a symmetric transition that has been commonly used in reservoir characterization [169–171]. The α_{ij} given by Eq. (10.51) is known to be relatively efficient at sampling the distribution compared to other possible choices [164, p. 114].

Numerical example

For a simple illustration, we consider the problem of generating normally distributed random numbers with mean 0 and variance 1 in the interval $[-3, 3]$. This is actually a truncated Gaussian distribution, although only a small fraction of samples would be generated outside this interval. We will use a Markov chain Monte Carlo approach to this problem, even though there are other methods that are far more efficient if the goal is only to generate normally distributed random numbers.

The probability density distribution that we wish to sample from is

$$\pi(x_j) = \left[\int_{-3}^{3} \exp(-x^2/2)\, dx\right]^{-1} \begin{cases} 0 & \text{for } x_j \leq -3, \\ \exp(-x^2/2) & \text{for } -3 < x_j \leq 3, \\ 0 & \text{for } x_j > 3, \end{cases} \tag{10.52}$$

where the normalization is required to make this a proper probability density function. One of the strengths of the McMC method, however, is that we do not need to know the normalizing constant, which in general is very difficult to approximate. To develop insight into the use of Markov chain Monte Carlo methods for sampling, and the effect of the choice of the proposal distribution on the performance, we will examine chains resulting from three different proposal distributions.

Uniform proposal distribution

For this example we will propose numbers from the uniform distribution on the interval $[-3, 3]$, i.e.

$$q_{ij} = \begin{cases} 0 & \text{for } x_j \leq -3, \\ 1/6 & \text{for } -3 < x_j \leq 3, \\ 0 & \text{for } x_j > 3. \end{cases} \tag{10.53}$$

Note that q_{ij} does not depend on the current state i in this example. Once we generate the proposed state from the uniform distribution on $[-3, 3]$, we must decide whether to accept the new state or to repeat the old state. The new state is accepted with probability

$$\begin{aligned} \alpha_{ij} &= \min\left[1, \frac{\pi(x_j)q_{ji}}{\pi(x_i)q_{ij}}\right] \\ &= \min\left[1, \frac{\pi(x_j)}{\pi(x_i)}\right] \\ &= \min\left[1, \frac{\exp(-x_j^2/2)}{\exp(-x_i^2/2)}\right]. \end{aligned} \tag{10.54}$$

The Markov chain Monte Carlo algorithm for this example is as follows.
- Propose a starting value x_{old}. This becomes the first element of the Markov chain.
- Set $i = 1$
- While $i \leq M$,
  1. propose a new random number $x_j \sim U[-3, 3]$,
  2. generate a random number $u \sim U[0, 1]$,
  3. if $\alpha(x_{old}, x_j) \geq u$ then set $x_{new} = x_j$, otherwise $x_{new} = x_{old}$,
  4. append x_{new} to the Markov chain.
  5. set $x_{old} = x_{new}$,
  6. increment the counter ($i = i + 1$).

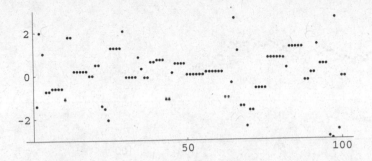

Figure 10.21. The first 100 elements of a Markov chain for sampling from a truncated Gaussian distribution.

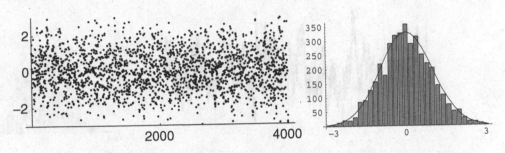

Figure 10.22. The first 4000 elements of a Markov chain for sampling from a truncated Gaussian distribution with mean 0 and variance 1.

We carried out this calculation for a chain length of 4000. Figure 10.21 shows the first 100 elements of the chain. Note that the first number proposed was approximately equal to -1.5. The next number proposed was approximately equal to 2. Although it is less probable than -1.5, it was accepted as the new state. After a few proposals, the chain remains at the same value or state for several iterations, then changes to another. Note that this generally happens when the value is relatively close to zero, because most proposed states will then have a lower probability than the current state.

Figure 10.22 shows all 4000 elements of the Markov chain for the same problem. Because of the scale, it is no longer possible to tell that the chain often stays at the same value for several iterations before changing to another value. The density of dots gives some indication, however, of the distribution of states. On the right-hand side of Fig. 10.22, the frequency of occurrence of values in the first 4000 elements is shown in a histogram. Note that the distribution is quite close to the expected distribution, even though the states were proposed from a uniform distribution.

Small perturbation

From Fig. 10.21, it appears that only about one half of the proposed realizations were accepted. We can generate a chain with a higher acceptance rate by modifying

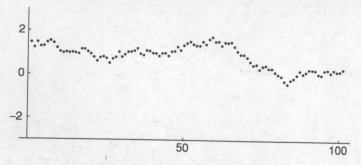

Figure 10.23. The first 100 elements of a Markov chain for sampling from a truncated Gaussian distribution with small perturbations.

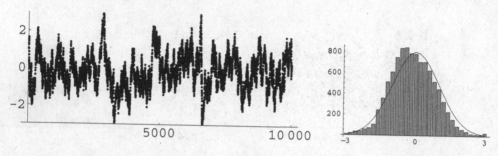

Figure 10.24. The first 10 000 elements of a Markov chain for sampling from a truncated Gaussian distribution with mean 0 and variance 1.

the proposal method. Consider the following probability density for proposing new realizations.

$$
q_{ij} = \begin{cases} 0 & \text{for } x_j \le x_i - 0.25, \\ 2 & \text{for } x_i - 0.25 < x_j \le x_i + 0.25, \\ 0 & \text{for } x_j > x_i + 0.25. \end{cases} \tag{10.55}
$$

In this example, the probability of proposing a new state x_j does depend on the current state x_i. Note, however, that $q_{ij} = q_{ji}$ so we can still use Eq. (10.54) to decide whether or not to accept the proposed state.

We carried out this calculation for a chain length of 10 000. Figure 10.23 shows the first 100 elements of the chain. Note how much different this chain is than the previous chain. While the acceptance rate is very high, the rate of mixing is quite low, meaning that it can take a long time to go from high values to low values.

Figure 10.24 shows all 10 000 elements of the Markov chain for this problem. At this scale, we can see that it typically takes at least 1000 perturbations (or steps) to go from high to low values. Or we could say that the realizations are correlated over distances on the order of 1000. On the right-hand side of Fig. 10.24, the frequency of occurrence of values in the 10 000 elements is shown in a histogram. The distribution

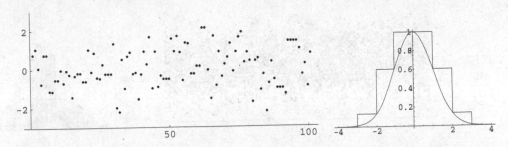

Figure 10.25. The first 100 elements of a Markov chain for sampling from a truncated Gaussian distribution with proposals from the piecewise constant PDF shown on the right-hand side.

seems to be somewhat skewed (more low values than high). If we made the chain much longer, the histogram would ultimately be a very good approximation to the Gaussian distribution.

Proposal from approximation to π_j

Instead of making small perturbations to the current state, we could try to get a high acceptance rate by making q_{ij} closer to π_j. We see from Eq. (10.51) that if $q_{ij} = \pi_j$ and $q_{ji} = \pi_i$, the acceptance probability is one. Consider the following proposal PDF,

$$
q_{ij} = \begin{cases}
0 & \text{for } x_j \leq -3, \\
\exp(-2^2/2) & \text{for } -3 < x_j \leq -2, \\
\exp(-1^2/2) & \text{for } -2 < x_j \leq -1, \\
\exp(-0^2/2) & \text{for } -1 < x_j \leq 1, \\
\exp(-1^2/2) & \text{for } 1 < x_j \leq 2, \\
\exp(-2^2/2) & \text{for } 2 < x_j \leq 3, \\
0 & \text{for } x_j > 3.
\end{cases}
\tag{10.56}
$$

In this example, the probability of proposing a new state x_j does not depend on the current state x_i, but $q_{ij} \neq q_{ji}$ so we must use Eq. (10.51) to decide whether or not to accept the proposed state.

We carried out this calculation for a chain length of 4000. Figure 10.25 shows the first 100 elements of the chain. Note that this chain looks similar to the chain in Fig. 10.21 except that it seems to get stuck less often. In this case we seem to have both a high acceptance rate and rapid mixing.

The left-hand side of Fig. 10.26 shows all 4000 elements of the Markov chain for this problem and on the right-hand side of Fig. 10.26, the frequency of occurrence of values in the first 4000 elements is shown in a histogram. The distribution seems to be a good approximation to the Gaussian distribution.

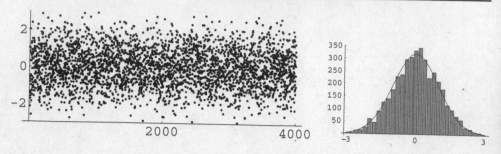

Figure 10.26. The first 4000 elements of a Markov chain for sampling from a truncated Gaussian distribution with mean 0 and variance 1.

10.4.3 Markov random fields

The approaches for simulation of the spatial distribution of geologic facies fall into two categories in general. The first category is object based, which parameterizes facies shapes on a background matrix facies. Typical applications for this type of approach are simulation of channel sand on a shale background [172, 173] and simulation of shales in a submarine fan environment [174]. The other category of approaches describes the facies field by pixels and a facies type is assigned to each gridblock without specification of matrix facies. One example for this type of approach is the truncated Gaussian method, which is introduced in the early part of this chapter. In this section, we introduce the Markov random field for simulating the facies distributions in petroleum reservoirs [175–179]. Although more general spatial arrangements are possible, we limit our discussion to Markov random fields on a rectangular lattice.

Neighborhood system and cliques

In a discrete 2D random field, a finite rectangular lattice is defined with the number of gridblocks (pixels) as $N = N_x \times N_y$. The set of coordinates of the gridblocks in the lattice is denoted as $L = \{(i, j), \text{ for } 1 \leq i \leq N_x \text{ and } 1 \leq j \leq N_y\}$. Any site (i, j) relates with other sites through a neighborhood system. The definition of the neighborhood system according to Derin and Elliott [180] is:

A collection of subsets of L

$$\eta = \{\eta_{ij}, \text{ such that } (i, j) \in L \text{ and } \eta_{ij} \subseteq L\} \tag{10.57}$$

is a neighborhood system on the grid (i, j) if and only if

1. the grid point (i, j) is not an element of the neighborhood η_{ij}: $(i, j) \notin \eta_{ij}$, and
2. the neighboring relationship is symmetric: $(k, l) \in \eta_{ij} \Leftrightarrow (i, j) \in \eta_{kl}$, for any $(i, j) \in L$ and $(k, l) \in L$.

The neighborhood system is ordered based on the Euclidean distance between the central location (i, j) and the grids within the neighborhood η_{ij}. Figure 10.27 shows a

5	4	3	4	5
4	2	1	2	4
3	1	X	1	3
4	2	1	2	4
5	4	3	4	5

Figure 10.27. The fifth-order neighborhood system.

neighborhood system for the central location denoted by "X." The four grids numbered 1 form the first-order neighborhood system, also known as the "nearest-neighborhood system." The second-order neighborhood system consists of all the grids with numbers 1 or 2. As there are eight such neighbors for every central location, it is also called the "eight-neighborhood system." The number n in each grid indicates the outermost neighbor grid in the nth-order neighborhood system. The largest system in Fig. 10.27 is a full fifth-order neighborhood system. For the locations at or near the boundaries of the lattice, there may not be complete neighborhoods, unless the reservoir is assumed to have periodic structure. For instance, grids at the corner of a lattice have two nearest neighbors instead of four. The neighborhood system has also been defined on irregular grid systems, such as the Delaunay triangulation [181, 182] and the Voronoi polygons [183].

A clique c is defined as a set of grids in the lattice L with the neighborhood system η, such that

1. c consists of a single pixel, or
2. each distinct pair of nodes are within the neighborhoods of each other:
   $$\forall (i, j), (k, l) \in c, \text{ and } (i, j) \neq (k, l) \Rightarrow (i, j) \in \eta_{kl}, \text{ and } (k, l) \in \eta_{ij}.$$

In Fig. 10.28, the subfigures (a)–(d) are a collection of all types of cliques for a second-order neighborhood. The subfigure (e) is not a clique for a second-order neighborhood, but a clique for fourth order and above.

Markov–Gibbs equivalence

Assume the number of facies in the lattice is K, the random field of facies are denoted as $\mathcal{F}$ for which any grid (i, j) in the lattice L is assigned a facies type: $\mathcal{F}_{ij} \in \{F_0, F_1, \ldots, F_{K-1}\}$. Let f denote a facies map realization over the entire lattice L:

$$f = \{\mathcal{F}_{11}, \mathcal{F}_{12}, \ldots, \mathcal{F}_{ij}, \ldots, \mathcal{F}_{N_x N_y}\} \in \Omega,$$

where Ω contains all the possible realizations of the facies distribution. There are $K^{N_x \times N_y}$ different combinations of the facies at all the gridblocks for the realizations of f. The stochastic field, $\mathcal{F}$, with joint distribution of the facies at every gridblock $\pi(f) = P(\mathcal{F} = f)$ is a Markov random field (MRF) if and only if it satisfies:

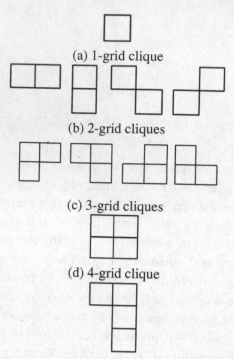

(a) 1-grid clique

(b) 2-grid cliques

(c) 3-grid cliques

(d) 4-grid clique

(e) Not a clique for a second-order neighborhood

Figure 10.28. Example of cliques for a second-order neighborhood.

1. $\pi(f) > 0$, for $\forall f \in \Omega$, and
2. $\pi(f_{ij}|f_{kl}) = \pi(f_{ij}|f_{\eta_{ij}})$ for all $(k, l) \neq (i, j)$ in the whole lattice, where f_{ij} is the facies type at the grid (i, j). The location (i, j) is random in the lattice L. $f_{\eta_{ij}}$ denotes the set of facies at the neighborhood system of (i, j). When the neighborhood system is as large as the lattice, the facies field is by default a Markov random field.

The random field $\mathcal{F} = \{\mathcal{F}_{ij}\}$ is said to be in a Gibbs distribution (GD) on L with respect to the neighborhood system η if and only if its joint distribution takes the form:

$$P(\mathcal{F} = f) = Z^{-1} \times e^{-U(f)},$$

(10.58)

for

$$Z = \sum_{f} e^{-U(f)},$$

(10.59)

and

$$U(f) = \sum_{c \in C} V_c(f_c),$$

(10.60)

where Z is called the partition function, and it is the sum over all possible facies distributions. The energy function $U(f)$ is the summation of the potentials $V_c(f_c)$ of each clique type c. C consists of all possible cliques for the neighborhood system η. The potentials $V_c(f_c)$ depend on the values assigned to the clique type c and the random variable $\mathcal{F}$ assigned to the grids in the clique c.

The Hammersley–Clifford theorem [175, 184] establishes the one-to-one equivalence of the MRF with the GRF. Therefore, the joint distribution of the MRF also takes the form:

$$\pi(\mathcal{F} = f) \propto \exp\left\{-\sum_{c \in C} V_c(f_c)\right\}. \tag{10.61}$$

The GRF is homogeneous if the value of potentials $V_c(f_c)$ is independent of the locations of the clique c in the lattice L. It is isotropic if the values of $V_c(f_c)$ are identical for all cliques with same shape but different orientation in L. For MRF, homogeneity is defined as the conditional probability $\pi(f_{ij}|f_{\eta_{ij}})$ being independent to the relative position of the location (i, j) in the lattice L. The homogeneity and the isotropy for GRF are equivalent to those for MRF.

Let M be the number of clique types in the specified neighborhood system η of a homogeneous MRF, and ξ_m is the assigned value to the mth clique type, which is proportional to the likelihood of the appearance of the mth clique for $m \in \{1, 2, \ldots, M\}$. Let L_m be the number of configurations of facies assignment in the mth type of clique, β_{lm} is the assigned value to the lth facies configuration in the clique m for $l \in \{1, 2, \ldots, L_m\}$. A general form of the definition of the potential functions is:

$$V_c(f_c) = \begin{cases} \beta_{lm} \cdot \xi_m & \text{for } c \subset \eta, \\ 0 & \text{otherwise.} \end{cases} \tag{10.62}$$

The joint distribution of the MRF is obtained by:

$$\pi(f) \propto \exp\left\{-\sum_{m=1}^{M}\sum_{l=1}^{L_m} \beta_{lm} \cdot \xi_m \cdot n_{lm}(f)\right\} \tag{10.63}$$

$$= \exp\left\{-\sum_{m=1}^{M}\sum_{l=1}^{L_m} V_{lm}(f)n_{lm}(f)\right\}, \tag{10.64}$$

where $n_{lm}(f)$ is the total number of the mth clique with configuration l, and V_{lm} denotes the potential of the mth clique with configuration l. The lower the potential value, the higher the probability $\pi(f)$ for the facies distribution f.

Simulation of MRF

The MRFs are generated iteratively with the evolution of the Markov chain. The procedure for generating a MRF is as follows.

- Propose a starting random facies field f_{old}. This becomes the first element of the Markov chain.
- Set $i = 1$.

- While $i \leq T$,
  1. propose a random location (i, j);
  2. replace the current facies value at (i, j) with a new one, which is uniformly sampled from the $K - 1$ available facies types; denote the new proposal as $f^\star$;
  3. compute the acceptance probability of $f^\star$:

  $$\alpha(f \to f^\star) = \min\left\{1, \frac{\pi(f^\star)}{\pi(f)}\right\}; \tag{10.65}$$

  4. generate a random number u from $U(0, 1)$;
  5. if $\alpha(f \to f^\star) \geq u$ then set $f_{new} = f^\star$, otherwise $f_{new} = f_{old}$;
  6. append f_{new} to the Markov chain;
  7. set $f_{old} = f_{new}$;
  8. increment the counter $(i = i + 1)$.

Example: generating 2D anisotropic Markov random fields

In this example, we first define the clique potentials associated with the second-order neighborhood in a three-facies MRF. Then we show the evolution of the MRF generated using the clique potential parameters. Finally we discuss the impacts of the parameters to the features and the stability of the MRF. For simplicity, here we only consider the cliques with up to two pixels, which is equivalent to defining $\xi_i = 0$ for the cliques with three or four pixels.

For the single pixel cliques, the clique potentials are

$$V_c(f_c) = \begin{cases} \beta_{11} \cdot \xi_1 \text{ for } f_c = F_1, \\ \beta_{21} \cdot \xi_1 \text{ for } f_c = F_2, \\ \beta_{31} \cdot \xi_1 \text{ for } f_c = F_3, \end{cases} \tag{10.66}$$

where the relative scales among β_{11}, β_{21}, and β_{31} affect the percentage of each facies type. In this example, all the three facies take the same percentage, i.e. 33% each, thus we set $\beta_{11} = \beta_{21} = \beta_{31} = 0$. So the potentials are $V_c(f_c) = 0$ for all single pixel cliques, which saves some computation cost.

All the cliques of the second-order neighborhood system with up to two pixels are listed and numbered as shown in Fig. 10.29. The total number of clique types is $M = 5$. In this anisotropic case, the cliques with identical shape but different directions are not assigned the same potential values. Therefore arrows are used in Fig. 10.29 to label out the direction of the facies transition scenarios. For instance, if we want to encourage

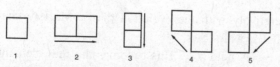

Figure 10.29. The cliques with up to two pixels are numbered 1–5. The arrows denote the positive direction for facies transitions.

the direction of the facies transition scenarios. For instance, if we want to encourage the facies transition $1 \to 2$ from left to right, the facies configuration $1 \to 2$ should be assigned a lower potential thus higher probability than the configuration $2 \to 1$. The number of facies configurations is $L_m = 9$ for each of the four two-pixel cliques in this example. Denote the facies at the first pixel along the positive direction as i, and the second as j, the facies configurations in any two-pixel clique are

$$(i, j) \in \{(1, 1), (1, 2), (1, 3), (2, 1), (2, 2), (2, 3), (3, 1), (3, 2), (3, 3)\},$$

for $i \in \{1, 2, 3\}$ and $j \in \{1, 2, 3\}$.

Previously we denoted the facies configuration factors as β_{lm}, where l is the facies configuration index $l \in \{1, 2, \ldots, L_m\}$. Here we let $\beta_{lm} = \beta_m(i, j)$ to reflect the facies configurations more clearly. The facies configuration factors on the two-pixel cliques are assigned values as

$$\begin{bmatrix} \beta_2(1, 1) \ \beta_2(1, 2) \ \beta_2(1, 3) \\ \beta_2(2, 1) \ \beta_2(2, 2) \ \beta_2(2, 3) \\ \beta_2(3, 1) \ \beta_2(3, 2) \ \beta_2(3, 3) \end{bmatrix} = \begin{bmatrix} -1.6 \ -1.0 \ 1.0 \\ 1.0 \ -1.6 \ -1.0 \\ -1.0 \ 1.0 \ -1.6 \end{bmatrix},$$

$$\begin{bmatrix} \beta_3(1, 1) \ \beta_3(1, 2) \ \beta_3(1, 3) \\ \beta_3(2, 1) \ \beta_3(2, 2) \ \beta_3(2, 3) \\ \beta_3(3, 1) \ \beta_3(3, 2) \ \beta_3(3, 3) \end{bmatrix} = \begin{bmatrix} -1.0 \ -1.0 \ 1.0 \\ 1.0 \ -1.0 \ -1.0 \\ -1.0 \ 1.0 \ -1.0 \end{bmatrix},$$

$$\begin{bmatrix} \beta_4(1, 1) \ \beta_4(1, 2) \ \beta_4(1, 3) \\ \beta_4(2, 1) \ \beta_4(2, 2) \ \beta_4(2, 3) \\ \beta_4(3, 1) \ \beta_4(3, 2) \ \beta_4(3, 3) \end{bmatrix} = \begin{bmatrix} 0.0 \ 0.0 \ 0.0 \\ 0.0 \ 0.0 \ 0.0 \\ 0.0 \ 0.0 \ 0.0 \end{bmatrix},$$

$$\begin{bmatrix} \beta_5(1, 1) \ \beta_5(1, 2) \ \beta_5(1, 3) \\ \beta_5(2, 1) \ \beta_5(2, 2) \ \beta_5(2, 3) \\ \beta_5(3, 1) \ \beta_5(3, 2) \ \beta_5(3, 3) \end{bmatrix} = \begin{bmatrix} -1.5 \ 1.0 \ -1.0 \\ -1.0 \ -1.5 \ 1.0 \\ 1.0 \ -1.0 \ -1.5 \end{bmatrix}.$$

For simplicity, assign unit value to each clique type:

$$\xi_i = 1, \text{ for } i = 1, \ldots, 5,$$

where i is the numbering of all the cliques. The parameters ξ_i for two-pixel cliques decide the shape, size, and anisotropy of the facies clusters. The potentials for each two-pixel cliques are computed using Eq. (10.62).

Figure 10.30 shows the evolution of a realization of a MRF generated using the parameters described above. At the beginning, all facies are assigned independently based on the marginal probability for the facies. So the initial facies distribution is spatially uncorrelated, but each facies honors the assigned proportions. By the 100th iteration, the basic properties of the field are apparent – long correlation length in the

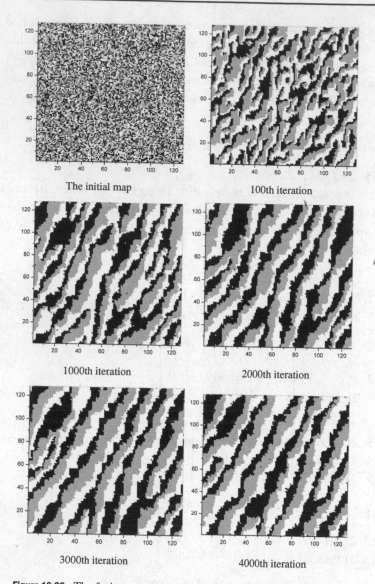

Figure 10.30. The facies maps generated with MRF algorithm at the initial, 100th, 1000th, 2000th, 3000th, and 4000th iterations. The black regions are populated by facies 1, gray for facies 2, and white for facies 3.

SW–NE direction, and transitions from facies 1 to 2 to 3 and back to facies 1 in the NW–SE direction.

Approximately one third of the pixels are populated by each facies type. The proportions of each facies are determined primarily by values for homogeneous cliques. The formation of large regions of continuous facies assignment was encouraged by assigning large negative values to the clique potentials for same facies. We also encouraged facies transitions of $1 \rightarrow 2 \rightarrow 3 \rightarrow 1$ in N$\rightarrow$S direction and in W$\rightarrow$E direction by assigning -1.0 to the corresponding facies configuration factors in $\beta_2(i, j)$ and $\beta_3(i, j)$. We assigned -1.5 to $\beta_5(1, 1)$, $\beta_5(2, 2)$, and $\beta_5(3, 3)$ to increase the width of the facies stripes as the number 5 clique extents NW–SE. The anisotropy angle of the MRF reflects the combined effect of the clique configuration factors.

The clique type parameters ξ_i for $i \in \{1, 2, \ldots, 5\}$ are positive numbers and control the appearance of random speckles in the MRFs. The inverse of $\max(\xi_i, i = 1, 2, \ldots, 5)$ is regarded as a temperature parameter. There are more appearances of random speckles for higher temperature. A few random speckles can be found in each iteration of the MRF in Fig. 10.30.

Although we are not sure that the Markov chain was sufficiently long to be completely independent of the starting model, it appears to have mixed rapidly and has a relatively high acceptance rate. The spatial correlation of the facies has been built within 100 iterations. The MRF at the 2000th iteration has nearly identical spatial features with the 4000th iteration. Therefore we believe that the MRF is stabilized after the 2000th iteration.

10.5   Simulation methods based on minimization

There are many methods that can be used to generate realizations from multi-normal distributions, or for independent stochastic variables. Of the many methods available for conditional simulation to exact data, two are particularly interesting because of the contrast in approaches. One approach is to generate a "rough" field (an unconditional simulation) with the same covariance as the true field, then to subtract a smooth correction that forces the simulated field to pass through the data [185]. A second approach is to calculate a "smooth" estimate that passes through the data, then use the LU decomposition of the estimation error covariance to add a stochastic component to the estimate. It is known that these methods are equivalent [156, 186].

The introduction of errors into the measurements requires a modification of both approaches. Clearly, it is not desirable to force the conditional simulation to honor inaccurate measurements exactly, but the ideas behind the two approaches generally are valid and can be shown to give equivalent results if carefully implemented.

Smooth plus rough – linear problem (Cholesky method)
In this algorithm, we first compute m_{map} and $C_{M'}$. The maximum a posteriori estimate is the model vector that minimizes

$$S(m) = (m - m_{\text{prior}})^{\text{T}} C_M^{-1} (m - m_{\text{prior}}) + (Gm - d_{\text{obs}})^{\text{T}} C_D^{-1} (Gm - d_{\text{obs}}), \quad (10.67)$$

or

$$m_{\text{map}} = m_{\text{prior}} + C_M G^{\text{T}} (G C_M G^{\text{T}} + C_D)^{-1} (d_{\text{obs}} - G m_{\text{prior}}). \quad (10.68)$$

Then, to generate N conditional realizations we use the following steps,
For i = 1 to N,

$$m_i = m_{\text{map}} + L_{M'} Z_M.$$

In this case, $L_{M'}$ is a square root of the posteriori covariance matrix, i.e. $L_{M'} L_{M'}^T = C_{M'}$, and

$$C_{M'} = \left(C_M^{-1} + G^T C_D^{-1} G \right)^{-1}.$$

Rough plus smooth – linear problem (RML method)

In this algorithm, we first generate unconditional realizations of the model (and data), then make a correction to the unconditional model. To generate N conditional realizations we use the following steps,
For $i = 1$ to N,
Generate an unconditional realization of the model variable $m_{u,i}$.
Generate an unconditional realization of the noise in the data and add the noise to the data to create $d_{u,i}$.
Compute the model $m_{c,i}$ that minimizes

$$S(m) = (m - m_{u,i})^T C_M^{-1} (m - m_{u,i}) + (Gm - d_{u,i})^T C_D^{-1} (Gm - d_{u,i}), \tag{10.69}$$

or

$$m_{c,i} = m_{u,i} - C_M G^T [G C_M G^T + C_D]^{-1} (Gm_{u,i} - d_{u,i}). \tag{10.70}$$

This algorithm is referred to as "randomized maximum likelihood (RML)."

Both of these algorithms are valid methods of generating realizations for linear problems with Gaussian measurement errors and a Gaussian prior PDF for the model variables. The analogous algorithms for nonlinear problems would be as follows.

Smooth plus rough – nonlinear problem (LMAP)

In this algorithm, we first compute the estimate of the model variables that maximize the probability density, m_{map}, and the corresponding estimate of the model covariance, $C_{M'}$. The maximum a posteriori estimate is the model vector that minimizes

$$S(m) = (m - m_{\text{prior}})^T C_M^{-1} (m - m_{\text{prior}}) + (g(m) - d_{\text{obs}})^T C_D^{-1} (g(m) - d_{\text{obs}}). \tag{10.71}$$

The linearized approximation to the a posteriori covariance matrix is constructed from the sensitivity matrix, G_∞, which is evaluated at the last iteration before obtaining m_{map}.

$$C_{M'} = \left(C_M^{-1} + G_\infty^T C_D^{-1} G_\infty \right)^{-1}.$$

Then, to generate N realizations we use the following steps,
For $i = 1-N$

$$m_i = m_{\text{map}} + L_{M'} Z_M, \quad \text{where } L_{M'} L_{M'}^T = C_{M'}.$$

Because this is only an approximation for nonlinear problems, there is no guarantee that the realizations generated by this algorithm are sampled correctly from the posterior PDF. The easiest test of the quality of the realizations is to see if the realizations generated by this algorithm approximately reproduce the data. As the approximation is based on a Taylor expansion in the neighborhood of the MAP estimate, this algorithm is often referred to as "linearization about the MAP (LMAP)."

Rough plus smooth – nonlinear problem (RML approximate)

In this algorithm, we first generate unconditional realizations of the model (and data), then make a correction to the unconditional model. To generate N conditional realizations we use the following steps.

For $i = 1$–N:

generate unconditional realization of model variables, $m_{u,i}$;

generate unconditional realization of measurement noise and add the noise to the data, $d_{u,i}$;

compute the model $m_{c,i}$ that minimizes

$$S(m) = (m - m_{u,i})^{\mathrm{T}} C_{\mathrm{M}}^{-1} (m - m_{u,i}) + (g(m) - d_{u,i})^{\mathrm{T}} C_{\mathrm{D}}^{-1} (g(m) - d_{u,i}). \qquad (10.72)$$

Note that this algorithm requires a great deal of additional work compared to the previous algorithm. For every realization, we must minimize the objective function given by Eq. (10.72). This is usually a difficult task, and quite expensive, for large problems. On the other hand, the realizations that are generated by this method will almost certainly approximately honor the data and look approximately correct. In that sense, this method is far more robust than the previous method. If we want to use the realizations to quantify uncertainty, however, we must examine the algorithm in more detail (see Section 10.5).

For variables that are multi-variate Gaussian, it is possible to generate realizations from the square root of the covariance [61]. This method has been applied to the problem of generating realizations of Gaussian random fields in petroleum engineering by Davis [187] and Alabert [155]. The Cholesky method is an exact method for generating conditional realizations from a linear inverse problem with a Gaussian prior probability and Gaussian noise in the data [188, p. 98]. Very similar to its application in generating unconditional realizations, the conditional realizations are generated by:

$$m_i = m_{\mathrm{map}} + L Z_i, \qquad (10.73)$$

where Z_i is a vector of independent normal random deviates $Z_i \rightarrow N[0, 1]$, L is a square root of the a posteriori covariance matrix $C_{\mathrm{M'}}$ ($C_{\mathrm{M'}} = LL^{\mathrm{T}}$), and m_{map} is the MAP estimate. The a posteriori covariance matrix $C_{\mathrm{M'}}$ of the model is:

$$C_{\mathrm{M'}} = \left(G^{\mathrm{T}} C_{\mathrm{D}}^{-1} G + C_{\mathrm{M}}^{-1} \right)^{-1}, \qquad (10.74)$$

where G is the sensitivity coefficient matrix, and C_{M} is the prior covariance matrix of model parameters.

Generation of other realizations is clearly inexpensive after the decomposition of $C_{M'}$ has been performed. This approach is valid only if the probability density of the model parameters is approximately Gaussian. This will always be the case if the model is linear and the prior probabilities are Gaussian. The relationships between drawdown and storativity, and drawdown and log transmissivity, are not linear, however, so the limitations on the use of the Cholesky method for conditional simulation are usually more complex for petroleum engineering problems.

When the relationship between data and model parameters is nonlinear, the square-root method for generating conditional realizations approximates the non-Gaussian PDF with a Gaussian PDF. The a posteriori PDF is generated by computing an approximation to the a posteriori covariance based on linearization of the data relationship at the maximum a posteriori point. This method is referenced as linearization about the MAP estimate (LMAP). Once the most probable model m_{map} is computed, the a posteriori covariance matrix $C_{M'}$ of the model can be approximated as:

$$C_{M'} = \left(G_{map}^{T} C_{D}^{-1} G_{map} + C_{M}^{-1}\right)^{-1}, \tag{10.75}$$

where G_{map} is the sensitivity coefficient corresponding to the MAP estimate of the model variables. Both the Cholesky method and the LMAP method sample from the PDF:

$$m_i \sim N(m_{map}, C_{M'}) \tag{10.76}$$

to obtain realizations conditional to data.

Beale [189] developed a numerical measure of the intrinsic nonlinearity of a model. His measure is approximately independent of the distance from the maximum likelihood estimate as long as that distance is not too great. Unfortunately the important question with conditional simulation of nonlinear models is knowing whether or not the distance is so great that the linearization is invalid.

The following algorithm fits both the Cholesky and the LMAP methods:
1. compute the maximum a posteriori model m_{map};
2. estimate the a posteriori covariance matrix $C_{M'}$;
3. decompose $C_{M'}$ to obtain L;
4. generate random normal deviates Z_i;
5. compute m_i using Eq. (10.73), if another realization is needed go to step 4.

In the remainder of this section, we will explore the randomized maximum likelihood method for generating conditional realizations to inaccurate data. The formulas for conditional simulation to inaccurate data have appeared in Oliver [190] but the connection to a minimization problem is not very clear in that paper. It was later proposed as a method of conditional simulation for nonlinear problems in the context of Markov chain Monte Carlo methods [191]. Kitanidis [192] described its use for quasi-linear problems. We begin by describing the algorithm, then show that it produces conditional realizations with the correct mean and covariance when the relationship between data and variables is linear.

10.5.1 Geostatistical approach

Geostatisticians often assume that the property field of interest $K_0(x)$ is a random stationary function and that the covariance is independent of location, i.e. $C(x, x') = C(x - x')$. Journel and Huijbregts [185, p. 495] then approach the problem of simulation as follows. Let $k_0(x)$ be the true random field, and x_α be any of the data locations. The simulated values must match experimental data values at data locations:

$$k_{sc}(x_\alpha) = k_0(x_\alpha), \tag{10.77}$$

where the possibility of measurement error is clearly not considered. At any location, the kriged (interpolated) values, $k_{0k}^*(x)$, differ from the true values, $k_0(x)$, by an unknown amount:

$$k_0(x) = k_{0k}^*(x) + \left[k_0(x) - k_{0k}^*(x) \right] \tag{10.78}$$

or in terms of random functions

$$K_0(x) = K_{0k}^*(x) + \left[K_0(x) - K_{0k}^*(x) \right]. \tag{10.79}$$

To obtain a conditional simulation we replace the error term in brackets, $[K_0(x) - K_{0k}^*(x)]$, by an isomorphic and independent kriging error, $\left[K_s(x) - K_{sk}^*(x) \right]$. Isomorphic in this context means that the two functions have the same expectation and the same second-order moment, $C(x, x')$. Procedurally, one would generate an unconditional simulation $k_s(x)$. This function would have the same covariance as the true field but would not necessarily honor the data at the data locations. The values of the simulated field at the data locations would then be kriged (interpolated). We call this field $k_{sk}^*(x)$. The required conditional simulation is then written as

$$k_{sc}(x) = k_{0k}^*(x) + \left[k_s(x) - k_{sk}^*(x) \right]. \tag{10.80}$$

Note that

$$\left[k_s(x_\alpha) - k_{sk}^*(x_\alpha) \right] = 0, \tag{10.81}$$

so the error term in Eq. (10.80) is zero at the data locations, as it should be.

10.5.2 RML for linear inverse problems

The algorithm of the linear RML method.

1. Generate an unconditional realization of m from the prior PDF for the model parameters, i.e.

$$m_{uc} = m_{prior} + L_M Z_M, \tag{10.82}$$

where Z_M is a vector of independent normal deviates with zero mean and unit variance, and L_M is a square root of the model covariance matrix, i.e. $L_M L_M^T = C_M$.

2. Generate an unconditional realization of the noise in the data from the prior PDF for the noise, i.e.

$$d_{uc} = d_{obs} - L_D Z_D, \tag{10.83}$$

where Z_D is a vector of independent normal deviates with zero mean and unit variance.

3. Find the model that minimizes the objective function

$$S(m) = (m - m_{uc})^T C_M^{-1}(m - m_{uc}) + (Gm - d_{uc})^T C_D^{-1}(Gm - d_{uc}). \tag{10.84}$$

Note that the model m_{cs} that minimizes Eq. (10.84) is given by

$$m_{cs} = m_{uc} - C_M G^T [GC_M G^T + C_D]^{-1}(Gm_{uc} - d_{uc}). \tag{10.85}$$

Note that a qualitative interpretation of the algorithm is that it finds a model that is as close as possible to the unconditional realization, while simultaneously minimizing the misfit to the realization of the data. The second step, adding errors to the observed data before minimizing the misfit, might seem strange, but we will show that it leads to correct results.

We begin by expanding the terms in Eq. (10.85):

$$
\begin{aligned}
m_{cs} &= m_{uc} - C_M G^T [GC_M G^T + C_D]^{-1}(Gm_{uc} - d_{uc}) \\
&= (m_{prior} + L_M Z_M) - C_M G^T [GC_M G^T + C_D]^{-1}[G(m_{prior} + L_M Z_M) \\
&\quad - (d_{obs} - L_D Z_D)] \\
&= m_{prior} - C_M G^T [GC_M G^T + C_D]^{-1}(Gm_{prior} - d_{obs}) \\
&\quad + L_M Z_M - C_M G^T [GC_M G^T + C_D]^{-1}(GL_M Z_M + L_D Z_D) \\
&= m_{map} + L_M Z_M - C_M G^T [GC_M G^T + C_D]^{-1}(GL_M Z_M + L_D Z_D) \\
&= m_{map} + \left([L_M\ 0] - C_M G^T(GC_M G^T + C_D)^{-1}[GL_M\ L_D]\right) \begin{bmatrix} Z_M \\ Z_D \end{bmatrix}.
\end{aligned} \tag{10.86}
$$

Clearly, the expectation of the random vectors generated by this method is m_{map}. To show that this realization is drawn from the a posteriori PDF, we also need to establish that the covariance of realizations generated by this method is equal to the a posteriori covariance matrix for linear problems with Gaussian additive noise. The covariance of m_{cs} is

$$
\begin{aligned}
E\{(m_{cs} - m_{map})(m_{cs} - m_{map})^T\} &= \left([L_M\ 0] - C_M G^T(GC_M G^T + C_D)^{-1}[GL_M\ L_D]\right) \\
&\quad E\left\{\begin{bmatrix} Z_M \\ Z_D \end{bmatrix} \begin{bmatrix} Z_M^T\ Z_D^T \end{bmatrix}\right\} \left([L_M\ 0] - C_M G^T(GC_M G^T + C_D)^{-1}[GL_M\ L_D]\right)^T.
\end{aligned}
$$

Because Z_M and Z_D are independent,

$$E\left\{\begin{bmatrix} Z_M \\ Z_D \end{bmatrix}\begin{bmatrix} Z_M^T & Z_D^T \end{bmatrix}\right\} = \begin{bmatrix} I_M & 0 \\ 0 & I_D \end{bmatrix}$$

and the covariance of the conditional simulation is

$$
\begin{aligned}
C_M^{cs} &= \left([L_M\ 0] - C_M G^T (G C_M G^T + C_D)^{-1}[G L_M\ L_D]\right) \\
&\qquad \left([L_M\ 0] - C_M G^T (G C_M G^T + C_D)^{-1}[G L_M\ L_D]\right)^T \\
&= \left([L_M\ 0] - C_M G^T (G C_M G^T + C_D)^{-1}[G L_M\ L_D]\right) \\
&\qquad \left(\begin{bmatrix} L_M^T \\ 0 \end{bmatrix} - \begin{bmatrix} L_M^T G^T \\ L_D^T \end{bmatrix}(G C_M G^T + C_D)^{-1} G C_M\right) \\
&= L_M L_M^T - C_M G^T (G C_M G^T + C_D)^{-1} G L_M L_M^T \\
&\quad - L_M L_M^T G^T (G C_M G^T + C_D)^{-1} G C_M \\
&\quad + C_M G^T (G C_M G^T + C_D)^{-1} \left(G L_M L_M^T G^T + L_D L_D^T\right)(G C_M G^T + C_D)^{-1} G C_M \\
&= C_M - C_M G^T (G C_M G^T + C_D)^{-1} G C_M, \tag{10.87}
\end{aligned}
$$

which we recognize as the a posteriori covariance for linear inverse problems with Gaussian prior probability density and additive Gaussian errors.

Note that Eq. (10.86) appears to be a "square-root" approach to conditional simulation. That is, we can interpret Eq. (10.86) as being of the form

$$m_{cs} = m_{map} + L'Z, \tag{10.88}$$

where

$$Z = [Z_M\ Z_D]^T \tag{10.89}$$

and

$$L' = [L_M\ 0] - C_M G^T (G C_M G^T + C_D)^{-1}[G L_M\ L_D]. \tag{10.90}$$

The advantage of Eq. 10.88 is that it uses the square root of the stationary a priori covariance to generate the square root of the a posteriori covariance. Because it is possible to factor a stationary covariance operator analytically (such as using the moving average method) for most common covariance models, it is not necessary to perform Cholesky decomposition to use Eq. 10.88 for conditional simulation.

It is interesting to note that the square root, L', given by Eq. 10.90 is not square. The dimensions of L' are $m \times (m + n)$, where m is the number of model parameters and n is the number of data. The L matrix does, however, satisfy $C_{M'} = L'L'^T$.

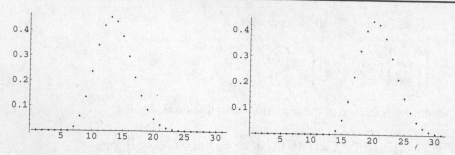

Figure 10.31. The seventh and 14th rows of the Cholesky (LL^T) decomposition of the covariance matrix.

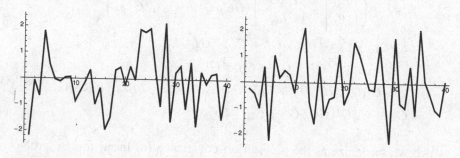

Figure 10.32. Two vectors of independent identically distributed (iid) random variables with mean 0 and standard deviation 1.

Example: application of RML to a simple linear problem

In this example, we applied RML in conditioning a 1D property field to inaccurate measurements. The field is evenly divided into 30 grids. Three measurements are made at grids 15, 21, and 25: $d_{\mathrm{obs}} = \{0.7, 0.3, -0.2\}$. The measurement error is in normal distribution $N(0, 0.1)$. The property field is known to have mean zero and a Gaussian covariance with range ten times that of a grid length. Figure 10.31 shows the seventh and the 14th rows of a "square root" of the prior model covariance matrix. Note that both curves have the same shape but shifted and recentered. Therefore, saving all the rows is sometimes unnecessary, especially when the covariance matrix is too large to carry out Cholesky decomposition. The method of moving average introduced in Section 10.3.3 is based on this idea of only saving one row. Figure 10.32 shows a vector of 30 independent identically distributed (iid) random variables with mean 0 and standard deviation 1, which is the Z vector in Eq. (10.82). Multiplying this iid vector with the L matrix, we obtain an unconditional realization, as shown on the left-hand side of Fig. 10.33. Another unconditional realization is generated by generating a different Z vector, which is shown on the right-hand side of Fig. 10.33. Two unconditional realizations take different shapes but have the same spatial correlation.

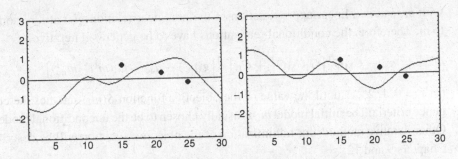

Figure 10.33. Two unconditional simulations of the parameter field m produced using $m_{uc} = m_{prior} + LZ$, with observed data plotted for comparison.

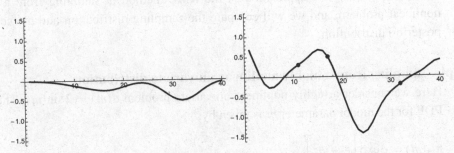

Figure 10.34. Left: A correction to the first unconditional simulation that will make it honor the data. Right: The conditional simulation that was generated from the first unconditional realization.

The unconditional realizations then are calibrated to honor the data following Eq. (10.85). The sensitivity coefficient matrix G in this case is 3×30 with all zeros except a "1" in each row at the grid corresponding to the measurement. Figure 10.34 shows the conditional realization modified from the first unconditional realization, and the correction made on the unconditional realization to honor the data. As the correction to unconditional realization is based on the data, grids far from the data (such as grids 1–5) are not changed in the conditional realization.

10.5.3   RML for nonlinear inverse problems

When the relation between the data and the model is nonlinear, the RML method samples approximately from the correct a posteriori distribution. The algorithm is similar with the one for the linear inverse problem except that the objective function to be minimized takes the form:

$$S(m) = (m - m_{uc})^T C_M^{-1} (m - m_{uc}) + (g(m) - d_{uc})^T C_D^{-1} (g(m) - d_{uc}), \qquad (10.91)$$

where $g(.)$ is the nonlinear operator representing the forward problem, and the simulated data is $g(m)$. The sensitivity of the data about model parameters $G(m) = \nabla_m g(m)$ varies with m, and is a linear approximation of the nonlinear relation $g(.)$ at a region close to m. Therefore, the conditional realizations have to be generated iteratively:

$$m^{i+1} = m_{uc} - C_M G_i^T [G_i C_M G_i^T + C_D]^{-1} [g(m^i) - d_{uc} - G_i(m^i - m_{uc})], \quad (10.92)$$

for $i = 0, 1, 2, \ldots$ until the value of the objective function $S(m_i)$ satisfies the convergence criteria. The initial model m_0 is usually chosen to be the unconditional model m_{uc}. Other iterative methods for minimization of the objective function will be described in Chapters 9 and 12.

As RML is an approximate sampling method for nonlinear problems, we must investigate the validity of the distribution of the model realizations. In the following example, we show the application of the RML method in sampling from a highly nonlinear problem, and we will compare the sampling distribution and the correct a posteriori distribution.

Example: sampling from a highly nonlinear univariate distribution

Here we consider a highly nonlinear univariate problem $d(m) = 2\sin(m)$. The prior PDF for the model parameter m is given by:

$$\pi_{pr}(m) = c \exp\left[-\frac{1}{2}m^2\right], \quad (10.93)$$

which is Gaussian with mean 0 and variance 1. The "likelihood" PDF is:

$$\pi_d(m|d_{obs}) \propto \exp\left[-\frac{1}{2}(g(m) - d_{obs})^T C_D^{-1}(g(m) - d_{obs})\right], \quad (10.94)$$

where the observation data is $d_{obs} = 0.5$, the variance of the data is 0.25, and $g(m) = 2\sin(m)$. Substituting these information into the likelihood equation,

$$\pi_d(m|d_{obs}) \propto \exp\left[-\frac{(2\sin(m) - 0.5)^2}{0.5}\right]. \quad (10.95)$$

Then from Bayes' rule, the a posteriori PDF for m is:

$$f(m) \propto \pi_{pr}(m)\pi_d(m|d_{obs})$$

$$\propto \exp\left[-\frac{1}{2}m^2\right] \exp\left[-\frac{(2\sin(m) - 0.5)^2}{0.5}\right]. \quad (10.96)$$

The procedure for generating a conditional realization using RML is as following:
1. generate an unconditional realization $m_{uc} \in N(0, 1)$;
2. add random measurement error to the observation $d_{uc} \in N(0.5, 0.25)$;
3. minimize the objective function

$$S(m) = \frac{1}{2}(m - m_{uc})^2 + \frac{(2\sin(m) - d_{uc})^2}{0.5}$$

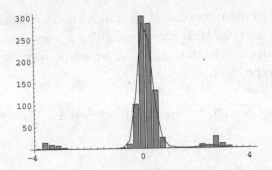

Figure 10.35. The histogram of the 1000 conditional realizations with the true a posteriori PDF.

by iteratively improving the model m:

$$m_{i+1} = m_{\mathrm{uc}} - 2\cos(m_i)[4\cos^2(m_i) + 0.25]^{-1}$$
$$\times [2\sin(m_i) - d_{\mathrm{uc}} - 2\cos(m_i)(m_i - m_{\mathrm{uc}})],$$

where $m_0 = m_{\mathrm{uc}}$;

4. stop the iterations and accept the model m_{i+1} when $\frac{S(m_i) - S(m_{i+1})}{S(m_{i+1})} < 10^{-2}$;

5. if the number of the conditional realizations is less than desired, go to step 1 and repeat the process.

The choice of the convergence criterion in step 4 is quite ad hoc. When the reduction of the objective function is less than 1% of the objective function value, the current realization is accepted. 1000 conditional realizations have been generated using this procedure. The histogram of the conditional realizations and the correct a posteriori PDF have been plotted together in Fig. 10.35. The realizations from the RML method seem to be in a satisfactory distribution in comparison to the true PDF, but it slightly overestimated the uncertainty.

The most important fact to remember about this method, is that it uses a minimization routine to generate a conditional simulation. Another way to say this is that it adds a smooth correction to an unconditional simulation. The primary advantage of the algorithm described in this section, however, is that it seems to be somewhat more robust for nonlinear models than other approximate methods. This feature will be examined more carefully in Section 10.8.

10.5.4 RML as Markov chain Monte Carlo

One way to ensure that the generated realizations are distributed correctly is to use the calibrated realizations from the Randomized Maximum Likelihood (RML) approximate method as trial states in a Markov chain Monte Carlo (McMC) method. In order, however, to use the Markov chain Monte Carlo method, we need to be able to calculate the probability of proposing the calibrated model. Recall that in the RML method we first generate a new unconditional realization of the permeability field m_u (it is not calibrated to the pressure data although it may be conditioned to other data) and a set of errors that are added to the observed data to generate d_u. The state m_{cal} that is proposed is the result of calibrating the unconditional realization to the simulated data (observed data plus noise) using Eq. (10.72). The joint probability density, $f(m_u, d_u)$, of proposing (m_u, d_u) is easily calculated because m_u and d_u are independent Gaussian random variables. Hence, for this problem,

$$f(m_u, d_u) \propto \exp\left[-\frac{1}{2}(m_u - m_{\text{prior}})^{\text{T}} C_{\text{M}}^{-1}(m_u - m_{\text{prior}}) - \frac{1}{2}(d_u - d_{\text{obs}})^{\text{T}} C_{\text{D}}^{-1}(d_u - d_{\text{obs}})\right].$$

(10.97)

The joint probability density, $h(m_{\text{cal}}, d_u)$, of proposing (m_{cal}, d_u) can, theoretically, be calculated if the functional relationship between (m_u, d_u) and (m_{cal}, d_u) is known.

The calibrated model m_{cal} is computed using a Gauss–Newton method to find the minimum of Eq. (10.72), given m_u and d_u. Reversing the procedure, we can instead solve for m_u as a function of m_{cal} and d_u. If we exclude the regions of the (m_{cal}, d_u) space that are inaccessible to the calibration routine, we obtain a unique one-to-one, invertible, relationship between (m_u, d_u) and (m_{cal}, d_u). The joint probability of proposing (m_{cal}, d_u) can then be calculated as follows [see 193, Section 2.8].

$$h(m_{\text{cal}}, d_u) = f(m_u, d_u)|J|,$$

(10.98)

where J is the Jacobian of the transformation, i.e.

$$J = \left|\frac{\partial(m_u)}{\partial(m_{\text{cal}})}\right|.$$

(10.99)

The probability of proposing m_{cal} is found by integrating $h(m_{\text{cal}}, d_u)$ over the data space,

$$q(m_{\text{cal}}) = \int_{\text{D}} h(m_{\text{cal}}, d_u)\, dd_u.$$

(10.100)

For most practical problems, evaluation of the integral in Eq. (10.100) is too difficult to attempt. We will show a one-dimensional example for which the calculation can be performed, and an approximation that seems to work well under a fairly broad range of conditions.

If the probability of proposing a transition to state m_j is independent of the current state, Hasting's rule for the acceptance of a proposed transition from state m_i to state m_j can be written as

$$\alpha_{ij} = \min\left\{1, \frac{\pi_j q_i}{\pi_i q_j}\right\}.$$

(10.101)

In our procedure, q_j is the probability of proposing the calibrated model and depends only on the proposed state. The probability density for the calibrated model, π_j, is

$$\pi_j \propto \exp\left[-\frac{1}{2}(m_j - \mu)^{\text{T}} C_{\text{M}}^{-1}(m_j - \mu) - \frac{1}{2}\left(g(m_j) - d_{\text{obs}}\right)^{\text{T}} C_{\text{D}}^{-1}\left(g(m_j) - d_{\text{obs}}\right)\right].$$

(10.102)

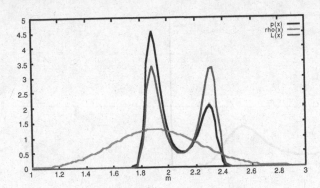

Figure 10.36. Prior probability, likelihood based on data, and a posteriori probability distributions for synthetic example [191].

Note that the probability is not based on the quality of the match obtained in the minimization, but on the quality of the match to the prior model and the observed data.

Synthetic 1D problem

Because it is difficult to evaluate sampling procedures on large multi-variate probability density functions, we first consider the simpler problem of sampling from the following univariate distribution

$$\pi(m) = a \exp\left[-\frac{(m - m_0)^2}{2\sigma_M^2} - \frac{(g(m) - d)^2}{2\sigma_d^2}\right],\tag{10.103}$$

where $m_0 = 1.9$, $d = 0.8$, $\sigma_M^2 = 0.1$, $\sigma_d^2 = 0.01$, $g(m) = 1 - \frac{9}{2}(m - 2\pi/3)^2$, and $a \approx 4.567$. The parameters in this example were chosen to make it as close as possible to the example in Bonet-Cunha *et al.* [194] without the additional complication of multiple minima outside the region of interest. The first term in Eq. (10.103) can be identified with an a priori probability distribution so we will use that terminology for it here. The other term is nonlinear and approximates the behavior of the likelihood term in the inverse problem. The a posteriori distribution, the a priori distribution, and the likelihood are all plotted in Fig. 10.36.

One approach to the problem of sampling from $\pi(m)$ is to linearize the model–data relationship about the maximum a posteriori (MAP) point, which occurs at $m \approx 1.8805$, then propose models for a Markov chain Monte Carlo method from the distribution based on the linearization, a Gaussian centered at the MAP point having a variance which satisfies the following relationship.

$$\sigma_{map}^{-2} = \frac{G_{map}^2}{\sigma_d^2} + \frac{1}{\sigma_M^2},\tag{10.104}$$

where $G_{map} \approx 1.893$ is the value of the derivative of $g(m)$, with respect to m, evaluated at the MAP value of m. Substituting numerical values into Eq. (10.104) gives $\sigma_{map}^2 \approx$

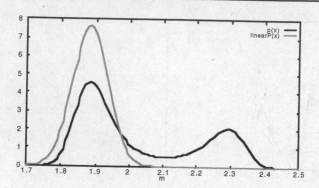

Figure 10.37. The a posteriori probability distributions for synthetic example and the linear approximation to the probability density based on the sensitivity at the maximum a posteriori point [191].

0.0027. The Gaussian approximation to the a posteriori probability density is shown in Fig. 10.37 along with the true probability density function.

In the method that we referred to as "global perturbations from the MAP," Oliver *et al.* [195] proposed transitions from the current state (state i) to another state (state j) based on the linearized approximation to the nonlinear probability distribution, that is, the Gaussian probability density of Fig. 10.37. Theory assures us that, even though the distribution that we sample from is quite unlike the true nonlinear distribution, the samples in the chain will (asymptotically) follow the true distribution as long as we use a correct acceptance criterion. Unfortunately, when this approach is used to generate five thousand states, the resulting distribution appears to be the result of sampling from a Gaussian probability density function centered on the MAP point. There is no hint of a secondary peak, or of any asymmetry to the distribution.

This result makes practical sense when we consider that the distribution we are proposing from is much narrower than the true distribution (see Fig. 10.37) so that the probability of proposing a transition to a value of m greater than 2.2, from a current realization in the region of the MAP point, is approximately

$$
q(m \geq 2.2) = c \int_{2.2}^{\infty} \exp\left(-\frac{(m - m_{\text{map}})^2}{2\sigma_{\text{map}}^2}\right) dm
$$

$$
\approx 6.6 \times 10^{-10}.
$$

(10.105)

Clearly, it is unlikely that any states in the neighborhood of $m = 2.3$ will be proposed in a chain of reasonable length. This seems inconsistent with the idea that the McMC method, if it is truly sampling the distribution $\pi(m)$, should get about 33% of its samples from the neighborhood of the second peak. The explanation is that, although there is a very low probability for proposing a transition to the neighborhood of the secondary

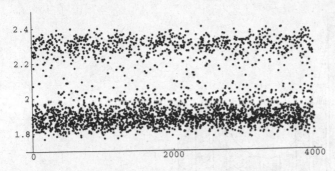

Figure 10.38. The Markov chain of states generated by "calibrating" unconditional simulations from the prior distribution to data [191].

maximum, once it is proposed it will be accepted and the chain will almost certainly remain stuck at that state for another 10^9 or so states. The chain will not just get stuck in the neighborhood of the secondary maximum, but rather it will remain stuck at a single state for a long period. Although many transitions to states in the neighborhood of the MAP will be proposed, the Metropolis–Hastings criterion gives a very low probability of acceptance to those transitions. While this behavior does serve to properly distribute the correct proportion of states within each interval, the distribution of realizations for chains of reasonable length will be very poor.

Calibrate the example

For this simple, but highly nonlinear, problem it seems that the method of conditional simulation that adds variablity to an optimal estimate is a poor method of proposing states. In order to evaluate the use of RML as a method for proposing transitions in a Markov chain Monte Carlo method, a simple routine was written to implement the conditional simulation procedure based on calibration of unconditional simulations of the model.

Figure 10.38 shows all samples generated in the first 4000 trials. The distribution of states looks qualitatively reasonable and succeeding states do not appear to be correlated. The histogram of the frequency of occurrence of values from a 10 000 element Markov chain shown in Fig. 10.39 appears to match the true distribution quite well with the exception of a slight undersampling of elements in the region between the peaks.

We quantitatively compared the distribution of realizations with the true distribution using a chi-square test on the number of realizations within each decile. To do this we first calculated the values of m corresponding to the decile points for the a posteriori distribution. Observed frequencies, by decile, are 1051, 1129, 1015, 1026, 958, 976, 943, 874, 1015, and 1013. Because the expected value within each decile is 1000, the value of χ^2 for this chain is 4.2. With nine degrees of freedom, the probability

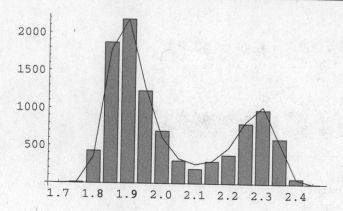

Figure 10.39. The frequency of occurrence of states within each frequency range seems to be a good approximation to the probability distribution that was to be sampled [191].

of obtaining a χ^2 value that large or larger is about 10%, which indicates that the calibration algorithm did a good job of sampling from the a posteriori distribution.

The only difficult part of this procedure is the computation of the marginal probability density as in Eq. (10.100). If it is done accurately, the resulting distribution of states is very close to the desired distribution. It seems hopeless, however, to be able to accurately calculate the probability $q(m_{cal})$ of proposing the calibrated states for large multi-variate problems. Instead, we evaluate the possibility that a simple approximation, not requiring accurate calculation of the Jacobian, might result in a reasonable distribution of states. If we assume that $q_j = \pi_j$ then we can simply accept all states that are the result of the calibration procedure. Figure 10.40 shows the distribution of states from the simple approximation of accepting all states. Note that the fraction of states in the two peaks seems to be approximately correct. The main deficiency is an undersampling of states in the low probability region between the peaks. Whether or not this is important in practice is unknown.

10.6 Conceptual model uncertainty

The traditional approach to assessment of uncertainty is to assume that the conceptual model is known, or can be determined and that uncertainty in predictions is only a result of uncertainty in the model parameters. In most cases, however, there are several levels of uncertainty, and uncertainty in the parameter values is only one of them [196]. Is the reservoir naturally fractured? Are channel sands present? Is compaction important, or can it be ignored? Is the reservoir highly compartmentalized by sealing faults? Højberg and Refsgaard [197] showed that uncertainty in the conceptual model structure could not be compensated for by assessment of parameter uncertainty within a single model.

In their study of solute transport in groundwater, the contribution to uncertainty from model uncertainty was always larger than the contribution from parameter uncertainty. As a result, ignoring the uncertainty in the conceptual model can result in bias and gross underestimation of uncertainty.

While some conceptual model uncertainties can be handled using the same approaches that are used for uncertainty in reservoir variables (e.g. a single porosity reservoir is the limiting case of a dual porosity reservoir as fracture porosity goes to zero), it is not clear that there can be a continuous transition between a channel sand environment and an aeolian sand environment, in which case it makes sense to speak of the relative probability of the two conceptual models being correct. We must distinguish, however, between the prior probability for the models, and the posterior probability for the conceptual models after data assimilation. Although defining the prior probability may be somewhat difficult, the technically challenging aspect is computation of the posterior probability.

If there are a finite number, K, of conceptual models, M_k, with prior probability $P(M_k)$, the posterior probability for model M_k is

$$P(M_k|d_{\text{obs}}) = \frac{P(d_{\text{obs}}|M_k)P(M_k)}{\sum_{j=1}^{K} P(d_{\text{obs}}|M_j)P(M_j)}, \tag{10.106}$$

where

$$P(d_{\text{obs}}|M_k) = \int P(d_{\text{obs}}|m_k, M_k)P(m_k, M_k)\,dm_k. \tag{10.107}$$

The goal is usually to assess uncertainty in predictions or outcomes to support policy decisions. In Bayesian Model Averaging [198], the posterior probability of a prediction

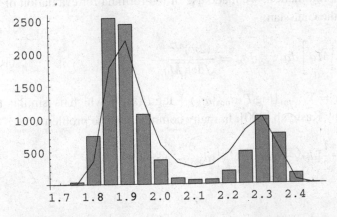

Figure 10.40. Approximate method for generating realizations from the non-Gaussian PDF. All transitions generated by the RML method were accepted [191].

Δ is a weighted average of predictions from all models, i.e.

$$P(\Delta|d_{\text{obs}}) = \sum_{k=1}^{K} P(\Delta|M_k, d_{\text{obs}})P(M_k|d_{\text{obs}}). \tag{10.108}$$

The number of model variables may be different for each conceptual model. The difficulty is clearly in evaluation of the integral in Eq. (10.107). The dimension of the space is almost always too large for regular quadrature methods to be useful.

Neuman [199] proposes that $P(d_{\text{obs}}|M_k)$ be evaluated using the maximum likelihood estimate of the model variables as described by Kashyap [200]. An expression similar to Kashyap's can be derived from a Taylor series expansion of the posterior probability in the neighborhood of the maximum likelihood estimate m_k for the model M_k.

$$\begin{aligned}
P(d_{\text{obs}}|M_k) &= \int P(d_{\text{obs}}|m_k)P(m_k)\,dm_k \\
&= \int \exp[\log P(d_{\text{obs}}|m_k) + \log P(m_k)]\,dm_k \\
&= \int \exp\Bigg[\log P(d_{\text{obs}}|\hat{m}_k) + \log P(\hat{m}_k) \\
&\quad -\frac{1}{2}(m_k - \hat{m}_k)^{\mathrm{T}}\nabla_{m_k,m_k}\big(\log P(d_{\text{obs}}|m_k)\big)(m_k - \hat{m}_k) \\
&\quad -\frac{1}{2}(m_k - \hat{m}_k)^{\mathrm{T}}\nabla_{m_k,m_k}\big(\log P(m_k)\big)(m_k - \hat{m}_k) + \text{h.o.t.}\Bigg]\,dm_k \\
&\approx P(d_{\text{obs}}|\hat{m}_k)P(\hat{m}_k)\int \exp\left[-\frac{1}{2}(m_k - \hat{m}_k)^{\mathrm{T}}A_k(m_k - \hat{m}_k)\right]\,dm_k \\
&= P(d_{\text{obs}}|\hat{m}_k)P(\hat{m}_k)\frac{(2\pi)^{K/2}}{\sqrt{\det(A_k)}}.
\end{aligned} \tag{10.109}$$

In evaluating the approximation, we made use of the formula for evaluation of multivariate integrals of the Gaussian:

$$\int \cdots \int \exp\left[-\frac{1}{2}q^{\mathrm{T}}Mq\right]dq_1\ldots dq_n = \frac{(2\pi)^{n/2}}{\sqrt{\det(M)}} \tag{10.110}$$

and we defined $A_k = \nabla_{m_k,m_k}\big(\log P(d_{\text{obs}}|m_k) + \log P(m_k)\big)$ which is similar to the expression derived by Kashyap [200]. In many common inverse problems,

$$P(d_{\text{obs}}|m) \propto \exp\left\{-\frac{1}{2}[d_{\text{obs}} - g(m)]^{\mathrm{T}}C_{\mathrm{D}}^{-1}[d_{\text{obs}} - g(m)]\right\}$$

and

$$P(m) \propto \exp\left\{-\frac{1}{2}[m - m_0]^{\mathrm{T}}C_M^{-1}[m - m_0]\right\}$$

in which case,

$$A_k = \nabla_{m_k, m_k} \left(\log P(d_{obs}|m_k) + \log P(m_k) \right)$$
$$= G^T C_D^{-1} G + C_M^{-1} \tag{10.111}$$

and

$$P(d_{obs}|M_k) \approx P(d_{obs}|\hat{m}_k) P(\hat{m}_k) \frac{(2\pi)^{K/2}}{\sqrt{\det \left(G^T C_D^{-1} G + C_M^{-1} \right)}}. \tag{10.112}$$

Evaluation of $P(d_{obs}|M_k)$ using Eq. (10.112) requires computation of the maximum a posteriori model variables for each conceptual model, and computation of the determinant of the Hessian (Eq. 10.111). For large models a Monte Carlo approach to the evaluation of the integral in Eq. (10.107), may be more efficient.

10.7 Other approximate methods

The sampling methods can also be categorized another way: those known to sample from the true a posteriori distribution and those only approximately correct. The degree of approximation can make a great difference, especially for highly nonlinear problems. The exact methods we discuss in this chapter are the rejection method and Markov chain Monte Carlo. Both of the two exact methods involve a criterion for accepting or rejecting realizations generated from a stochastic process. We also investigate four approximate methods: linearization about the maximum a posteriori model (LMAP), randomized maximum likelihood (RML), pilot-point methods (PP) and gradual deformation (GD) method.

In addition to the methods described above, several other optimization methods such as simulated annealing and the genetic algorithm have been used to generate conditional realizations. The simulated annealing method has been discussed by Alabert [201], Deutsch [202], Hird and Kelkar [203], and Holden et al. [204]. The genetic algorithm has been discussed by Sen et al. [205] and Romero et al. [206] among others. Unfortunately, these methods are usually far too computationally expensive and require too many iterations to be useful for generating realizations matching reservoir production data.

10.7.1 Pilot point

As described in Section 9.8.1, the pilot-point methods are actually methods of parameterization of the reservoir property field. They were developed in order to reduce the dimension of the history-matching problem. In pilot-point methods, reservoir properties are calculated in a small number of locations and then interpolation is used to assign values to the remaining gridblocks.

When applied to the problem of simulation, the pilot-point method can be considered an approximation to the RML method. In the pilot-point method, perturbations to the model parameters are made only at selected locations (the pilot points). The system of equations to be solved in an iteration step for minimization is typically smaller than in the randomized maximum likelihood method because the number of parameters is reduced. In the pilot-point method, the columns of C_{M}, associated with pilot-point locations, are used as basis vectors for computation of corrections to the model parameters. One widely used implementation of the algorithm is as follows:

1. generate an unconditional realization of the model parameters, $m_{\mathrm{u}} \leftarrow N[m_{\mathrm{pr}}, C_{\mathrm{M}}]$;
2. generate an unconditional realization of the noise and add it to the data, $d_{\mathrm{u}} \leftarrow N[d_{\mathrm{obs}}, C_{\mathrm{D}}]$;
3. minimize the function:

$$J_{\mathrm{D}}(\delta m) = \left[d_{\mathrm{u}} - g(m_{\mathrm{u}} + \delta m)\right]^{\mathrm{T}} C_{\mathrm{D}}^{-1}\left[d_{\mathrm{u}} - g(m_{\mathrm{u}} + \delta m)\right], \tag{10.113}$$

where $\delta m = \sum C_{\mathrm{M},i}\alpha_i$ and the summation is only over the values of i corresponding to gridblocks containing pilot points.

Pilot points, and the related master points, have been used in history matching by de Marsily *et al.* [123], Bréfort and Pelcé [124], and LaVenue and Pickens [125]. The application to conditional simulation was made by RamaRao *et al.* [126] and Gómez-Hernández *et al.* [131] among others. As Oliver *et al.* [207] point out, the use of Eq. (10.113) seems to be based on a mistaken belief that the regularization term (model mismatch squared) is not necessary when using the pilot-point method. In a Bayesian context, the correct alternative is to minimize

$$J(\delta m) = \left[d_{\mathrm{u}} - g(m_{\mathrm{u}} + \delta m)\right]^{\mathrm{T}} C_{\mathrm{D}}^{-1}\left[d_{\mathrm{u}} - g(m_{\mathrm{u}} + \delta m)\right] + (\delta m)^{\mathrm{T}} C_{\mathrm{M}}^{-1} \delta m. \tag{10.114}$$

10.7.2 Gradual deformation

The method of gradual deformation [208] is an approximate method for gradually deforming continuous geostatistical models to generate reservoir models which honor historic production data. This algorithm has been used by Hu *et al.* [209] to incorporate historic production data to reduce uncertainty in production forecasts. Because the gradual deformation algorithm attempts to honor the geostatistical parameters (covariance model and range) while deforming the model to honor the data, it seemed intuitive that it might generate realizations from the PDF for model variables conditioned to data.

The principal idea of the gradual deformation method is that new realizations of a random field Z with mean μ and covariance C_Z can be written as the linear combination of a set of independent random Gaussian fields with expected mean μ and covariance C_Z, i.e.

$$Z(K) = \sum_{i=1}^{n} k_i(Z_i - \mu) + \mu. \tag{10.115}$$

If the coefficients k_i satisfy:

$$\sum_{i=1}^{n} k_i^2 = 1 \tag{10.116}$$

then it is easy to show that the expected mean and covariance of the random vector Z are also μ and C_Z, assuming that k_i does not depend on Z_i:

$$E[Z] = E\left[\sum_{i=1}^{n} k_i(Z_i - \mu) + \mu\right] \tag{10.117}$$

$$= \mu + \sum_{i=1}^{n} k_i E[Z_i - \mu] \tag{10.118}$$

$$= \mu, \tag{10.119}$$

$$\text{cov}[Z] = E\left[\left(\sum_{i=1}^{n} k_i(Z_i - \mu)\right)\left(\sum_{i=1}^{n} k_i(Z_i - \mu)\right)^{\text{T}}\right] \tag{10.120}$$

$$= \sum_{i=1}^{n}\sum_{j=1}^{n} k_i k_j E\left[(Z_i - \mu)(Z_j - \mu)^{\text{T}}\right] \tag{10.121}$$

$$= E\left[\sum_{i=1}^{n} k_i^2 (Z_i - \mu)(Z_i - \mu)^{\text{T}}\right] \tag{10.122}$$

$$= \sum_{i=1}^{n} k_i^2 C_Z \tag{10.123}$$

$$= C_Z. \tag{10.124}$$

However, in practice, k_i is the optimum weight for each Z_i such that Eq. (10.115) has the best combination of the random vectors to minimize the objective function. Therefore the algorithm of the gradual deformation method is not an exact method and it is difficult to quantify its sampling distribution.

The most basic form of the gradual deformation algorithm has pairs of vectors combined:

$$Z(\rho) = Z_1 \cos(\pi\rho) + Z_2 \sin(\pi\rho), \tag{10.125}$$

where ρ is the deformation parameter with the range from 0 to 2. Each of the Z_i is a vector of independent deviates from the same Gaussian distribution with expectation 0 and variance 1. The procedure for generating a realization conditional to d_{obs} is as follows:

1. generate an initial vector Z_1 of independent normal deviates;
2. generate a second vector of independent normal deviates Z_2;

3. search the ρ value which minimizes the objective function S_d;

$$S_d(\rho) = \frac{1}{2}[g(m(\rho)) - d_{\text{obs}}]^T C_D^{-1}[g(m(\rho)) - d_{\text{obs}}], \tag{10.126}$$

where reservoir model realizations are calculated by:

$$m(\rho) = m_{\text{prior}} + LZ(\rho) \tag{10.127}$$

where

$$LL^T = C_M; \tag{10.128}$$

note that the objective function to be minimized contains only the squared data mismatch;

4. if the minimum value of the objective function is sufficiently small, then stop the procedure; otherwise, replace Z_1 with the optimal $Z(\rho)$ and return to step 2.

Local perturbation

When the historic production data are scattered spatially in the reservoir, adding an independent vector Z is likely to improve the fit in some gridblocks while deteriorating the fit in other locations. This led Hu *et al.* [209] to develop a procedure for modifying the values of Z only within a limited region for which the data mismatch was large. In cases for which it is not clear how to choose a limited region for interference tests, the region of change can be limited to a single gridblock. The location of the gridblock to be modified could be chosen randomly in each iteration. In this case, the Z vector in Eq. (10.127) is calculated as:

$$z_i(\rho) = \begin{cases} z_{1,k} & \text{for } i \neq k, \\ \cos(\pi\rho)z_{1,i} + \sin(\pi\rho)z_{2,i} & \text{for } i = k, \end{cases} \tag{10.129}$$

where k is a randomly selected perturbation location. $z_{2,i}$ is a realization of a random variable sampled from the Gaussian distribution with mean 0 and variance 1. $z_{1,i}$ is the ith element of the vector Z_1.

10.8 Comparison of uncertainty quantification methods

Four comparative investigations of the validity of sampling algorithms for quantifying uncertainty of reservoir performance conditional to flow data have been reported in the literature. Zimmerman *et al.* [210] compared several geostatistical inverse techniques to determine which is better suited for making probabilistic forecasts of solute transport in an aquifer. The comparison criteria were the predicted traveltimes and the travel paths taken by conservative radioactive tracers if accidentally released from storage at the Waste Isolation Pilot Plant (WIPP) site in New Mexico. The main conclusion achieved by this study was the importance of the appropriate selection of the variogram and "the time and experience devoted by the user of the method."

In a large investigation of uncertainty quantification supported by the European Commission, Floris et al. [71] applied several methods of evaluating uncertainty to a synthetic reservoir characterization study based on a real field case. Participants received reservoir parameters only at well locations, "historic" production data (with noise), and a general geologic description of the reservoir. Nine different techniques for conditioning of the reservoir models to the production data were evaluated, and results (production forecast for a certain period) were compared in the form of a cumulative distribution function. Variation in the parameterization of the problem was identified as the main discriminating factor in this study. The differences in the quality of the history matching and the production forecast caused by the distinct approaches to the problem also resulted in major differences in the resulting cumulative distribution functions.

Barker et al. [133] used the same synthetic test problem as Floris et al. [71], but focussed their investigation of sampling on three methods: history matching of multiple realizations using a pilot-point approach, rejection sampling, and Markov chain Monte Carlo. They obtained very different distributions of realizations from rejection and Markov chain Monte Carlo methods. The difference was attributed to variations in the prior information used by the participants, but this made evaluation of the results difficult.

Liu and Oliver [211] evaluated the ability of the various sampling methods to correctly assess the uncertainty in reservoir predictions by comparing the distribution of realizations with a distribution from a long Markov chain Monte Carlo. The ensemble of realizations from five sampling algorithms for a synthetic, one-dimensional, single-phase flow problem were compared in order to establish the best algorithm under controlled conditions. Five thousand realizations were generated from each of the approximate sampling algorithms. The distributions of realizations from the approximate methods were compared to the distributions from McMC. The following material is a summary of that paper.

10.8.1 Test problem

Because our objective is to compare the distribution of the realizations of reservoir predictions generated by approximate sampling methods with the distribution generated by methods that are known to assess uncertainty correctly, it was important to choose the test problem carefully – we know that some of the approximate methods sample correctly when the relationships between the conditioning data and the model parameters are linear, so we designed our problem to be highly nonlinear. We also needed to be able to generate large numbers of realizations so that the resulting distributions do not depend significantly on the random seed. By choosing a single-phase transient flow problem with highly accurate pressure measurements, fairly large uncertainty in

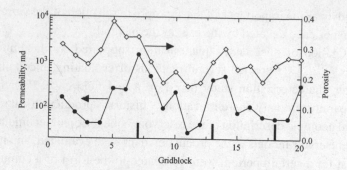

Figure 10.41. The true synthetic permeability and porosity fields, used to generate pressure data to test the sampling algorithms. Well locations are shown by solid bars along the base of the figure. (Adapted from Liu and Oliver [211], copyright SPE.)

the property field, and a short correlation length, we were able to obtain a problem with multiple local maxima in the likelihood function, yet for which a flow simulation required only 0.02 seconds.

Our test problem is a one-dimensional heterogeneous reservoir whose permeability and porosity fields are shown in Fig. 10.41. The reservoir is discretized into 20 grid-blocks, each of which is 50 feet in length. Both the log-permeability ($\ln k$) and the porosity fields were assumed to be multi-variate Gaussian with exponential covariance and a range of 175 ft. The prior means for porosity and log-permeability are 0.25 and 4.5, respectively. The standard deviation of the porosity field is 0.05 and the standard deviation of the log-permeability field is 1.0. The correlation coefficient between porosity and log-permeability is 0.5. The flow is single phase with an oil viscosity of 2 cp and a total compressibility of 4×10^{-6} psi^{-1}. The initial reservoir pressure is 3500 psi.

The well located in gridblock 13 produces at a constant rate. Observation wells are located in gridblocks 7 and 18. Gaussian random noise with a standard deviation of 0.5 psi has been added to the data generated from the true reservoir model. The observed pressure data for all three wells are shown in Fig. 10.42. Although there are ten measurements of pressure drop at each well, the first three measurements at the observation wells are below the noise level. Porosity measurements at well locations were not included in this study as their introduction would have made the a posteriori (conditional) PDF for model variables more nearly Gaussian.

10.8.2 Results analysis

Each of the approximate sampling methods, except the gradual deformation, was used to generate 5000 conditional reservoir realizations. We ran the gradual deformation algorithm until the objective function was reduced to 50 or less. If the objective function

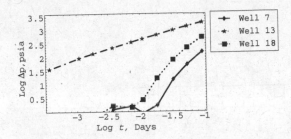

Figure 10.42. The observed pressure drop data at all wells. Random noise added to the true pressure drop causes the nonphysical appearance at low values of Δp. (Adapted from Liu and Oliver [211], copyright SPE.)

was not reduced to 50 by the 10 000th iteration, it was discarded. For comparison, over 99% of the McMC realizations have squared data mismatch values less than 50, so 50 is a relatively loose tolerance on the data mismatch. 86 realizations were generated out of 1000 gradual deformation sequences using the global perturbation methods. The local perturbation method performed slightly better with 11% of the sequences reaching the convergence criterion before the 10 000th iteration. The 86 realizations from global perturbation were used for comparison of the distributions with those from other methods. The average porosity and the effective permeability ($k_{\mathrm{eff}} = 20(\sum_{i=1}^{20} k^{-1})^{-1}$) for each realization were then computed as prediction of the uncertainty in these quantities would be important for assessing oil in place and recovery. We also computed the maximum permeability for each realization because it had been argued previously by Oliver *et al.* [207] that the pilot-point method tends to produce extreme values in the property fields and this would provide a check on the validity of that conjecture. Finally, we computed the squared data mismatch and the squared model mismatch. The model mismatch provides an indication of the probability that the realization could be a sample from the prior distribution (a Gaussian random field with known covariance) and the data mismatch provides a measure of the likelihood of the model given the data. If the realizations do not approximately honor the data they are unlikely to be valid samples from the conditional distribution and little confidence could be placed in predictions of future performance.

The distributions of realizations are summarized in the form of box plots. The method of generation of realizations is identified by an acronym along the bottom axis beneath each box. The meaning of the acronyms should be obvious with the possible exception that the several pilot-point methods are identified by number of pilot-point locations (6 or 9) and by the objective function that was minimized (D if the objective function minimized the data mismatch only).

Figure 10.43 shows comparisons of the distributions of summary reservoir properties from each of the methods. All of the approximate methods in this case seem to generate reservoir realizations whose effective permeability distributions have been shifted to

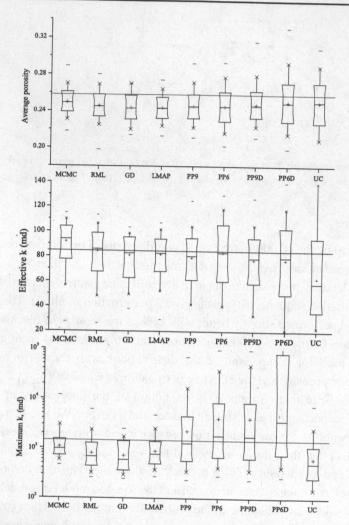

Figure 10.43. Distribution of conditional realizations of average porosity, effective permeability, and maximum permeability from the approximate sampling algorithms and from the very long McMC. The unconditional distribution and the true values are shown for comparison. (Adapted from Liu and Oliver [211], copyright SPE.)

lower values than the distribution obtained from McMC (top of Fig. 10.43). The differences among the methods do not seem to be large, except that the pilot-point methods that used six locations tend to predict greater uncertainty in effective permeability than the other methods. All sampling methods also gave similar results for the distribution of average reservoir porosity.

Interestingly, the average reservoir porosity of the true reservoir was 25.8% while the median values from the ensembles of realizations were all between 24.5 and 25.0% (middle of Fig. 10.43). One might be tempted to conclude that any method would be acceptable for assessing uncertainty in average porosity, but the good agreement

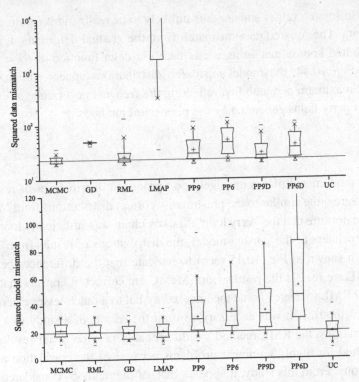

Figure 10.44. Distribution of conditional realizations of squared data mismatch and squared model mismatch from the approximate sampling algorithms and from the very long McMC. The unconditional distribution and the true values are shown for comparison. (Adapted from Liu and Oliver [211], copyright SPE.)

between methods seen here may be due to the fact that the conditional mean is nearly the same as the unconditional mean. Again, the pilot-point methods with small numbers of pilot points tend to substantially overestimate the uncertainty in average reservoir porosity.

The distributions of maximum reservoir permeability in the bottom plot of Fig. 10.43 show considerably greater variability. Randomized maximum likelihood and linearization about the MAP both give distributions of realizations that are wider and slightly shifted, but otherwise similar, compared to McMC. The pilot-point methods, on the other hand, generate reservoir realizations for which very large permeability values occur frequently. This tendency to produce realizations with extreme values has been one of the primary objections to the use of pilot points in reservoir characterization [212].

Because the square of the mismatch between data computed from a reservoir model and the observed data is an indication of the likelihood of a model being correct, it is extremely important that the squared data mismatch values be relatively small. The upper plot in Fig. 10.44 shows that realizations generated by the LMAP method have very large data mismatch values and are thus unlikely to be realizations conditioned to the observed data. The squared data mismatch from the gradual deformation, instead of being distributed approximately as χ^2, is nearly a delta function at $S_d = 50$. At the bottom of Fig. 10.44, the model mismatch distributions appear similar to the distributions of maximum permeability, reflecting the frequent occurrence of extreme values in the property fields generated by the pilot-point methods.

10.8.3 Discussion

The method of randomized maximum likelihood produced distributions of reservoir properties that were quite similar to the presumably correct distributions from McMC. Although we are not sure that the "very long" Markov chain was sufficiently long to be completely independent of the starting model, the distributions of results from chains of varying length shown in Fig. 10.45 seem to indicate that the differences between McMC and RML are real. If the results from McMC are correct, it appears that there is a tendency for RML to overestimate the uncertainty but to a much lesser extent than the pilot-point methods. If it was necessary to limit the results of a study to a small number of realizations, the RML method would probably provide better results than the McMC method for highly nonlinear problems because of the correlation among McMC realizations. From this study, it appears that, of the methods considered here, generating realizations using the randomized maximum likelihood method is the only practical alternative that provides acceptable assessment of uncertainty.

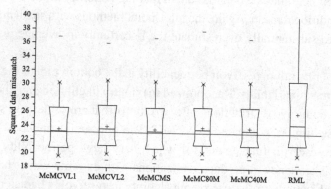

Figure 10.45. Distributions of conditional realizations of squared data mismatch from four Markov chains of different lengths. The distribution from randomized maximum likelihood is shown for comparison. McMCVL means one very long chain (320 million iterations); McMCMS means multiple short chains (each of which is 2 million iterations in length); McMC80M means one chain of 80 million iterations; McMC40M means one chain of 40 million iterations. (Adapted from Liu and Oliver [211], copyright SPE.)

11 Recursive methods

This chapter reviews methods for updating estimates of reservoir properties and states when new data are available for assimilation. These methods are called recursive because the new estimate is a function of the previous estimate. The Kalman filter is the classical approach to data assimilation for linear problems, and the extended Kalman filter is an adaptation for nonlinear problems. Neither method is well suited for the large problems that typify reservoir history matching or characterization. In the final sections of this chapter, the ensemble Kalman filter (EnKF) method is introduced as a viable alternative to classical recursive methods and to traditional history matching.

11.1 Basic concepts of data assimilation

In reservoir history-matching problems, the assumption is often made that the initial state of the reservoir is known (pressure, saturations, and concentrations are at equilibrium), and that the joint probability of the reservoir parameters before assimilation of data can be characterized. In many cases, the permeability and porosity are assumed to be realizations of a random process whose characteristics are known. It is often valid to also assume that the conditional probability distribution of future states of the reservoir, given the present state and all past states, depends only upon the current state and not on any past states. The data (or observations) at time t depend only on the state of the reservoir at time t, and not on previous states. If these assumptions are valid, the probability density for the parameters and the state variables can be defined recursively, given the following information.

initial probability density for model variables $\qquad$ $p(m_0)$
probability of current state given probability of prior state $\quad$ $p(m_k|m_{k-1})$
probability of observations given model parameters $\qquad$ $p(d_{\text{obs},k}|m_k)$.

Denote the observations up to time t by $D_k = \{d_{\text{obs},1}, \ldots, d_{\text{obs},k}\}$. Our aim is to estimate the posterior distribution $p(m_k|D_k)$.

The posterior probability distribution for the model parameters and state variables is given by Bayes' theorem,

$$p(m_k|D_k) = \frac{p(D_k|m_k)p(m_k)}{\int p(D_k|m_k)p(m_k)\,dm_k},$$ (11.1)

which can be rewritten to emphasize the recursive relationship between the probability density for model variables at time $t_k + 1$ and time t_k:

$$p(m_{k+1}|D_{k+1}) = p(m_k|D_k)\frac{p(d_{\mathrm{obs},k+1}|m_{k+1})p(m_{k+1}|m_k)}{p(d_{\mathrm{obs},k+1}|D_k)}.$$ (11.2)

Assume that at time t_k, we have an ensemble of samples of state and model vectors from the posterior $p(y_k, m_k|D_k)$ where $D_k = \{D_{k-1}, d_{\mathrm{obs},k}\}$ is the collection of all data through time t_k. Bayes' theorem relates the probability density for the state variables, f_k, and model variables, m_k, after assimilation of data $d_{\mathrm{obs},k}$ at time t_k to the prior probability density at t_k as follows:

$$\begin{aligned}
p(f_k, m_k|D_k) &\propto p(d_{\mathrm{obs},k}|f_k, m_k)p(f_k, m_k|D_{k-1})\\
&\propto p(d_{\mathrm{obs},k}|f_k, m_k)p(f_k|m_k, D_{k-1})p(m_k|D_{k-1})\\
&\propto p(d_{\mathrm{obs},k}|m_k)p(m_k|D_{k-1}).
\end{aligned}$$ (11.3)

The equivalence is a result of the fact that the state variables f_k can be computed from the model variables m_k. Although the first line of Eq. (11.3) is fundamental for the traditional Kalman filter because it explains how to update the model and state variables directly, the third line simply points out the possibility of using the Kalman filter to update only the model variables (and initial conditions), then computing the state from the model variables.

The posterior PDF from which samples should be drawn (Eq. 11.3) is the product of two terms. The prior is represented by the ensemble of model and state vectors. As in the traditional EnKF, for purposes of updating the variables, we approximate the prior by a Gaussian whose mean and covariance are estimated from the ensemble. The other term is the likelihood; when both terms are Gaussian, the product is Gaussian and the Kalman filter can be used to compute the updated model and state variables. When the likelihood is not Gaussian (e.g. when the relationship between model variables and observation variables is nonlinear), the product is not a Gaussian, but sampling from an approximation to the posterior can still be accomplished fairly efficiently.

11.2 Theoretical framework

Consider the sequential assimilation of data for model variables with a Gaussian prior probability density and a linear relationship between data and model variables. Let

$$Gm = \begin{bmatrix} G_1 \\ G_2 \end{bmatrix} m \quad \text{and} \quad d_{\mathrm{obs}} = \begin{bmatrix} d_1 \\ d_2 \end{bmatrix},$$ (11.4)

where the data have been partitioned into two sets with independent errors so that

$$C_D = \begin{bmatrix} C_{D_1} & 0 \\ 0 & C_{D_2} \end{bmatrix}. \tag{11.5}$$

If two data sets are assimilated sequentially: d_1 followed by d_2, then after assimilation of the data d_1, the posterior estimate of the model variables, m_1, is

$$m_{1,\text{post}} = m_{\text{pr}} + (G_1^T C_{D_1}^{-1} G_1 + C_M^{-1})^{-1} G_1^T C_{D_1}^{-1}(d_1 - G_1 m_{\text{pr}}) \tag{11.6}$$
$$= m_{\text{pr}} + C_M G_1^T (G_1 C_M G_1^T + C_{D1})^{-1}(d_1 - G_1 m_{\text{pr}})$$

and the posterior estimate of the model covariance is

$$C_{M'} = (G_1^T C_{D_1}^{-1} G_1 + C_M^{-1})^{-1} \tag{11.7}$$
$$= C_M - C_M G_1^T (G_1 C_M G_1^T + C_{D_1})^{-1} G_1 C_M.$$

Define

$$S_0 = C_M^{-1} \tag{11.8}$$
$$S_1 = G_1^T C_{D_1}^{-1} G_1 \tag{11.9}$$
$$S_2 = G_2^T C_{D_2}^{-1} G_2 \tag{11.10}$$

and rewrite Eqs. (11.6) and (11.7) in terms of S_0 and S_1 as

$$m_{1,\text{post}} = m_{\text{pr}} + (S_1 + S_0)^{-1} G_1^T C_{D_1}^{-1}(d_1 - G_1 m_{\text{pr}}) \tag{11.11}$$
$$C_{M'} = (S_1 + S_0)^{-1}. \tag{11.12}$$

When incorporating the new data, d_2, we use the posterior values of the estimate and the covariance as prior values:

$$
\begin{aligned}
m_2 &= m_{1,\text{post}} + (G_2^T C_{D_2}^{-1} G_2 + C_{M'}^{-1})^{-1} G_2^T C_{D_2}^{-1}(d_2 - G_2 m_{1,\text{post}}) \\
&= m_{1,\text{post}} + (S_2 + S_1 + S_0)^{-1} G_2^T C_{D_2}^{-1}(d_2 - G_2 m_{1,\text{post}}) \\
&= [m_{\text{pr}} + (S_1 + S_0)^{-1} G_1^T C_{D_1}^{-1}(d_1 - G_1 m_{\text{pr}})] \\
&\quad + (S_2 + S_1 + S_0)^{-1} G_2^T C_{D_2}^{-1} d_2 \\
&\quad - (S_2 + S_1 + S_0)^{-1} S_2 m_{1,\text{post}} \\
&= [m_{\text{pr}} + (S_1 + S_0)^{-1} G_1^T C_{D_1}^{-1}(d_1 - G_1 m_{\text{pr}})] \\
&\quad + (S_2 + S_1 + S_0)^{-1} G_2^T C_{D_2}^{-1} d_2 \\
&\quad - (S_2 + S_1 + S_0)^{-1} S_2 [m_{\text{pr}} + (S_1 + S_0)^{-1} G_1^T C_{D_1}^{-1}(d_1 - G_1 m_{\text{pr}})] \\
&= [m_{\text{pr}} + (S_1 + S_0)^{-1} G_1^T C_{D_1}^{-1}(d_1 - G_1 m_{\text{pr}})] \\
&\quad + (S_2 + S_1 + S_0)^{-1} G_2^T C_{D_2}^{-1}[d_2 - G_2[m_{\text{pr}}] \\
&\quad - (S_2 + S_1 + S_0)^{-1} S_2 (S_1 + S_0)^{-1} G_1^T C_{D_1}^{-1}(d_1 - G_1 m_{\text{pr}}) \\
&\quad - m_{\text{pr}} + [(S_1 + S_0)^{-1} \quad (S_2 + S_1 + S_0)^{-1} S_2 (S_1 + S_0)^{-1}] G_1^T C_{D_1}^{-1}(d_1 - G_1 m_{\text{pr}}) \\
&\quad + (S_2 + S_1 + S_0)^{-1} G_2^T C_{D_2}^{-1}[d_2 - G_2[m_{\text{pr}}] \\
&= m_{\text{pr}} + (S_2 + S_1 + S_0)^{-1}[G_2^T C_{D_2}^{-1}(d_2 - G_2 m_{\text{pr}}) + G_1^T C_{D_1}^{-1}(d_1 - G_1 m_{\text{pr}})].
\end{aligned}
\tag{11.13}
$$

This is exactly the same result that is obtained by assimilating all of the data simultaneously:

$$
\begin{aligned}
\hat{m} &= m_{\mathrm{pr}} + \left[G^{\mathrm{T}} C_D^{-1} G + C_M^{-1} \right)^{-1} G^{\mathrm{T}} C_D^{-1} (d - G m_{\mathrm{pr}}) \\
&= m_{\mathrm{pr}} + \left(G_1^{\mathrm{T}} C_{D_1}^{-1} G_1 + G_2^{\mathrm{T}} C_{D_2}^{-1} G_2 + C_M^{-1} \right)^{-1} \\
&\quad \times \left[G_2^{\mathrm{T}} C_{D_2}^{-1} (d_2 - G_2 m_{\mathrm{pr}}) + G_1^{\mathrm{T}} C_{D_1}^{-1} (d_1 - G_1 m_{\mathrm{pr}}) \right] \\
&= m_{\mathrm{pr}} + (S_2 + S_1 + S_0)^{-1} \left[G_2^{\mathrm{T}} C_{D_2}^{-1} (d_2 - G_2 m_{\mathrm{pr}}) + G_1^{\mathrm{T}} C_{D_1}^{-1} (d_1 - G_1 m_{\mathrm{pr}}) \right].
\end{aligned}
\tag{11.14}
$$

Hence, for linear problems with Gaussian errors the results for sequential assimilation of all data are identical to the results from sequential assimilation of independent data sets.

There are two steps required in the sequential data assimilation algorithm. The first is the computation of the posterior estimate (Eq. 11.6) and the second is the computation of the posterior covariance (Eq. 11.7). At the next data assimilation, these values become the prior values.

Note that there is nothing in this approach that dictates the order of data assimilation – the results are the same even if the last data to be acquired are assimilated first. Much of the computational effort, however, is typically expended in the computation of the theoretical data, Gm. This effort can be minimized in dynamical systems if the data are assimilated in temporal order.

11.3   Kalman filter and extended Kalman filter

The Kalman filter [213] specifically addresses the problem of estimating the current state of the system that is described by a linear stochastic differential equation of the form,

$$
x_k = A_k x_{k-1} + w_{k-1}
\tag{11.15}
$$

with measurements, $d_{\mathrm{obs},k}$, that are related to the state variables through a linear relationship

$$
d_{\mathrm{obs},k} = G_k x_k + \epsilon_k.
\tag{11.16}
$$

The noise in the process and the noise in the measurements are assumed to be independent and Gaussian with zero mean. The linearity of the data relationship and the assumptions of normality are identical to the assumptions made for linear inverse problems in Chapter 8.

The Kalman filter provides an estimate of the posterior probability density for the state when the data and the dynamical relationships are both linear, and the prior is Gaussian. In a short expository article, Meinhold and Singpurwalla [214] summarize the meaning of the Kalman filter from a Bayesian standpoint. Consider the following statement of knowledge of the state at time t_{k-1}:

$$p(x_{k-1}|D_{k-1}) \sim N(\hat{x}_{k-1}, C_{x,k-1}), \tag{11.17}$$

where $\hat{x}_{k-1}$ is the expectation of x_{k-1} and $C_{x,k-1}$ is the variance, conditional to all data through time t_{k-1}, $D_{k-1} = \{D_{k-2}, d_{k-1}\}$. We would like to compute a similar expression for the posterior probability at a later time t_k, after assimilation of data d_{t_k}. Because the prior is assumed to be Gaussian (and the relationships are linear), the posterior is also Gaussian, and the probability density is completely characterized by the expectation and the covariance.

The Kalman filter method consists of two sequential stages: computation of $p(x_k|D_{k-1})$, followed by computation of $p(x_k|D_k)$. The first is the probability forecast based on advancing the solution of the dynamical equations for flow and transport in the reservoir from time t_{k-1} to time t_k. The second is the conditioning of the prior (forecast) estimate to the observations at t_k.

Stage 1 (forecast stage): Given the estimate $\hat{x}_{k-1}$ at time t_{k-1} the best estimate of x_k (before conditioning to data) can be obtained directly from the dynamic relationship (Eq. 11.15), and the best estimate of the covariance is obtained from the identity (Eq. 4.35), which shows how to relate the covariance of two variables that are linearly related.

$$x_k^f = A_k \hat{x}_{k-1} \tag{11.18}$$

and

$$C_k^f = A_k C_{k-1} A_k^T. \tag{11.19}$$

Stage 2 (analysis or assimilation stage): Given all the data through time t_k, we compute the posterior for x_k using Bayes' rule.

$$\begin{aligned} p(x_k|D_k) &= p(x_k|d_{\text{obs},k}, D_{k-1}) \\ &\propto p(d_{\text{obs},k}|x_k, D_{k-1}) p(x_k|D_{k-1}). \end{aligned} \tag{11.20}$$

Since $d_{\text{obs},k} = G_k x_k + \epsilon_k$, where ϵ_k is a vector of unknown observation errors, and $E[\epsilon_k \epsilon_k^T] = C_{D,k}$, the likelihood is

$$p(d_{\text{obs},k}|x_k, D_{k-1}) \propto \exp\left[-\frac{1}{2}(G_k x_k - d_{\text{obs},k})^T C_{D,k}^{-1}(G_k x_k - d_{\text{obs},k})\right]. \tag{11.21}$$

The second term in the computation of the posterior from Eq. (11.20) is the prior for Stage 2 (the posterior for Stage 1):

$$p(x_k|D_{k-1}) \propto \exp\left[-\frac{1}{2}(x_k - x_k^f)^T (C_k^f)^{-1}(x_k - x_k^f)\right]. \tag{11.22}$$

The best estimate of x_k at the analysis or assimilation stage is the state vector that maximizes the product of the expressions in Eqs. (11.21) and (11.22), or equivalently, minimizes the argument of the exponential function:

$$S(x) = \frac{1}{2}(G_k x - d_{\text{obs},k})^{\text{T}} C_{D,k}^{-1}(G_k x - d_{\text{obs},k}) + \frac{1}{2}(x - x_k^f)^{\text{T}}(C_k^f)^{-1}(x - x_k^f). \quad (11.23)$$

Taking the derivative of $S(x)$ with respect to x, and setting it equal to zero, the best estimate of x_k is obtained.

$$\hat{x}_k = x_k^f + C_k^f G_k^{\text{T}}(G_k C_k^f G_k^{\text{T}} + C_{D,k})^{-1}(d_{\text{obs},k} - G_k x_k^f)$$
$$= x_k^f + K(d_{\text{obs},k} - G_k x_k^f), \quad (11.24)$$

where

$$K = C_k^f G_k^{\text{T}}(G_k C_k^f G_k^{\text{T}} + C_{D,k})^{-1} \quad (11.25)$$

is called the Kalman gain. The posterior covariance at time t_k after data assimilation is

$$C_k = ((C_k^f)^{-1} + G_k^{\text{T}} C_{D,k}^{-1} G_k)^{-1}$$
$$= C_k^f - C_k^f G_k^{\text{T}}(C_{D,k} + G_k C_k^f G_k^{\text{T}})^{-1} G_k C_k^f. \quad (11.26)$$

Although Eqs. (11.24) and (11.26) are identical to Eqs. (7.19) and (7.21), the Kalman filter requires repeated computation of the covariance, while the posterior covariance is not required when all data are assimilated simultaneously (except to characterize the uncertainty). The Kalman filter allows one to continuously update the variable estimates as new observations are acquired, but at the cost of computing the posterior covariance for each model update. The covariance matrix for reservoir flow problems is typically very large and the computational effort required makes this method impractical for history matching. Additionally, the assumption of linearity of the model dynamics is violated for multi-phase reservoir flow.

The *extended Kalman filter* [215] is an ad hoc extension of the Kalman filter methodology to problems with nonlinear dynamics. For nonlinear problems, the dynamical relationship expressing the time evolution of the system can often be written as

$$x_k = h_k(x_{k-1}) + w_k, \quad (11.27)$$

which, if the relationship is only weakly nonlinear, can be approximated as

$$x_k \approx A_k x_{k-1} + w_k, \quad (11.28)$$

where A_k is the tangent linear operator (Jacobian) of $h_k(x_{k-1})$. Similarly, the data relationship,

$$d_{\text{obs},k} = g_k(x_k) + \epsilon_k, \quad (11.29)$$

may be approximated by the linearized relationship

$$d_{\text{obs},k} \approx G_k x_k + \epsilon_k. \tag{11.30}$$

Eqs. (11.28) and (11.30) are of the same form as the relationships assumed for the Kalman filter, and the resulting estimates are identical. In addition to the computational expense of computing the posterior covariance at every analysis stage, use of the extended Kalman filter also requires computation of G_k, which as we have seen (Section 9.5) is also prohibitively expensive when there are many data.

11.4   The ensemble Kalman filter

The ensemble Kalman filter (EnKF) method is a Monte Carlo implementation of the Kalman filter in which the mean of an ensemble of realizations provides the best estimate of the population mean and the ensemble itself provides an empirical estimate of the probability density. The method was first introduced by Evensen [216], and much of the development has taken place in the field of weather prediction. Applications to petroleum engineering began with Nævdal *et al.* [217] and have expanded rapidly since that time [218–221]. The EnKF avoids several of the limitations of the Kalman filter and the extended Kalman filter. In particular, there is no need to linearize the dynamical equations or the relationship between the state variables and the data. There is also no need to compute and update the estimate of the covariance, so the method can be practical for very large models.

Implementation of the ensemble Kalman filter begins with the generation of an ensemble of N_e initial models (typically 40–100) consistent with prior knowledge of the initial state and its probability distribution. We denote the ensemble of augmented state vectors at time t_k by Ψ_k:

$$\Psi_k = \{y_{k,1}, y_{k,2}, \ldots, y_{k,N_e}\},$$

where N_e is the number of ensemble members; $y_{k,i}$ for $i = 1, \ldots, N_e$ are state vectors. Each of the state vectors in the ensemble Kalman filter contains all the uncertain and dynamic variables that define the state of the system. For a reservoir flow problem, the state vector could be a reservoir simulation restart file with all gridblock permeabilities, porosities, saturations, pressures, and PVT properties. It is useful to augment the state vector with a vector of predicted or theoretical data. These data may in general be nonlinear functions of the state variables.

At time t_k, the ith state vector for the reservoir model is expressed as:

$$y_{k,i} = \left[(x_{k,i})^{\mathrm{T}}, g(x_{k,i})^{\mathrm{T}} \right]^{\mathrm{T}},$$

where $x_{k,i}$ consists of variables for rock properties and flow system in every gridblock, $g(x_{k,i})$ is the simulated data from the model state $x_{k,i}$.

Like the Kalman filter, the ensemble Kalman filter for data assimilation consists of two sequential stages. The first is the forecast forward in time based on solution of the dynamical equations for flow and transport in the reservoir:

$$y^f_{k,i} = f(y_{k-1,i}), \text{ for } i = 1, N_e,$$

where $f(\cdot)$ is the reservoir simulator. The superscript f indicates the "forecast" state. This stage does not usually modify the rock properties (unless they happen to be time dependent), but replaces the pressure, saturation, and simulated data in the predicted state vector with new values appropriate for time t_k. The initial ensemble for $k = 0$ refers to the collection of initial state vectors, which are sampled from the prior probability density function of the state vector before any data assimilation.

The second stage is the analysis or data assimilation stage in which the variables describing the state of the system are updated to honor the observations. The update to each ensemble member is made using an ensemble approximation to the Kalman gain formula:

$$y_j = y^f_j + K_e(d_j - Hy^f_j), \text{ for } j = 1, \dots, N_e,$$

where K_e is the ensemble Kalman gain, and H is the measurement operator that extracts the simulated data from the state vector y^f:

$$H_k = \begin{bmatrix} 0 & I \end{bmatrix}.$$

d_j is a vector of observational data that has been perturbed with random errors from the same distribution as the measurement error:

$$d_j = d_{\text{obs}} + \epsilon_j, \text{ for } j = 1, \dots, N_e.$$

The ensemble Kalman gain is identical in form to Eq. (11.25):

$$K_e = C_{\Psi,e} H^{\text{T}} (H C_{\Psi,e} H^{\text{T}} + C_D)^{-1},$$

but in this case, the covariance matrix of the state vectors $C_{\Psi,e}$ at any time is approximated from the ensemble members:

$$C_{\Psi,e} = \frac{1}{N_e - 1} \sum_{i,j=1}^{N_e} (y^f_i - \bar{y}^f)(y^f_j - \bar{y}^f)^{\text{T}},$$

where the indices i and j refer to ensemble members. $\bar{y}^f$ is the mean of the N_e ensemble members at the current data assimilation step. If the size of each state vector is N_y, the covariance matrix $C_{\Psi,e}$ is $N_y \times N_y$. It is not practical to compute or store $C_{\Psi,e}$ except for problems that are quite small. Fortunately, computation of $C_{\Psi,e}$ is almost never necessary. Instead, we compute and store a square root of $C_{\Psi,e}$:

$$L_e = \frac{1}{\sqrt{N_e - 1}} (\Psi - \bar{y}^f) \tag{11.31}$$

in which case, a computationally efficient form of the Kalman gain is

$$K_e = L_e(HL_e)^{\mathrm{T}}\big[(HL_e)(HL_e)^{\mathrm{T}} + C_D\big]^{-1}.$$

Note that HL_e is simply the last N_d rows of L_e and that its dimension is $N_d \times N_e$.

11.5 Application of EnKF to strongly nonlinear problems

The dynamical equations for multi-phase flow in porous media are highly nonlinear and although the ensemble Kalman filter handles the nonlinear dynamics correctly during the forecast step, it sometimes fails badly in the analysis (or updating) of saturations [221]. In those cases, an iterative ensemble Kalman filter for data assimilation can be used to enforce constraints and to ensure that the resulting ensemble is representative of the conditional PDF (i.e. that the uncertainty quantification is correct).

For applications to nonlinear assimilation problems, the iterative ensemble Kalman filter can be thought of as a least-squares method in which an average gradient for minimization is obtained not from a variational approach, but from an empirical correlation between model variables [222, 223].

For a problem in which the relationship between the state variables, the model parameters, and the data is linear, both the model parameters and the state variables can be updated simultaneously using the Kalman filter. The result is an improved estimate of the (nonvarying) model parameters and also an improved estimate of the current value of the state variables. For a nonlinear problem, however, it may be impossible to update the state variables to be consistent with the updated model parameters without resolving the nonlinear forward problem to obtain state variables. In many applications of the ensemble Kalman filter, the objective is primarily to estimate the current state of the system. For petroleum reservoir applications, however, it is generally important to estimate not only the current state of the system (the pressures and saturations), but also the values of permeability, porosity, and fault transmissibility. Knowledge of these variables is sufficient to determine the state of the system.

Consistent with previous sections, we denote the static model variables by the symbol m, the dynamic state variables at time t_k by $f_k(m)$, and the predictions of observations at time t_k by $g_k(m)$. In this nomenclature, the augmented state vector at time t_k is

$$Y_k = [m, f_k(m), g_k(m)]^{\mathrm{T}}. \tag{11.32}$$

Because the primary focus of this section is on the assimilation step for highly nonlinear data relationships, we will omit the time subscript, understanding that the functions $f(\cdot)$ and $g(\cdot)$ are generally functions of time. The relationship between the observations and the true static model variables is

$$d_{\mathrm{obs}} = g(m_{\mathrm{true}}) + \epsilon, \tag{11.33}$$

where ϵ is the measurement error assumed to be Gaussian and $E[\epsilon\epsilon^T] = C_D$. The relationship between the observations and the true augmented state vector can also be written as

$$d_{\text{obs}} = HY_{\text{true}} + \epsilon, \tag{11.34}$$

where H can be represented as a matrix whose elements are all ones or zeros.

11.5.1 Nonlinear dynamic system

Let the state vector, Y_k, at time t_k, consist of model variables, m_k, whose estimate changes with time, but whose true value is constant, state variables, $f_k(m_k)$, and simulated data, $g_k(m_k)$. Suppose that $g(m) \approx g(m_\ell) + G_\ell(m - m_\ell)$ and $f(m) \approx f(m_\ell) + F_\ell(m - m_\ell)$, so the approximation to the covariance based on linearization at m_ℓ is

$$C_Y \approx \begin{bmatrix} C_M & C_M F_\ell^T & C_M G_\ell^T \\ F_\ell C_M & F_\ell C_M F_\ell^T & F_\ell C_M G_\ell^T \\ G_\ell C_M & G_\ell C_M F_\ell^T & G_\ell C_M G_\ell^T \end{bmatrix}. \tag{11.35}$$

Although the relationship between the observations and the model variables is generally nonlinear, i.e. $d_{\text{obs}} = g(m) + \epsilon$, the relationship between the observations and the augmented state vector is linear, i.e.

$$d_{\text{obs}} = HY_{\text{true}} + \epsilon \tag{11.36}$$

with $E[\epsilon\epsilon^T] = C_D$.

The vector of model variables that maximizes the conditional probability density also minimizes the following objective function.

$$S(m) = \frac{1}{2}\big(g(m) - d\big)^T C_D^{-1}\big(g(m) - d\big) + \frac{1}{2}(m - m_{\text{pr}})^T C_M^{-1}(m - m_{\text{pr}}). \tag{11.37}$$

In this notation, m_{pr} denotes the estimate of m at the end of the forecast step (before the assimilation or analysis step). When the number of data is smaller than the number of model variables, an appropriate iterative form of the Gauss–Newton method for finding the model vector m that minimizes the objective function in Eq. (11.37) is

$$m^{\ell+1} = m_{\text{pr}} - C_M G_\ell^T \big(C_D + G_\ell C_M G_\ell^T\big)^{-1}\big[g(m^\ell) - d_{\text{obs}} - G_\ell(m^\ell - m_{\text{pr}})\big]. \tag{11.38}$$

On the other hand, if the problem is sufficiently nonlinear that a reduced step length is required, a Gauss–Newton formula for iteration is

$$m^{\ell+1} = \beta_\ell m_{\text{pr}} + (1 - \beta_\ell)m^\ell$$
$$- \beta_\ell C_M G_\ell^T \big(C_D + G_\ell C_M G_\ell^T\big)^{-1}\big[g(m^\ell) - d_{\text{obs}} - G_\ell(m^\ell - m_{\text{pr}})\big], \tag{11.39}$$

where β_ℓ is an adjustment to the step length whose optimal value can be determined by standard methods [66].

Note that C_M in Eqs. (11.38) and (11.39) is the model covariance at the end of the forecast step, prior to assimilation of the current data but after assimilation of all data before the current time. C_M should not be changed during the Gauss–Newton iteration, although the linear approximation, G, to the measurement operator, g, may change with each iteration. Because the relationship between the data and the model variables is assumed to be nonlinear, the value of G should also be different for each ensemble member, but in many cases it seems to work quite well to use a single average value for all vectors [224]. As a result, the computation of the product $G_\ell C_M G_\ell^T$ is not as straightforward in an iterative formulation, as in the EnKF where it is only necessary to compute $(H \Delta Y)(H \Delta Y)^T$.

11.5.2 Implementation of the EnRML

Let m_{pr} denote the ensemble of model vectors at the end of the forecast step, after assimilation of all previous data. There are N_e of these vectors, although for convenience, the index has been neglected here. Denote the vector of means of the prior variables by $\bar{m}_{pr}$ and the vector of deviations from the means by Δm_{pr}. The ensemble estimate of the prior model variable covariance (after assimilation of all previous data) is $C_M = \Delta m_{pr} \Delta m_{pr}^T / (N_e - 1)$. One feature that makes the implementation of the traditional ensemble Kalman filter so efficient, however, is that it is never necessary to compute C_M, only the products $H C_M H^T$ and $C_M H^T$. This computation is not as straightforward in an iterative filter, because it is important to maintain the distinction between the model covariance matrix estimate, which should be based on the prior models, and the sensitivity matrix, which should be based on the current values. At the ℓth iteration, let Δd^ℓ represent the deviation of each vector of computed data from the mean vector of computed data and let Δm^ℓ represent the deviation of each vector of model variables from the current mean. The sensitivity matrix G_ℓ is the coefficient matrix relating the changes in model parameters to the changes in computed data,

$$\Delta d^\ell = G_\ell \Delta m^\ell, \tag{11.40}$$

where Δm^ℓ is $N_M \times N_e$; Δd^ℓ is $N_D \times N_e$; G_ℓ is $N_D \times N_M$. As Δm^ℓ is not generally invertible (or even square), we compute a pseudoinverse.

Except for the iterative aspects, the EnRML procedure is very similar to the EnKF procedure and the basic procedure is as follows.

1. Generate an ensemble of random vectors from the prior probability density for the model variables. Compute the mean and the deviation from the mean for each vector. From this, it is possible to compute an approximation of the prior covariance. Store the prior deviations (not the estimate of the covariance).

2. Generate an approximation of the data sensitivity matrix by "solving" the equation $\Delta d^\ell = G_\ell \Delta m^\ell$ from the ensemble for G_ℓ.

3. Apply the Gauss–Newton update formula Eq. (11.39) to compute improved estimates of model variables, $m^{\ell+1}$.

4. Recompute theoretical data $g(m^{\ell+1})$ using the updated model variables.

5. Check to see if the estimates have converged. If not, return to step 2.

Many steps in the iterative method are identical to the steps in the standard method. A comprehensive description of the noniterative procedure, including refinements that can be used to decrease the effects of small sample size can be found in Evensen [225].

11.6   1D example with nonlinear dynamics and observation operator

This example deals with data assimilation for 1D, two-phase, immiscible flow without capillary pressure. The problem is chosen because the saturation shock results in a bimodal probability density for saturation that is difficult for the standard Kalman filters to handle. The test case has 32 grid cells in a 1D grid (see Table 11.1). Water is injected at a constant rate in grid 1 and fluid is produced at a constant pressure from grid 32. Water saturations are measured at the observation well in grid 16.

A correlated exponential covariance model is used to generate 65 initial realizations of porosity and log-permeability. The practical correlation range of the covariance model is approximately 15 grids. The mean and standard deviation of porosity fields are 0.2 and 0.04, respectively. The mean and standard deviation of $\ln k$ fields are 5.5 and 0.7. The correlation coefficient between porosity and $\ln k$ is 0.6.

In a previous investigation, it was found that the analysis step in the ensemble Kalman filter worked fairly well for water saturation field when corrections to the saturation were small, but that the updated values exceeded physical bounds when the observed values of saturation were far from the forecast values. For this test, water saturation observations at days 40, 90, 140, and 190 are used to refine the reservoir models. None of the models have breakthrough at the measurement location before 190 days so there are no changes to the model or state variables until 190 days. At that time, the correction is fairly large.

Figure 11.1 shows the ensemble of water saturation profiles after the assimilation of data at day 190. The vertical line indicates the observation location (grid 16). The

Table 11.1. *Schematic grid setup for the 1D problem.*

1	... ...	16	... ...	32
Injector		Obs. Well		Producer

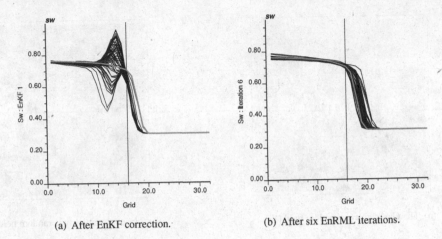

(a) After EnKF correction.   (b) After six EnRML iterations.

Figure 11.1. Assimilation of Sw observation at 190 days. (Reproduced from Gu and Oliver [226], copyright SPE.)

gray curve is the true saturation profile at this time. The black curves are saturation profiles from the ensemble of realizations. The analysis step of the ensemble Kalman filter gives saturation profiles that are completely implausible; in some cases water saturations exceed $1 - S_{or}$ and in others the saturation profiles oscillate between high and low values (Fig. 11.1a). Results from the iterative EnKF are both plausible and consistent with the observed saturation at the measurement location (Fig. 11.1b).

In each of these methods, both the model variables (porosity and log-permeability) and the state variables (saturation and pressure) are updated to be consistent with observations. The saturation profiles in Fig. 11.1 show, however, that the errors in saturation in the neighborhood of the front can be quite large for EnKF. The errors in the porosity are generally smaller and more consistent between the methods.

11.7 Example – geologic facies

The problem of estimating the boundaries of facies, or regions of relatively uniform properties, from two-phase flow data is highly nonlinear and non-Gaussian. For that reason, it provides a good test case for the application of the ensemble Kalman filter to problems for which it does not seem appropriate. The geologic facies in this section are modeled using the truncated pluri-Gaussian method [227]. Observations consist of facies at well locations, and production data such as phase production or injection rates and down-hole pressures.

Figure 11.2 illustrates graphically, the procedure for generating random facies on a 2D region using the truncated pluri-Gaussian method. In this application, the truncation regions for simulation of geologic facies from two Gaussian random fields are described by three intersecting threshold lines [220], although this is not the standard method.

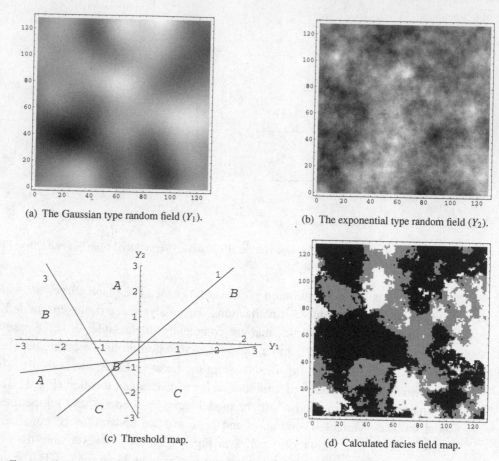

(a) The Gaussian type random field (Y_1).

(b) The exponential type random field (Y_2).

(c) Threshold map.

(d) Calculated facies field map.

Figure 11.2. Simulation of lithofacies distribution in the field by truncation of random Gaussian fields Y_1 and Y_2 using intersecting line thresholds. (Reproduced from Liu and Oliver [104], copyright SPE.)

Three lines intersecting each other without all passing through the same point divide the plane into seven regions. A facies type can be attributed to each region, so up to seven different facies types can be modeled. Two Gaussian random fields Y_1 and Y_2 are required over the region of interest. In this example, the Gaussian random field Y_1 (Fig. 11.2a) has Gaussian covariance and Y_2 (Fig. 11.2b) has exponential covariance. The coordinates of the threshold map (Fig. 11.2c) are Y_1 and Y_2 respectively. Three facies or lithotypes, A, B, and C, are assigned to the seven regions in the threshold map. The facies type at any gridblock in the field is determined by the location of the Y_1 and Y_2 values for each gridblock on the threshold map.

In this study, the geostatistical parameters for generating the two Gaussian fields are assumed to be known, so the only variables to be modified in data assimilation are the random Gaussian fields Y_1 and Y_2. As the hard data measurements do not

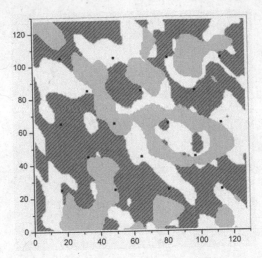

Figure 11.3. The true facies map with all the well locations denoted by black dots. The wells are numbered 1–18 from the lower left-hand corner to upper right-hand corner by rows [220].

depend on the dynamic states of the reservoir fluid flow, the state vector for cases with only facies measurements is $y_j = \{Y_1, Y_2, d_{sim}\}$ where d_{sim} is a vector of predicted facies at observation locations. When there are production data in d_{sim}, the state vector includes the pressure and the saturation in every gridblock, $y_j = \{Y_1, Y_2, P, S, d_{sim}\}$. Both Gaussian fields have the same size as the reservoir grid, therefore the size of the state vector is $N_y = 4 \times n_{grid} + n_d$, where n_d is the number of data obtained at each observation time.

11.7.1 Matching facies observations at wells

In the first application of EnKF to facies estimation, eighteen facies observations are made over a 128×128 grid. The Gaussian field Y_1 is anisotropic with the principal direction in NW 30°. The range in the principal direction (approximately 20 grid-blocks) is twice the range in the perpendicular direction. The second Gaussian field Y_2 is isotropic with range of approximately 20 gridblocks. The true facies field with observation locations is shown in Fig. 11.3.

The facies variable is not continuous nor necessarily numerical, so quantifying the data mismatch is not straightforward. Note for example that the difference between facies 1 and 3, may not be larger than the difference between facies 1 and 2. One solution is to define a facies mismatch f instead of using the difference between simulated and observed facies values for updating the state vectors. The facies mismatch can be

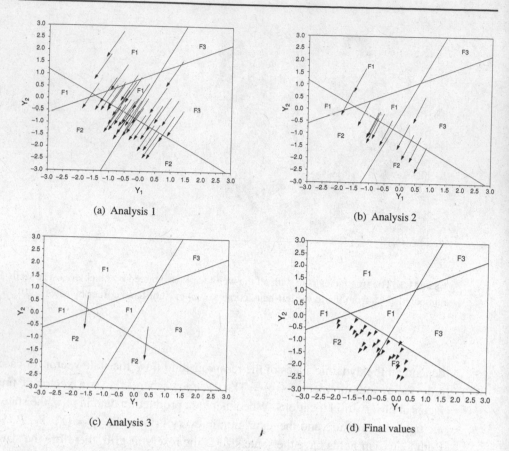

(a) Analysis 1

(b) Analysis 2

(c) Analysis 3

(d) Final values

Figure 11.4. After replacing the simulated facies F_{sim} with the facies mismatch f in the state vectors, only realizations with facies mismatch at the well are updated. The thick lines in each plot are the threshold lines. The arrows point from the values of Y_1 and Y_2 before update to the locations after update [220].

defined as:

$$f = \begin{cases} 0, & F_{\text{sim}} = F_{\text{obs}} \\ 1, & F_{\text{sim}} \neq F_{\text{obs}}. \end{cases}$$

Consequently, the state vector update step becomes:

$$y_j^u = y_j^p - K_e f.$$

Figure 11.4 shows a sequence of updates to the values of Y_1 and Y_2 at well 2. Only the realizations of Y_1 and Y_2 that do not match the facies observation before the update are shown in each subfigure. Figure 11.4(a) shows the modification in the first iteration to the Gaussian variables at well 2 from each of the ensemble members that does not match the observed facies. The facies assigned to each region have been labeled as F_1,

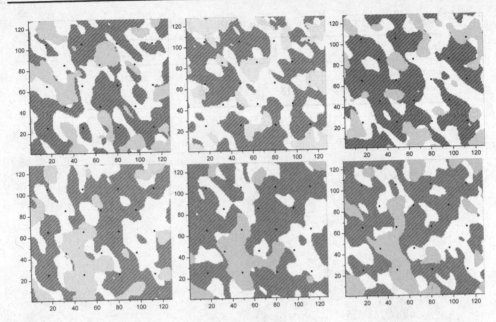

Figure 11.5. Three initial (top row) and corresponding final (bottom row) facies realizations that match the facies observations from all 18 wells [220].

F_2, and F_3. Note that because of the way facies mismatch is defined (either 0 or 1), all updates with nonzero mismatch have the same magnitude and direction. Because the prior probability for Y_1 and Y_2 to fall into facies 2 regions is relatively small, 45 of the 50 initial ensemble members have nonzero facies mismatch. In the first step of model update, 31 ensemble members are corrected to simulate facies 2 at the observation grid (Fig. 11.4a). Three iterations are required for all the ensemble members to simulate facies 2 at grid (49, 25). The final locations of all Y_1 and Y_2 pairs for well 2 is shown in Fig. 11.4(d).

Figure 11.5 shows the first three facies realizations from an initial ensemble of 120, and the corresponding final facies realizations that matched all 18 facies observations. The resulting realizations appear to be drawn from the posterior PDF, as they all honor facies observations at wells, and show appropriate dimensions and structure.

11.7.2 Matching production data

The second example demonstrates the application of the ensemble Kalman filter to the problem of history matching production data for a reservoir with unknown facies boundaries. The true reservoir model in this case is 50 gridblocks by 50 gridblocks with four producers and one injector as shown in Fig. 11.6. Facies 1 is dark grey, facies 2 is light grey, facies 3 is white in the figures. The rock properties for each facies

Table 11.2. *Properties of the lithofacies in the synthetic problem.*

Index	Facies 1	Facies 2	Facies 3
Permeability (md)	174.0	80.0	372.0
Porosity	0.18	0.146	0.25

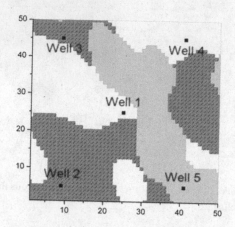

Figure 11.6. The true facies map for the 2D case study of matching both the facies observations and the production data. This facies map is a 50×50 square taken from the 128×128 true facies map in the case study of matching 18 facies observations [220].

are shown in Table 11.2. Bottomhole pressures are fixed at 5000 psia for the injector (well 1) and at 1500 psia for producers. The field is produced for 195 days and the rates are observed and assimilated every 15 days beginning at day 15. Well 3 has water breakthrough on day 183. In this case, there are 14 data for each assimilation step, including 1 injection rate, eight production rates, and five facies observations. Every member of the initial ensemble honors facies observations at the well locations. Three of the 100 initial facies realizations for matching production data are shown in the top row of Fig. 11.7.

Although the initial realizations all honor the facies observations, the theoretical facies observations are kept in the state vector for all assimilation times because updating of Y_1 and Y_2 from matching production data may change the facies type at well locations. When the facies type at a well location is wrong, correction of the Gaussian variables Y_1 and Y_2 using production data can be difficult. An iteration step over the facies observations is made after each model update to ensure the rock properties at well locations are always correct.

The bottom row of Fig. 11.7 shows three facies maps from the ensemble after assimilating all the production data. Each of the final facies maps shown has kept

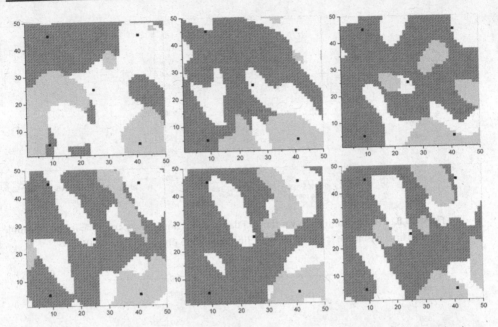

Figure 11.7. The initial facies maps (top row) and final facies maps (bottom row) from realizations 20, 40, and 60 after matching both the production data and the facies observations. Well locations are denoted by the black dots [220].

some features from the initial state, but has developed features that are common among the ensemble members. Some of the common features do not exist in the true facies map.

The variability of the final ensemble has obviously been reduced and, for this relatively small ensemble, the subspace spanned by the ensemble members may not include features that are in the true facies map. After matching the facies observations, the average variances for the two Gaussian fields have decreased from 1 to less than 0.4, and 0.7, respectively. After assimilating the production data, the variances decreased to less than 0.3. The reduction of the average variance for both Gaussian fields in all the ensemble members indicates that the variability among the ensemble members has reduced and the ensemble members become more and more similar with data assimilations.

Although uncertainty quantification for geologic facies models from an ensemble of 100 realizations may not be feasible, the ensemble Kalman filter is able to assimilate production data for this strongly nonlinear problem. The simulated rates and the observed data for two wells are plotted in Figs. 11.8 and 11.9. The box plot on the left-hand side of Fig. 11.8 shows the injection rates from the initial ensemble conditional only to facies observations. The observed injection rate is plotted as the thick line. The distribution of injection rates from the 100 final reservoir models

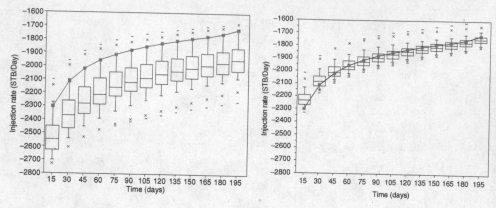

Figure 11.8. The injection rate over the 195 days production history from the initial ensemble (left) and the final ensemble (right). The thick line shows the observed data [220].

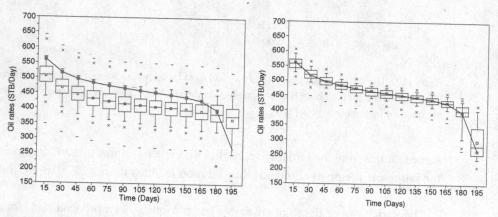

Figure 11.9. The oil rate of well 3 over the 195 days production history from the initial ensemble (left) and the final ensemble (right). The thick line shows the observed data [220].

is much narrower than the initial distribution and almost centered at the observed data (Fig. 11.8, right).

The distribution of the oil rates for well 3 from the final ensemble is shown in Fig. 11.9; the variability is considerably reduced and the ensemble mean is much closer to the observed data than the mean of the initial realization. The rapid drop in oil production rate near the end of history is due to water breakthrough at well 3. Note that only a few of the realizations in the initial ensemble have water breakthrough within 195 days, but after data assimilation almost all the reservoir models have breakthrough in 195 days.

Index

References

[1] D. D. Jackson. Interpretation of inaccurate, insufficient, and inconsistent data. *Geophys. J. R. Astron. Soc.*, **28**, 97–109, 1972.

[2] W. Menke. *Geophysical Data Analysis: Discrete Inverse Theory*. San Diego, CA: Academic Press (revised edition), 1989.

[3] R. L. Parker. *Geophysical Inverse Theory*. Princeton, NJ: Princeton University Press, 1994.

[4] A. Tarantola. *Inverse Problem Theory: Methods for Data Fitting and Model Parameter Estimation*. Amsterdam: Elsevier, 1987.

[5] N.-Z. Sun. *Inverse problems in Groundwater Modeling*. Dordrecht: Kluwer Academic Publishers, 1994.

[6] J. Scales and M. Smith. *Introductory Geophysical Inverse Theory*. Golden, CO: Samizdat Press, 1997.

[7] J. Hadamard. Sur les problémes aux dérivées partielles et leur signification physique. *Princeton Univ. Bull.*, **13**, 49–52, 1902.

[8] W. H. Press, S. A. Teukolsky, W. T. Vetterling, and B. P. Flannery. *Numerical Recipes in FORTRAN: the Art of Scientific Computing*. Cambridge: Cambridge University Press, 1992.

[9] R. L. Burden, J. D. Faires, and A. C. Reynolds. *Numerical Analysis*. Boston, MA: Prindle, Weber and Schmidt, 1978.

[10] R. S. Anderssen, F. R. de Hoog, and M. A. Lucus (eds). *The Application and Numerical Solution of Integral Equations*. Alphen aan den Rijn: Sijthoff & Noordhoff, 1980.

[11] D. W. Peaceman. Interpretation of well-block pressures in numerical reservoir simulation with non-square grid blocks and anisotropic permeability. *SPE J.*, **23**(6), 531–543, 1983.

[12] G. Strang. *Linear Algebra and Its Applications*. New York: Academic Press, 1976.

[13] A. N. Tikhonov. Regularization of incorrectly formulated problems and the regularization method. *Dokl. Akad. Nauk SSSR*, **151**, 501–504, 1963.

[14] G. E. Backus and J. F. Gilbert. Numerical applications of a formalism for geophysical inverse problems. *Geophys. J. R. Astron. Soc.*, **13**, 247–276, 1967.

[15] D. S. Oliver. Estimation of radial permeability distribution from well test data. *SPE Form. Eval.*, **7**(4), 290–296, 1992.

[16] D. S. Oliver. The averaging process in permeability estimation from well-test data. *SPE Form. Eval.*, **5**(3), 319–324, 1990.

[17] D. S. Oliver. The influence of nonuniform transmissivity and storativity on drawdown. *Water Resour. Res.*, **29**(1), 169–178, 1993.

[18] J. Hooke. River meander behaviour and instability: a framework for analysis. *Trans. Inst. Brit. Geog.*, **28**, 238–253, 2003.

[19] S. Bhattacharya, A. P. Byrnes, and P. M. Gerlach. Cost-effective integration of geologic and petrophysical characterization with material balance and decline curve analysis to

develop a 3D reservoir model for PC-based reservoir simulation to design a waterflood in a mature Mississippian carbonate field with limited log data, 2003. Also available as www.kgs.ku.edu/PRS/publication/2003/ofr2003-31/P2-03.html.

[20] E. C. Capen. The difficulty of assessing uncertainty. *J. Petrol. Technol.*, **28**(8), 843–850, 1976.

[21] D. S. Oliver. Multiple realizations of the permeability field from well-test data. *SPE J.*, **1**(2), 145–154, 1996.

[22] P. K. Kitanidis. *Introduction to Geostatistics: Applications in Hydrogeology.* Cambridge: Cambridge University Press, 1997.

[23] G. E. P. Box and D. R. Cox. An analysis of transformations. *J. Roy. Stat. Soc. B*, **26**, 211–246, 1964.

[24] R. A. Freeze. A stochastic–conceptual analysis of one-dimensional groundwater flow in a non-uniform, homogeneous media. *Water Resour. Res.*, **11**(5), 725–741, 1975.

[25] G. Christakos. *Random Field Models in Earth Sciences.* San Diego, CA: Academic Press, 1992.

[26] C. Lanczos. *Linear Differential Operators.* New York: Van Nostrand–Reinhold, 1961.

[27] S. Vela and R. M. McKinley. How areal heterogeneities affect pulse-test results. *SPE J.*, **10**(6), 181–191, 1970.

[28] R. C. Earlougher. *Advances in Well Test Analysis.* New York: Society of Petroleum Engineers of AIME, 1977.

[29] D. Bourdet, T. M. Whittle, A. A. Douglas, and Y. M. Pirard. A new set of type curves simplifies well test analysis. *World Oil*, pp. 95–101, May 1983.

[30] C. V. Theis. The relationship between the lowering of piezometric surface and rate and duration of discharge of wells using groundwater storage. In *Trans. AGU 16th Annual Meeting, pt. 2,* pp. 519–524, 1935.

[31] C. R. Johnson, R. A. Greenkorn, and E. G. Woods. Pulse-testing: a new method for describing reservoir flow properties between wells. *J. Petrol. Technol.*, **Dec**, 1599–1604, 1966.

[32] R. M. McKinley, S. Vela, and L. A. Carlton. A field application of pulse-testing for detailed reservoir description. *J. Petrol. Technol.*, **March**, 313–321, 1968.

[33] A. Mercado and E. Halevy. Determining the average porosity and permeability of a stratified aquifer with the aid of radioactive tracers. *Water Resour. Res.*, **2**(3), 525–531, 1966.

[34] D. B. Grove and W. A. Beetem. Porosity and dispersion constant calculations for a fractured carbonate aquifer using the two well tracer method. *Water Resour. Res.*, **7**(1), 128–134, 1971.

[35] M. Ivanovich and D. B. Smith. Determination of aquifer parameters by a two-well pulsed method using radioactive tracers. *J. Hydrol.*, **36**, 35–45, 1978.

[36] J. Hagoort. The response of interwell tracer tests in watered-out reservoirs. *SPE 11131*, p. 21, 1982.

[37] W. E. Brigham and M. Abbaszadeh-Dehghani. Tracer testing for reservoir description. *J. Petrol. Technol.*, **39**(5), 519–527, 1987.

[38] D. W. Hyndman, J. M. Harris, and S. M. Gorelick. Coupled seismic and tracer test inversion for aquifer property characterization. *Water Resour. Res.*, **30**(7), 1965–1977, 1994.

[39] D. S. Oliver. The sensitivity of tracer concentration to nonuniform permeability and porosity. *Transport Porous Media*, **30**(2), 155–175, 1998.

[40] D. W. Vasco, S. Yoon, and A. Datta-Gupta. Integrating dynamic data into high-resolution reservoir models using streamline-based analytic sensitivity coefficients. *SPE J.*, **4**(4), 389–399, 1999.

[41] Z. Wu, A. C. Reynolds, and D. S. Oliver. Conditioning geostatistical models to two-phase production data. *SPE J.*, **4**(2), 142–155, 1999.

[42] R. Li, A. C. Reynolds, and D. S. Oliver. Sensitivity coefficients for three-phase flow history matching. *J. Can. Petrol. Technol.*, **42**(4), 70–77, 2003.

[43] A. F. Veneruso, C. Ehlig-Economides, and L. Petitjean. Pressure gauge specification considerations in practical well testing. In *Proc. 66th Annual Tech. Conf. of SPE, Prod. Eng., Soc. Petr. Eng.*, Richardson, TX, pp. 865–870, 1991.

[44] K. Aki and P. G. Richards. *Quantitative Seismology*. San Francisco, CA: W. H. Freeman, 1980.

[45] Ö. Yilmaz. *Seismic Data Processing*. Tulsa, OK: Society of Exploration Geophysicists, 1987.

[46] M. B. Dobrin and C. H. Savit. *Introduction to Geophysical Prospecting*. New York: McGraw-Hill, 1988.

[47] B. J. Evans. *A Handbook for Seismic Data Acquisition in Exploration. Geophysical Monograph Series*. Tulsa, OK: Society of Exploration Geophysicists, 1997.

[48] M. Bacon, R. Simm, and T. Redshaw. *3-D Seismic Interpretation*. New York: Cambridge University Press, 2003.

[49] A. Buland and H. Omre. Bayesian linearized AVO inversion. *Geophysics*, **68**, 185–198, 2003.

[50] Y. Dong and D. S. Oliver. Quantitative use of 4D seismic data for reservoir description. *SPE J.*, **10**(1), 91–99, 2005.

[51] J.-A. Skjervheim, G. Evensen, S. I. Aanonsen, B. O. Ruud, and T. A. Johansen. Incorporating 4D seismic data in reservoir simulation models using ensemble Kalman filter (SPE 95789). In *SPE Annual Technical Conference and Exhibition, 9–12 October*, Dallas, TX, 2005.

[52] D. S. Oliver. A comparison of the value of interference and well-test data for mapping permeability and porosity. *In Situ*, **20**(1), 41–59, 1996.

[53] L. Chu, A. C. Reynolds, and D. S. Oliver. Computation of sensitivity coefficients for conditioning the permeability field to well-test data. *In Situ*, **19**(2), 179–223, 1995.

[54] R. D. Carter, L. F. Kemp, and A. C. Pierce. Discussion of comparison of sensitivity coefficient calculation methods in automatic history matching. *SPE J.*, **22**, 205–208, 1982.

[55] P. Jacquard. Théorie de l'interprétation des mesures de pression. *Rev. L'Institut Français Pétrole*, **19**(3), 297–334, 1964.

[56] L. Chu and A. C. Reynolds. Determination of permeability distributions for heterogeneous reservoirs with variable porosity and permeability. TUPREP report, The University of Tulsa, 1993.

[57] H. Tjelmeland, H. Omre, and B. Kåre Hegstad. Sampling from Bayesian models in reservoir characterization. Technical Report Statistics No. 2/1994, Norwegian Institute of Technology, Trondheim, Norway, 1994.

[58] P. K. Kitanidis and E. G. Vomvoris. A geostatistical approach to the inverse problem in groundwater modeling (steady state) and one-dimensional simulations. *Water Resour. Res.*, **19**(3), 677–690, 1983.

[59] P. K. Kitanidis. Parameter uncertainty in estimation of spatial functions: Bayesian estimation. *Water Resour. Res.*, **22**(4), 499–507, 1986.

[60] F. Zhang, A. C. Reynolds, and D. S. Oliver. The impact of upscaling errors on conditioning a stochastic channel to pressure data. *SPE J.*, **8**(1), 13–21, 2003.

[61] C. R. Rao. *Linear Statistical Inference and Its Applications*, 2nd edn. New York: John Wiley & Sons, 1973.

[62] H. Omre and O. P. Lødøen. Improved production forecasts and history matching using approximate fluid-flow simulators. *SPE J.*, **9**(3), 339–351, 2004.

[63] R. Li, A. C. Reynolds, and D. S. Oliver. History matching of three-phase flow production data. *SPE J.*, **8**(4), 328–340, 2003.

[64] R. Fletcher. *Practical Methods of Optimization*, 2nd edn. New York: John Wiley & Sons, 1987.

[65] Y. Abacioglu. The use of subspace methods for efficient conditioning of reservoir models to production data. Ph.D. thesis, University of Tulsa, Tulsa, OK, 2001.

[66] J. E. Dennis and R. B. Schnabel. *Numerical Methods for Unconstrained Optimization and Nonlinear Equations*. Englewood Cliffs, NJ: Prentice-Hall, 1983.

[67] J. Nocedal and S. J. Wright. *Numerical Optimization*. Berlin: Springer-Verlag, 1999.

[68] G. H. Golub and C. F. van Loan. *Matrix Computations*, 2nd edn. Baltimore, MD: The Johns Hopkins University Press, 1989.

[69] R. Li. Conditioning geostatistical models to three-dimensional three-phase flow production data by automatic history matching. Ph.D. thesis, University of Tulsa, Tulsa, OK, 2001.

[70] S. I. Aanonsen, I. Aavatsmark, T. Barkve, *et al.* Effect of scale dependent data correlations in an integrated history matching loop combining production data and 4D seismic data (SPE 79665). *Proc. 2003 SPE Reservoir Simulation Symposium*, 2003.

[71] F. J. T. Floris, M. D. Bush, M. Cuypers, F. Roggero, and A.-R. Syversveen. Methods for quantifying the uncertainty of production forecasts: A comparative study. *Petrol. Geosci.*, **7**(SUPP), 87–96, 2001.

[72] R. Bissell. Calculating optimal parameters for history matching. In *4th European Conf. on the Mathematics of Oil Recovery*, 1994.

[73] G. Gao. Data integration and uncertainty evaluation for large scale automatic history matching problems. Ph.D. thesis, University of Tulsa, Tulsa, OK, 2005.

[74] P. E. Gill, W. Murray, and M. H. Wright. *Practical Optimization*. San Diego, CA: Academic Press, 1981.

[75] D. G. Luenberger. *Linear and Nonlinear Programming*. New York: Addison-Wesley, 1984.

[76] G. Gao and A. C. Reynolds. An improved implementation of the LBFGS algorithm for automatic history matching. *SPE J.*, **11**(1), 5–17, 2006.

[77] F. Zhang and A. C. Reynolds. Optimization algorithms for automatic history matching of production data. In *Proc. 8th European Conf. on the Mathematics of Oil Recovery*, 2002.

[78] G. Gao, M. Zafari, and A. C. Reynolds. Quantifying uncertainty for the PUNQ-S3 problem in a Bayesian setting with RML and EnKF. *SPE J.*, **11**(4), 506–515, 2006.

[79] T. G. Kolda, D. P. O'Leary, and L. Nazareth. BFGS with update skipping and varying memory. *SIAM J. Optimiz.*, **8**(4), 1060–1083, 1998.

[80] F. Zhang. Automatic history matching of production data for large scale problems. Ph.D. thesis, University of Tulsa, Tulsa, OK, 2002.

[81] C. G. Broyden, J. E. Dennis, and J. J. Moré. On the local and superlinear convergence of quasi-Newton methods. *J. Inst. Math. Appl.*, **12**, 223–245, 1973.

[82] J. E. Dennis and Jorge J. More. Quasi-Newton methods, motivation and theory. *SIAM Rev.*, **19**, 46–89, 1977.

[83] S. S. Oren and E. Spedicato. Optimal conditioning of self-scaling variable metric algorithms. *Math. Program.*, **10**, 70–90, 1976.

[84] S. S. Oren and D. G. Luenberger. Self-scaling variable metric (SSVM) algorithms I: Criteria and sufficient conditions for scaling a class of algorithms. *Manag. Sci.*, **20**, 845–862, 1974.

[85] S. S. Oren. Self-scaling variable metric (SSVM) algorithms II: Implementation and experiments. *Manag. Sci.*, **20**, 863–874, 1974.

[86] D. F. Shanno and K.-H. Phua. Matrix conditioning and nonlinear optimization. *Math. Program.*, **14**, 149–160, 1978.

[87] J. Nocedal. Updating quasi-Newton matrices with limited storage. *Math. Comp.*, **35**, 773–782, 1980.

[88] D. Liu and J. Nocedal. On the limited memory BFGS method for large scale optimization. *Math. Program.*, **45**, 503–528, 1989.

[89] F. Zhang, J. A. Skjervheim, and A. C. Reynolds. An automatic history matching example. *EAGE 65th Conf. and Exhibition*, p. 4, 2003.

[90] E. M. Makhlouf, W. H. Chen, M. L. Wasserman, and J. H. Seinfeld. A general history matching algorithm for three-phase, three-dimensional petroleum reservoirs. *SPE Advan. Technol. Series*, **1**(2), 83–91, 1993.

[91] E. M. Oblow. Sensitivity theory for reactor thermal-hydraulics problems. *Nucl. Sci. Eng.*, **68**, 322–337, 1978.

[92] J. F. Sykes, J. L. Wilson, and R. W. Andrews. Sensitivity analysis for steady-state groundwater-flow using adjoint operators. *Water Resour. Res.*, **21**(3), 359–371, 1985.

[93] F. Anterion, B. Karcher, and R. Eymard. Use of parameter gradients for reservoir history matching, SPE 18433. In *10th SPE Reservoir Simulation Symp.*, pp. 339–354, 1989.

[94] J. E. Killough, Y. Sharma, A. Dupuy, R. Bissell, and J. Wallis. A multiple right hand side iterative solver for history matching, SPE 29119. In *Proc. 13th SPE Symp. on Reservoir Simulation*, pp. 249–255, 1995.

[95] N. He. Three dimensional reservoir description by inverse theory using well-test pressure and geostatistical data. PhD thesis, University of Tulsa, Tulsa, OK, 1997.

[96] R. D. Carter, L. F. Kemp, A. C. Pierce, and D. L. Williams. Performance matching with constraints. *Soc. Petrol. Eng. J.*, **14**(4), 187–196, 1974.

[97] P. Jacquard and C. Jain. Permeability distribution from field pressure data. *Soc. Petrol. Eng. J.*, **5**(4), 281–294, 1965.

[98] W. H. Chen, G. R. Gavalas, J. H. Seinfeld, and M. L. Wasserman. A new algorithm for automatic history matching. *Soc. Petrol. Eng. J.*, **14**, 593–608, 1974.

[99] G. M. Chavent, M. Dupuy, and P. Lemonnier. History matching by use of optimal control theory. *Soc. Petrol. Eng. J.*, **15**(1), 74–86, 1975.

[100] N. He, A. C. Reynolds, and D. S. Oliver. Three-dimensional reservoir description from multi-well pressure data and prior information. *Soc. Petrol. Eng. J.*, **2**(3), 312–327, 1997.

[101] A. C. Reynolds, R. Li, and D. S. Oliver. Simultaneous estimation of absolute and relative permeability by automatic history matching of three-phase flow production data. *J. Can. Petrol. Technol.*, **43**(3), 37–46, 2004.

[102] K. N. Kulkarni and A. Datta-Gupta. Estimating relative permeability from production data: a streamline approach. *SPE J.*, **5**(4), 402–411, 2000.

[103] F. Zhang, A. C. Reynolds, and D. S. Oliver. Evaluation of the reduction in uncertainty obtained by conditioning a 3D stochastic channel to multiwell pressure data. *Math. Geol.*, **34**(6), 715–742, 2002.

[104] N. Liu and D. S. Oliver. Automatic history matching of geologic facies. *Soc. Petrol. Eng. J.*, **9**(4), 188–195, 2004.

[105] K. Aziz and A. Settari. *Petroleum Reservoir Simulation*. London: Elsevier Applied Science Publishers, 1979.

[106] C. C. Paige and M. A. Saunders. Algorithm 583, LSQR: sparse linear equations and least squares problems. *ACM Trans. Math. Software*, **8**(2), 195–209, 1982.

[107] L. Chu, M. Komara, and R. A. Schatzinger. An efficient technique for inversion of reservoir properties using iteration method. *SPE J.*, **5**(1), 71–81, 2000.

[108] J. R. P. Rodrigues. Calculating derivatives for history matching in reservoir simulators (SPE 93445). In *Proc. SPE Reservoir Simulation Symposium*, 31 January–2 Feburary 2005.

[109] J. R. P. Rodrigues. Calculating derivatives for automatic history matching. *Comput. Geosci.*, **10**(1), 119–136, 2006.

[110] J. F. B. M. Kraaijevanger, P. J. P. Egberts, J. R. Valstar, and H. W. Buurman. Optimal waterflood design using the adjoint method (SPE 105764). In *SPE Reservoir Simulation Symp.*, 26–28 February 2007.

[111] Y. Dong. Integration of time-lapse seismic data into automatic history matching. Ph.D. thesis, University of Oklahoma, Tulsa, OK, 2005.

[112] H. O. Jahns. A rapid method for obtaining a two-dimensional reservoir description from well pressure response data. *SPE J.*, **6**(12), 315–327, 1966.

[113] A.-A. Grimstad and T. Mannseth. Nonlinearity, scale, and sensitivity for parameter estimation problems. *SIAM J. Sci. Comput.*, **21**(6), 2096–2113, 2000.

[114] A.-A. Grimstad, T. Mannseth, S. Aanonsen, *et al.* Identification of unknown permeability trends from history matching of production data. *SPE J.*, **9**(4), 419–428, 2004.

[115] P. C. Shah, G. R. Gavalas, and J. H. Seinfeld. Error analysis in history matching: the optimum level of parameterization. *Soc. Petrol. Eng. J.*, **18**(6), 219–228, 1978.

[116] G. R. Gavalas, P. C. Shah, and J. H. Seinfeld. Reservoir history matching by Bayesian estimation. *Soc. Petrol. Eng. J.*, **16**(6), 337–350, 1976.

[117] A. C. Reynolds, N. He, L. Chu, and D. S. Oliver. Reparameterization techniques for generating reservoir descriptions conditioned to variograms and well-test pressure data. *SPE J.*, **1**(4), 413–426, 1996.

[118] B. L. N. Kennett and P. R. Williamson. Subspace methods for large-scale nonlinear inversion. In *Mathematical Geophysics*. Dordrecht: Reidel, pp. 139–154, 1988.

[119] D. W. Oldenburg, P. R. McGillivray, and R. G. Ellis. Generalized subspace methods for large-scale inverse problems. *Geophys. J. Int.*, **114**(1), 12–20, 1993.

[120] D. W. Oldenburg and Y. Li. Subspace linear inverse method. *Inverse Problems*, **10**, 915–935, 1994.

[121] Y. Abacioglu, D. S. Oliver, and A. C. Reynolds. Efficient reservoir history matching using subspace vectors. *Comput. Geosci.*, **5**(2), 151–172, 2001.

[122] C. R. Vogel and J. G. Wade. Iterative SVD-based methods for ill-posed problems. *SIAM J. Sci. Comput.*, **15**(3), 736–754, 1994.

[123] G. de Marsily, G. Lavedan, M. Boucher, and G. Fasanino. Interpretation of interference tests in a well field using geostatistical techniques to fit the permeability distribution in a reservoir model. In *Geostatistics for Natural Resources Characterization, Part 2*. Dordrecht: Reidel, pp. 831–849, 1984.

[124] B. Bréfort and V. Pelcé. Inverse modeling for compressible flow. Application to gas reservoirs. In *2nd European Conf. on the Mathematics of Oil Recovery*, pp. 331–334, 1990.

[125] A. Marsh LaVenue and J. F. Pickens. Application of a coupled adjoint sensitivity and kriging approach to calibrate a groundwater flow model. *Water Resour. Res.*, **28**(6), 1543–1569, 1992.

[126] B. S. RamaRao, A. Marsh LaVenue, G. de Marsily, and M. G. Marietta. Pilot point methodology for automated calibration of an ensemble of conditionally simulated transmissivity fields, 1. Theory and computational experiments. *Water Resour. Res.*, **31**(3), 475–493, 1995.

[127] A. Marsh LaVenue, B. S. RamaRao, G. de Marsily, and M. G. Marietta. Pilot point methodology for automated calibration of an ensemble of conditionally simulated transmissivity fields, 2. Application. *Water Resour. Res.*, **31**(3), 495–516, 1995.

[128] D. McLaughlin and L. R. Townley. A reassessment of the groundwater inverse problem. *Water Resour. Res.*, **32**(5), 1131–1161, 1996.

[129] A. Datta-Gupta, D. W. Vasco, J. C. S. Long, P. S. D'Onfro, and W. D. Rizer. Detailed characterization of a fractured limestone formation by use of stochastic inverse approaches. *SPE Formation Eval.*, **10**(3), 133–140, 1995.

[130] A. Datta-Gupta, D. W. Vasco, and J. C. S. Long. Sensitivity and spatial resolution of transient pressure and tracer data for heterogeneity characterization, SPE 30589. *Proc. of SPE Annual Tech Conf., Formation Evaluation*, pp. 625–637, 1995.

[131] J. J. Gómez-Hernández, A. Sahuquillo, and J. E. Capilla. Stochastic simulation of transmissivity fields conditional to both transmissivity and piezometric data. 1. Theory. *J. Hydrology*, **203**, 162–174, 1997.

[132] Z. Wu. Conditioning geostatistical models to two-phase flow production data. Ph.D. thesis, University of Tulsa, Tulsa, OK, 1999.

[133] J. W. Barker, M. Cuypers, and L. Holden. Quantifying uncertainty in production forecasts: another look at the PUNQ-S3 problem. *SPE J.*, **6**(4), 433–441, 2001.

[134] R. Li, A. C. Reynolds, and D. S. Oliver. History matching of three-phase flow production data, SPE 66351. In *Proc. 2001 SPE Reservoir Simulation Symposium*, 2001.

[135] J. V. Beck and K. J. Arnold. *Parameter Estimation in Engineering and Science*. Chichester: John Wiley & Sons, 1977.

[136] R. Plackett and J. Burman. The design of optimum multifactorial experiments. *Biometrika*, **33** (4), 305–325, 1946.

[137] B. J. Winer. *Statistical Principles in Experimental Design*. New York: McGraw-Hill, 1962.

[138] G. E. P. Box, W. G. Hunter, and J. S. Hunter. *Statistics for Experimenters. Wiley Series in Probability and Mathematical Statistics*. New York: Wiley, 1978.

[139] M. D. McKay, R. J. Beckman, and W. J. Conover. A comparison of three methods for selecting values of input variables in the analysis of output from a computer code. *Technometrics*, **21**, 239–245, 1979.

[140] G. E. P. Box and N. R. Draper. *Empirical Model Building and Response Surfaces*. New York: John Wiley & Sons, 1987.

[141] R. H. Myers and D. C. Montgomery. *Response Surface Methodology*. New York: Wiley & Sons, 1995.

[142] A. L. Eide, L. Holden, E. Reiso, and S. I. Aanonsen. Automatic history matching by use of response surfaces and experimental design. In *4th European Conf. on the Mathematics of Oil Recovery*, 1994.

[143] C. D. White, B. J. Willis, K. Narayanan, and S. P. Dutton. Identifying and estimating significant geologic parameters with experimental design. *SPE J.*, **6**(3), 311–324, 2001.

[144] C. S. Kabir and N. J. Young. Handling production-data uncertainty in history matching: The Meren reservoir case study SPE 71621. In *Proc. 2001 SPE Annual Technical Conf. and Exhibition*, 2001.

[145] L. Alessio, S. Coca, and L. Bourdon. Experimental design as a framework for multiple realization history matching: F6 further development studies (SPE 93164). In *Proc. 2005 Asia Pacific Oil and Gas Conf. and Exhibition*, 2005.

[146] W. T. Peake, M. Abadah, and L. Skander. Uncertainty assessment using experimental design: Minagish Oolite reservoir (SPE 91820). In *Proc. 2005 SPE Reservoir Simulation Symp.*, 2005.

[147] G. R. King, S. Lee, P. Alexandre, *et al*. Probabilistic forcasting for matural fields with significant production histtory: a Nemba field case study (SPE 95869). In *Proc. 2005 SPE Annual Technical Conf. and Exhibition*, 2005.

[148] D. C. Montgomery. *Design and Analysis of Experiments*. New York: John Wiley and Sons, 2001.

[149] A. I. Khuri and J. A. Cornell. *Empirical Model Building and Response Surfaces*. New York: Marcel Dekker, 1987.

[150] G. A. Lewis, D. Mathieu, and R. Phan-Tan-Luu. *Pharmaceutical Experimental Design*. New York: Marcel Dekker, 1998.

[151] R. W. Mee and M. Peralta. Semifolding 2^{k-p} designs. *Technometrics*, **42**(2), 122–134, 2000.

[152] P. W. M. John. *Statistical Design and Analysis of Experiments. SIAM Classics in Applied Mathematics*. Philadelphia, PA: SIAM, 1971.

[153] G. R. Luster. Raw materials for Portland cement: applications of conditional simulation of coregionalization. Ph.D. thesis, Stanford University, Stanford, CA, 1985.

[154] M. Davis. Generating large stochastic simulations – the matrix polynomial approximation method. *Math. Geol.*, **19**(2), 99–107, 1987.

[155] F. Alabert. The practice of fast conditional simulations through the LU decomposition of the covariance matrix. *Math. Geol.*, **19**(5), 369–386, 1987.

[156] C. R. Dietrich and G. N. Newsam. Efficient generation of conditional simulations by Chebyshev matrix polynomial approximations to the symmetric square root of the covariance matrix. *Math. Geol.*, **27**(2), 207–228, 1995.

[157] T. C. Black and D. L. Freyberg. Simulation of one-dimensional correlated fields using a matrix-factorization moving average approach. *Math. Geol.*, **22**(1), 39–62, 1990.

[158] D. S. Oliver. Moving averages for Gaussian simulation in two and three dimensions. *Math. Geol.*, **27**(8), 939–960, 1995.

[159] G. Le Loc'h, H. Beucher, A. Galli, B. Doligez, and Heresim Group. Improvement in the truncated Gaussian method: combining several Gaussian Functions. In *Proc. ECMOR IV, 4th European Conf. on the Mathematics of Oil Recovery*, 1994.

[160] N. Liu and D. S. Oliver. Automatic history matching of geologic facies, SPE 84594. *Proc. 2003 SPE Annual Technical Conf. and Exhibition*, pp. 1–15, 2003.

[161] B. D. Ripley. Simulating spatial patterns: dependent samples from a multivariate density (algorithm as 137). *Appl. Statist.*, **28**, 109–102, 1979.

[162] R. Y. Rubinstein. *Simulation and the Monte Carlo Method*. New York: John Wiley & Sons, 1981.

[163] M. H. Kalos and P. A. Whitlock. *Monte Carlo Methods. Volume I: Basics*. New York: John Wiley & Sons, 1986.

[164] B. D. Ripley. *Stochastic Simulation*. New York: John Wiley & Sons, 1987.

[165] W. Feller. *An Introduction to Probability Theory and Its Applications*, vol. I, 3rd edn, New York: John Wiley & Sons, 1968.

[166] J. M. Hammersley and D. C. Handscomb. *Monte Carlo Methods*. New York: John Wiley & Sons, 1964.

[167] N. Metropolis, A. W. Rosenbluth, M. N. Rosenbluth, A. H. Teller, and E. Teller. Equations of state calculations by fast computing machines. *J. Chem. Phys.*, **21**, 1087–1092, 1953.

[168] W. K. Hastings. Monte Carlo sampling methods using Markov chains and their applications. *Biometrika*, **57**(1), 97–109, 1970.

[169] C. L. Farmer. Numerical rocks. In *Mathematics of Oil Recovery*. Oxford: Clarendon Press, pp. 437–447, 1992.

[170] C. V. Deutsch and A. G. Journel. *GSLIB: Geostatistical Software Library and User's Guide*. New York: Oxford University Press, 1992.

[171] R. K. Sagar, B. G. Kelkar, and L. G. Thompson. Reservoir description by integration of well test data and spatial statistics, SPE 26462. In *Proc. 68th Annual Technical Conf. of the SPE*, pp. 475–489, 1993.

[172] F. Georgsen and H. Omre. Combining fibre processes and Gaussian random functions for modelling fluvial reservoirs. In *Proc. Geostatistics Tróia '92*, ed. A. Soares, pp. 425–439. Dordrecht: Kluwer, 1993.

[173] Z. Bi, D. S. Oliver, and A. C. Reynolds. Conditioning 3D stochastic channels to pressure data. *SPE J.*, **5**(4), 474–484, 2000.

[174] H. Haldorsen and L. Lake. A new approach to shale management in field-scale models. *Soc. Petrol. Eng. J.*, **24**, 447–457, 1984.

[175] J. Besag. Spatial interaction and the statistical analysis of lattice systems. *J. Roy. Statist. Soc. B*, **36**(2), 192–236, 1974.

[176] S. Geman and D. Geman. Stochastic relaxation, Gibbs distributions, and Bayesian restoration of images. *IEEE Trans. Patt. Anal. Machine Intell.*, **6**(6), 721–741, 1984.

[177] J. Besag and P. J. Green. Spatial statistics and bayesian computation. *J. Roy. Statist. Soc. B*, **55**(1), 25–37, 1993.

[178] N. A. C. Cressie. *Statistics for Spatial Data*. New York: John Wiley & Sons, Inc., 1993.

[179] H. Tjelmeland.Modeling of the spatial facies distribution by Markov random fields. In *Geostatistics Wollongong '96, Proc. 5th International Geostatistical Congress*, Wollongong, Australia, September 22–27. Dordrecht: Kluwer, pp. 512–523, 1996.

[180] H. Derin and H. Elliott. Modeling and segmentation of noisy and textured images using Gibbs random fields. *IEEE Trans. Patt. Anal. Machine Intell.*, **9**(1), 39–55, 1987.

[181] A. Bowyer. Computing Dirichlet tessellations. *Computer J.*, **24**(2), 162–166, 1981.

[182] D. F. Watson. Computing the n-dimensional Delaunay tessellation with application to Voronoi polytopes. *Computer J.*, **24**, 167–171, 1981.

[183] J. Besag. Statistical analysis of non-lattice data. *Statistician*, **24**, 179–195, 1975.

[184] D. Griffeath. Introduction to random fields. In *Denumerable Markov Chains*, ed. J. G. Kemeny, J. L. Snell, and A. W. Knapp. Berlin: Springer-Verlag, 1976.

[185] A. Journel and C. J. Huijbregts. *Mining Geostatistics*. New York: Academic Press, 1978.

[186] W. J. Krzanowski. *Principles of Multivariate Analysis: a User's Perspective*. Oxford: Clarendon Press, 1988.

[187] M. Davis. Production of conditional simulations via the LU decomposition of the covariance matrix. *Math. Geol.*, **19**(2), 91–98, 1987.

[188] B. D. Ripley. *Spatial Statistics*. New York: John Wiley & Sons, 1981.

[189] E. M. L. Beale. Confidence regions in non-linear estimation. *J. R. Stat. Soc. B*, **22**, 41–76, 1960.

[190] D. S. Oliver. On conditional simulation to inaccurate data. *Math. Geol.*, **28**(6), 811–817, 1996.

[191] D. S. Oliver, N. He, and A. C. Reynolds. Conditioning permeability fields to pressure data. In *European Conf. for the Mathematics of Oil Recovery, V*, pp. 1–11, 1996.

[192] P. K. Kitanidis. Quasi-linear geostatistical theory for inversing. *Water Resour. Res.*, **31**(10), 2411–2419, 1995.

[193] S. S. Wilks. *Mathematical Statistics*. New York: John Wiley & Sons, 1962.

[194] L. Bonet-Cunha, D. S. Oliver, R. A. Rednar, and A. C. Reynolds. A hybrid Markov chain Monte Carlo method for generating permeability fields conditioned to multiwell pressure data and prior information. *SPE J.*, **3**(3), 261–271, 1998.

[195] D. S. Oliver, L. B. Cunha, and A. C. Reynolds. Markov chain Monte Carlo methods for conditioning a permeability field to pressure data. *Math. Geol.*, **29**(1), 61–91, 1997.

[196] W. E. Walker, P. Harremoës, J. Rotmans, *et al.* Defining uncertainty: a conceptual basis for uncertainty management in model-based decision support. *Integr. Assess.*, **4**(1), 5–17, 2003.

[197] A. L. Højberg and J. C. Refsgaard. Model uncertainty – parameter uncertainty versus conceptual models. *Water Sci. Technol.*, **52**(6), 177–186, 2005.

[198] J. A. Hoeting, D. Madigan, A. E. Raftery, and C. T. Volinsky. Bayesian model averaging: a tutorial. *Statist. Sci.*, **14**(4), 382–417, 1999.

[199] S. P. Neuman. Maximum likelihood Bayesian averaging of uncertain model predictions. *Stochast. Environ. Rese. Risk Assess.*, **17**(5), 291–305, 2003.

[200] R. L. Kashyap. Optimal choice of AR and MA parts in autoregressive moving average models. *IEEE Trans. Patt. Anal. Machine Intel.*, **4**(2), 99–104, 1982.

[201] F. G. Alabert. Constraining description of randomly heterogeneous reservoirs to pressure test data: a Monte Carlo study, SPE 19600. In *64th Annual SPE Technical Conf.*, pp. 307–321, 1989.

[202] C. V. Deutsch. Conditioning reservoir models to well test information. In *Geostatistics Tróia '92*. Dordrecht: Kluwer, pp. 505–518, 1993.

[203] K. B. Hird and M. G. Kelkar. Conditional simulation method for reservoir description using spatial and well performance constraints (SPE 24750). In *1992 SPE Annual Technical Conf. and Exhibition*, pp. 887–902, 1992.

[204] L. Holden, R. Madsen, A. Skorstad, *et al.* Use of well test data in stochastic reservoir modelling, SPE 30591. In *Proc. SPE Annual Technical Conf. and Exhibition*, 1995.

[205] M. K. Sen, A. D. Gupta, P. L. Stoffa, L. W. Lake, and G. A. Pope. Stochastic reservoir modeling using simulated annealing and genetic algorithm. In *Proc. 67th Annual Technical Conf. and Exhibition of the Society of Petroleum Engineers*, pp. 939–950, 1992.

[206] C. E. Romero, A. C. Gringarten, J. N. Carter, and R. W. Zimmerman. A modified genetic algorithm for reservoir characterisation. *SPE 64765*, 2000.

[207] D. S. Oliver, A. C. Reynolds, Z. Bi, and Y. Abacioglu. Integration of production data into reservoir models. *Petrol. Geosci.*, **7**(SUPP), 65–73, 2001.

[208] F. Roggero and L. Y. Hu. Gradual deformation of continuous geostatistical models for history matching. In *SPE 49004, Annual Technical Conf.*, 1998.

[209] L. Y. Hu, M. Le Ravalec, G. Blanc, *et al.* Reducing uncertainties in production forecasts by constraining geological modeling to dynmaic data. In *Proc. 1999 SPE Annual Technical Conf. and Exhibition*, pp. 1–8, 1999.

[210] D. A. Zimmerman, G. de Marsily, C. A. Gotway, *et al.* A comparison of seven geostatistically based inverse approaches to estimate transmissivities for modeling advective transport by groundwater flow. *Water Resour. Res.*, **34**(6), 1373–1413, 1998.

[211] N. Liu and D. S. Oliver. Evaluation of Monte Carlo methods for assessing uncertainty. *SPE J.*, **8**(2), 188–195, 2003.

[212] D. S. Oliver. Five things I dislike about pilot points. (unpublished talk). *Statoil Research Summit, Conditioning Reservoir Models and Forecasts to Dynamic Data*, Trondheim, Norway, September 13–15, 1999.

[213] R. E. Kalman. A new approach to linear filtering and prediction problems. *Trans. ASME, J. Basic Eng.*, **82**, 35–45, 1960.

[214] R. J. Meinhold and N. D. Singpurwalla. Understanding the Kalman filter. *Amer. Statist.*, **37**(2), 123–127, 1983.

[215] A. H. Jazwinski. *Stochastic Processes and Filtering Theory*. New York: Academic Press, 1970.

[216] G. Evensen. Sequential data assimilation with a nonlinear quasi-geostrophic model using Monte Carlo methods to forecast error statistics. *J. Geophys. Res.*, **99**(C5), 10 143–10 162, 1994.

[217] G. Nævdal, T. Mannseth, and E. H. Vefring. Near-well reservoir monitoring through ensemble Kalman filter, SPE 75235. In *Proc. SPE/DOE Improved Oil Recovery Symp.*, April 13–17, 2002.

[218] R. J. Lorentzen, G. Nævdal, B. Vàlles, A. M. Berg, and A. A. Grimstad. Analysis of the ensemble Kalman filter for estimation of permeability and porosity in reservoir models, SPE 96 375. In *SPE Annual Technical Conf. and Exhibition*, 2005.

[219] Y. Gu and D. S. Oliver. History matching of the PUNQ-S3 reservoir model using the ensemble Kalman filter. *SPE J.*, **10**(2), 51–65, 2005.

[220] N. Liu and D. S. Oliver. Ensemble Kalman filter for automatic history matching of geologic facies. *J. Petrol. Sci. Eng.*, **47**(3–4), 147–161, 2005.

[221] Y. Gu and D. S. Oliver. The ensemble Kalman filter for continuous updating of reservoir simulation models. *J. Energy Resour. Technol.*, **128**(1), 79–87, 2006.

[222] J. L. Anderson. A local least squares framework for ensemble filtering. *Mon. Weather Rev.*, **131**(4), 634–642, 2003.

[223] M. Zafari, G. Li, and A. C. Reynolds. Iterative forms of the ensemble Kalman filter. In *Proc. 10th European Conf. on the Mathematics of Oil Recovery*, Amsterdam, p. A030, 2006.

[224] J. D. Annan and J. C. Hargreaves. Efficient parameter estimation for a highly chaotic system. *Tellus A*, **56**(5), 520–526, 2004.

[225] G. Evensen. *Data Assimilation: the Ensemble Kalman Filter*. Berlin: Springer-Verlag, 2006.

[226] Y. Gu and D. S. Oliver. An iterative ensemble Kalman filter for multiphase fluid flow data assimilation. *SPE J.*, **12**(4), 438–446, 2007.

[227] A. Galli, H. Beucher, G. Le Loc'h, B. Doligez, and Heresim Group. The pros and cons of the truncated Gaussian method. In *Geostatistical Simulations*. Dordrecht: Kluwer, pp. 217–233, 1994.

Printed in the United States
By Bookmasters